Photoionization of Atoms

Photoionization of Atoms

Guest Editors

Sultana N. Nahar
Guillermo Hinojosa

Basel • Beijing • Wuhan • Barcelona • Belgrade • Novi Sad • Cluj • Manchester

Guest Editors

Sultana N. Nahar
The Ohio State University
Columbus, OH
USA

Guillermo Hinojosa
Universidad Nacional Autónoma de México
Cuernavaca
Mexico

Editorial Office
MDPI AG
Grosspeteranlage 5
4052 Basel, Switzerland

This is a reprint of the Special Issue, published open access by the journal *Atoms* (ISSN 2218-2004), freely accessible at: https://www.mdpi.com/journal/atoms/special_issues/photoionization_atoms.

For citation purposes, cite each article independently as indicated on the article page online and as indicated below:

Lastname, A.A.; Lastname, B.B. Article Title. *Journal Name* **Year**, *Volume Number*, Page Range.

ISBN 978-3-7258-2671-1 (Hbk)
ISBN 978-3-7258-2672-8 (PDF)
https://doi.org/10.3390/books978-3-7258-2672-8

Cover image courtesy of NASA, JPL-Caltech, J. Rho (SSC/Caltech)
Trifid nebula, infrared image was taken by NASA's Spitzer Space Telescope

© 2024 by the authors. Articles in this book are Open Access and distributed under the Creative Commons Attribution (CC BY) license. The book as a whole is distributed by MDPI under the terms and conditions of the Creative Commons Attribution-NonCommercial-NoDerivs (CC BY-NC-ND) license (https://creativecommons.org/licenses/by-nc-nd/4.0/).

Contents

About the Editors vii

Preface ix

Sultana N. Nahar, Edgar Hernández, David Kilcoyne, Armando Antillón, Aaron Covington, Olmo González-Magaña, et al.
Experimental and Theoretical Study of Photoionization of Cl III
Reprinted from: *Atoms* 2023, 11, 28, https://doi.org/10.3390/atoms11020028 1

Jean-Paul Mosnier, Eugene T. Kennedy, Jean-Marc Bizau, Denis Cubaynes, Ségolène Guilbaud, Christophe Blancard, et al.
L-Shell Photoionization of Magnesium-like Ions with New Results for Cl^{5+}
Reprinted from: *Atoms* 2023, 11, 66, https://doi.org/10.3390/atoms11040066 14

Muhammad Aslam Baig
Measurement of Photoionization Cross-Section for the Excited States of Atoms: A Review
Reprinted from: *Atoms* 2022, 10, 39, https://doi.org/10.3390/atoms10020039 34

Momar Talla Gning and Ibrahima Sakho
Photoionization Study of Neutral Chlorine Atom
Reprinted from: *Atoms* 2023, 11, 152, https://doi.org/10.3390/atoms11120152 82

Anand K. Bhatia
Photoejection from Various Systems and Radiative-Rate Coefficients
Reprinted from: *Atoms* 2022, 10, 9, https://doi.org/10.3390/atoms10010009 105

Sultana N. Nahar
Theoretical Spectra of Lanthanides for Kilonovae Events: Ho I-III, Er I-IV, Tm I-V, Yb I-VI, Lu I-VII
Reprinted from: *Atoms* 2024, 12, 24, https://doi.org/10.3390/atoms12040024 126

Anil Pradhan
Photoionization and Opacity
Reprinted from: *Atoms* 2023, 11, 52, https://doi.org/10.3390/atoms11030052 163

D. John Hillier
Photoionization and Electron–Ion Recombination in Astrophysical Plasmas
Reprinted from: *Atoms* 2023, 11, 54, https://doi.org/10.3390/atoms11030054 175

Tu-Nan Chang, Te-Kuei Fang, Chensheng Wu and Xiang Gao
Atomic Processes, Including Photoabsorption, Subject to Outside Charge-Neutral Plasma
Reprinted from: *Atoms* 2022, 10, 16, https://doi.org/10.3390/atoms10010016 202

Carlos Allende Prieto
The Shapes of Stellar Spectra
Reprinted from: *Atoms* 2023, 11, 61, https://doi.org/10.3390/atoms11030061 221

About the Editors

Sultana N. Nahar

Prof. Sultana N. Nahar is an atomic astrophysicist in the Department of Astronomy of the Ohio State University, USA. She studies the radiative and collisional atomic processes of photo-excitation, photoionization, electron-ion-recombination, and electron impact excitation that are dominant in astrophysical plasma. She uses Breit–Pauli R-matrix method to reveal the details of complex resonances in the photoionization process that can interpret spectroscopic phenomena. This requires large-scale computations using the high-performance computers of the Ohio Supercomputer Center. Her extensive research has identified various characteristic features in the resonances of photoionization of ground and excited, low and high states and focused on the high-energy region of the ion. These features are demonstrated in the book "Atomic Astrophysics and Spectroscopy" (A.K. Pradhan and S.N. Nahar, Cambridge University Press, 2011). She is a member of the International Opacity Project and the Iron Project, which study atomic processes in astrophysical modeling with high accuracy. She and her group at Ohio State University have been studying the Sun and its plasma opacity, which determines radiation absorption, elemental abundances, etc., making considerable advances. Her study also includes the biosignature spectroscopy of exoplanets and interpretation of the heavy-element spectra of kilonova observed following the detection of gravitational waves in the merger of two neutron stars. She and her team are also studying biomedical applications of high-energy x-rays that are similar to the study of black holes. The extensive atomic data that she has computed are available at the NORAD-Atomic-Data database at Ohio State University. She is actively engaged globally in various STEM education and research programs, as well as delivering online astrophysics course with computational workshops. She is a founder of the International Society of Muslim Women in Science, which has members in over 30 countries.

Guillermo Hinojosa

Dr. Hinojosa is an atomic and molecular physicist at the Institute of Physical Sciences of the National Autonomous University of Mexico, where he also undertakes vigorous teaching duties. Dr. Hinojosa participated in the early stages of the photoionization experiment installed in the 10.0.1 end-station at the Advanced Light Source (ALS) synchrotron at Lawrence Berkeley National Laboratory. This international collaboration was followed by the publication of many referential datasets in the field and gave Dr. Hinojosa the opportunity to collaborate with some of the world's top researchers. After winning "beam-time" by applying to the ALS's general users proposal open competitive examination system, he became the first Mexican principal investigator, where he and his international team of scientists carried out several photoionization experimental campaigns for three years. The complexity of the interpretation required by the phenomenon of photoionization lead Dr. Hinojosa to use Dr. S. N. Nahar's method, which is based on first principles and large numbers of configurations, leading into a successful collaboration detailed in this Special Issue. More recently, Dr. Hinojosa made important contributions in the field of simple negative ions. For example, he discovered o the negative ion of methane and realized its quantum structure, a $CH_2^-:H_2$ molecular exciplex. Also, the development of a new method to measure the lifetimes of excited state anions at the nanoseconds scale yielded a lifetime of about 30 ns for the $(2p^2)^3P$ state of H^-.

Preface

Photoionization, a process occurring in the presence of a radiative source, is vital for studying astronomical objects, such as, stars, nebular plasma. Sultana N. Nahar from Ohio State University and Guillermo Hinojosa from the National Autonomous University of Mexico have co-edited a Special Issue titled "Photoionization of Atoms" for the journal *Atoms*. They provide a comprehensive overview of photoionization, covering its characteristics, methodologies, data availability, and applications in astrophysics.

The editors emphasize the complexity of photoionization, particularly due to resonances that can complicate its study. Close coupling (CC) approximation is known as a theoretical approach that calculates both background cross-sections and resonances. Earlier theoretical studies focused on background cross-sections and calculated resonances separately. The International Opacity Project conducted the first systematic studies incorporating resonances using CC approximation and the R-matrix.

This Special Issue includes significant contributions, such as Nahar and Hinojosa's paper presenting R-matrix photoionization cross-sections for Cl III, benchmarked against experimental data from the Advanced Light Source. Another paper discusses the photoionization spectrum of Cl VI measured at SOLEIL, revealing resonance structures related to the ground and excited states, with calculated cross-sections using the MCDF method showing better agreement with experiments than z-scaled values from the R-matrix results.

M. A. Baig reviews the measured photoionization of excited states of alkali atoms, employing a non-synchrotron multi-step laser excitation technique to study Rydberg resonances. Theoretical predictions for higher effective quantum number resonances using R-matrix methods face challenges due to their density and overlap. Gning and Sakhoa present high-n resonances of Cl I using a semi-empirical method that they have developed.

Bhatia introduces a hybrid method for more accurate background cross-sections, which can be applied to various processes, including the photoionization of neutral helium. Chang et al. explore photoabsorption in a plasma environment, yielding good agreement with measured values.

Nahar's paper presents the astrophysical spectra of lanthanide ions, relevant for phenomena like neutron star mergers. The interplay between photoionization and electron-ion recombination is crucial for modeling nebular atmospheres, as discussed by D. J. Hillier. A. Pradhan highlights the importance of photoionization in solar opacity, particularly for addressing discrepancies between observed and predicted opacities.

C. Prieto quantitatively evaluates photoionization's role in stellar atmospheres. Detailed photoionization cross-sections with resonances are accessible through databases like TOPbase, TIPbase, and NORAD-Atomic-Data. The papers collectively provide valuable insights into photoionization, contributing to a broader understanding of this fundamental process in nature.

Sultana N. Nahar and Guillermo Hinojosa
Guest Editors

Article

Experimental and Theoretical Study of Photoionization of Cl III

Sultana N. Nahar [1], Edgar M. Hernández [2], David Kilcoyne [3,†], Armando Antillón [4], Aaron M. Covington [5], Olmo González-Magaña [4], Lorenzo Hernández [4], Vernon Davis [5], Dominic Calabrese [6], Alejandro Morales-Mori [4], Dag Hanstorp [7], Antonio M. Juárez [4] and Guillermo Hinojosa [4,*]

1 Department of Astronomy, The Ohio State University, Columbus, OH 43210, USA
2 Centro de Investigación en Ciencias-IICBA, Universidad Autónoma del Estado de Morelos, Cuernavaca 62209, Mexico
3 The Advanced Light Source, Lawrence Berkeley National Laboratory, CA 94720, USA
4 Instituto de Ciencias Físicas, Universidad Nacional Autónoma de México, A. P. 48-3, Cuernavaca 62251, Mexico
5 Physics Department, University of Nevada, Reno, NV 89557-0220, USA
6 Department of Physics, Sierra College, Rocklin, CA 95677, USA
7 Department of Physics, University of Gothenburg, SE-412 96 Gothenburg, Sweden
* Correspondence: hinojosa@icf.unam.mx
† Deceased.

Abstract: Photoionization of Cl III ions into Cl IV was studied theoretically using the ab initio relativistic Breit–Pauli R-matrix (BPRM) method and experimentally at the Advanced Light Source (ALS) synchrotron at the Lawrence Berkeley National Laboratory. A relative-ion-yield spectrum of Cl IV was measured with a photon energy resolution of 10 meV. The theoretical study was carried out using a large wave-function expansion of 45 levels of configurations $3s^23p^2$, $3s3p^3$, $3s^23p3d$, $3s^23p4s$, $3s3p^23d$, and $3p^4$. The resulting spectra are complex. We have compared the observed spectrum with photoionization cross sections (σ_{PI}) of the ground state $3s^23p^3(^4S^o_{3/2})$ and the seven lowest excited levels $3s^23p^3(^2D^o_{5/2})$, $3s^23p^3(^2D^o_{3/2})$, $3s^23p^3(^2P^o_{3/2})$, $3s^23p^3(^2P^o_{1/2})$, $3s3p^4(^4P_{5/2})$, $3s3p^4(^4P_{3/2})$ and $3s3p^4(^4P_{1/2})$ of Cl III, as these can generate resonances within the energy range of the experiment. We were able to identify most of the resonances as belonging to various specific initial levels within the primary Cl III ion beam. Compared to the first five levels, resonant structures in the σ_{PI} of excited levels of $3s3p^4$ appear to have a weaker presence. We have also produced combined theoretical spectra of the levels by convolving the cross sections with a Gaussian profile of experimental width and summing them using statistical weight factors. The theoretical and experimental features show good agreement with the first five levels of Cl III. These features are also expected to elucidate the recent observed spectra of Cl III by Sloan Digital Scan Survey project.

Keywords: relativistic multi-configurations interaction; R-matrix method; photoionization; phosphorus isoelectronic ion Cl IV; merged-beams technique

1. Introduction

The Io satellite plasma torus in Jupiter is a ring-shaped cloud of plasma in which Cl III ions were discovered by the Far Ultraviolet Spectroscopic Explorer (FUSE) [1,2] and also by the Galileo probe [3]. In addition, the CORONAS-F spacecraft has detected high-charge states of chlorine in coronal solar flares [4]. The apparently large abundance of chlorine in the interstellar medium is an unresolved question in astrophysics [5] prompting the need for more investigations into the fundamental properties of this element. In general, measurements of photoionization spectra of ions of chlorine [6] have helped in benchmarking theoretical efforts such as those of the OPACITY [7–9], IRON [10], and NORAD [11] projects that have for several decades focused on understanding the interaction of photons with ions.

Photoionization and spectroscopic data on chlorine ions remain rare. Experimental efforts have provided lifetimes and oscillator strengths for specific transitions observed by FUSE [12] for Cl II ions. A multi-configuration Hartree–Fock approach [13] showed a strong correlation between Rydberg series and perturbed states for the case of singly-ionized chlorine. An experimental effort investigating the photoionization of Cl II from threshold to 28 eV offered Rydberg series resonance energies and cross sections [6]. This investigation was followed by large-scale Dirac Coulomb R-matrix calculations [7,14], rendering an identification for the observed spectrum.

In the present study, we present a measurement of the relative ion yield resulting from the single-photon photoionization of Cl III. The data consists of the effective current of the Cl IV ions as a function of the photon beam energy. Photoionization can proceed via a direct process as in the following,

$$h\nu + Cl\ III \rightarrow e^- + Cl\ IV, \quad (1)$$

where $h\nu$ represents the photon (giving the background cross section), or through an intermediate autoionizing state at an energy belonging to a Rydberg series state,

$$Cl\ III + h\nu \rightleftharpoons (Cl\ III)^{**} \rightleftharpoons Cl\ IV + e^-, \quad (2)$$

which introduces a resonance in the cross section. The introduction of an intermediate state for resonances can be studied naturally by the close-coupling (CC) approximation and the R-matrix method. This report presents resonant features in the measured photoionization cross section of Cl III, studied theoretically using the relativistic Breit–Pauli R-matrix (BPRM) method; the experimental and theoretical results are each benchmarked against the other. Cl III is isoelectronic to phosphorous. For this reason, the photoionization study of atomic phosphorus by Berkowitz and collaborators [15] is relevant to the present work. We point out that the spectroscopy of elements isoelectronic to phosphorus is also relevant in astrobiology. Because phosphorus is one of the basic elements of life, its spectroscopic features are used to label phosphorus-containing exoplanetary atmospheres as possible life sustaining regions (e.g., Ref. [16]).

2. Theoretical Study with Breit–Pauli R-Matrix Method

The theoretical study was carried out using the relativistic Breit–Pauli R-matrix (BPRM) method with a large close-coupling (CC) wavefunction expansion that includes 45 levels of the core ion Cl IV. It is the core ion excitation that enables autoionizing resonances in photoionization cross sections. A brief outline of the method is given below.

In the CC approximation, the atomic system has (N+1) electrons and the core (or 'target') ion is an N-electron system interacting with the (N+1)th electron. The (N+1)th electron is bound or in the continuum depending on whether its energy E is negative or positive, respectively. The total wave function Ψ_E, of $SL\pi J$ symmetry, is expressed by an expansion of the form (e.g., Ref. [17])

$$\Psi_E(e^- + ion) = A \sum_i \chi_i(ion)\theta_i + \sum_j c_j \Phi_j, \quad (3)$$

where the core ion eigenfunction χ_i represents the ground and various excited states. The sum is over the number of states considered. The core is coupled with the (N+1)th electron function θ_i. The (N+1)th electron (of kinetic energy k_i^2) is in a channel labeled as $S_i L_i \pi_i J_i k_i^2 \ell_i (SL\pi J)$. A is the anti-symmetrization operator. In the second sum, the Φ_j are bound-channel functions of the (N+1)-electron system that provides the orthogonality between the continuum and the bound electron orbitals, and accounts for short-range correlations.

Substitution of $\Psi_E(e^- + ion)$ in the Schrödinger equation

$$H_{N+1}\Psi_E = E\Psi_E, \quad (4)$$

introduces a set of coupled equations that are solved using the R-matrix approach. This approach divides the space into an inner space of radius r_a that includes all short-range interactions, and an outer space up to infinity where the wavefunction is treated as Coulombic (e.g., Refs. [8,17–20]). The relativistic effects are included through Breit–Pauli approximation (e.g., Ref. [17]), in which the Hamiltonian is given by

$$H^{BP}_{N+1} = \sum_{i=1}^{N+1}\left\{-\nabla_i^2 - \frac{2Z}{r_i} + \sum_{j>i}^{N+1}\frac{2}{r_{ij}}\right\} + H^{mass}_{N+1} + H^{Dar}_{N+1} + H^{so}_{N+1}, \quad (5)$$

in Rydberg units. The relativistic correction terms are corrected by a mass term, $H^{mass} = -\frac{\alpha^2}{4}\sum_i p_i^4$, the Darwin term, $H^{Dar} = \frac{Z\alpha^2}{4}\sum_i \nabla^2(\frac{1}{r_i})$, and a spin-orbit interaction term, $H^{so} = Z\alpha^2 \sum_i \frac{1}{r_i^3}\mathbf{l}_i.\mathbf{s}_i$. The R-matrix Breit–Pauli (BPRM) approximation also includes parts of several two-body interaction terms, such as those without the momentum operators [17].

The solution of the BRPM approximation is a continuum wave function Ψ_F for an electron with positive energies (E > 0), or a bound state Ψ_B at a negative total energy (E ≤ 0). The complex resonant structures in the photoionization spectra are produced from channel couplings between continuum channels that are open ($k_i^2 > 0$), and ones that are closed ($k_i^2 < 0$), at electron energies k_i^2 corresponding to autoionizing states of the Rydberg series $S_i L_i J_i \pi_i \nu \ell$. ν is the effective quantum number of the series converging to excited core thresholds.

The photoionization cross section (σ_{PI}) is given by (e.g., Ref. [17])

$$\sigma_{PI} = \frac{4\pi^2}{3c}\frac{1}{g_i}\omega \mathbf{S}, \quad (6)$$

where g_i is the statistical weight of the bound state, ω is the incident photon energy, and $\mathbf{S}$ is the generalized line strength

$$\mathbf{S} = |<\Psi_j||\mathbf{D}_L||\Psi_i>|^2 = \left|\left\langle\psi_f\left|\sum_{j=1}^{N+1}r_j\right|\psi_i\right\rangle\right|^2. \quad (7)$$

Here Ψ_i and Ψ_f are the initial and final state wave functions, respectively, and $\mathbf{D}_L$ is the dipole operator in the length form.

3. Computation of Photoionization Cross Sections

BPRM computations were carried out through the package of R-matrix codes [20–22], which consists of several stages. The computation starts with the wave-function expansion of the core ion as the initial input. For this wave function, an atomic structure calculation was carried out using the code SUPERSTRUCTURE (SS) [23,24]. SS uses a Thomas–Fermi–Dirac–Amaldi potential and includes relativistic contributions in a Breit–Pauli approximation in which a full Breit interaction and parts of two-body correction terms are included. A set of 13 configurations of the core ion Cl IV was optimized to obtain the wave-function expansion for Cl III. The configurations, which include orbitals up to the $4d$, are $3s^23p^2$, $3s3p^3$, $3s^23p3d$, $3s^23p4s$, $3s^23p4p$, $3s^23p4d$, $3s3p^23d$, $3s3p^24s$, $3s3p^24p$, $3p^4$, $3p^33d$, $3p^34s$, and $3p^34p$. Table 1 presents 45 (the ground and 44 excited) fine-structure levels of Cl IV included in the wave-function expansion of Cl III. The calculated energies from SS are compared with observed values [25] (listed in the NIST [26] compilation). A comparison shows the agreement between SS and the observed values to within a few percent to a larger number, and some, e.g., $3p^4(^3P_{0,1,2})$, are over 10%. SS is not adjusted for any observed values. The optimization for energies and wavefunctions is carried out through the number of configurations and by varying the orbital wavefunctions with Thomas–Fermi scaling parameters. SS computes the radial wavefunction of all orbitals which remain the same while the energies change with the angular momenta of the states. The present set of energies is the resultant of efforts to reach an overall agreement between the calculated and measured values instead of focusing to improve any small set of energies, and the optimized set of configurations is selected for manageable computation by the R-matrix codes, which will retain all important physical characteristics of the atomic process. The accuracy of results provided by the R-matrix method is much improved over those of core ion energies since the method can handle a much larger set of configurations, e.g., 47 configurations of the core ion and interacting electron were used to compute the photoionization cross sections of Cl III. Hence, the effect of uncertainties in the core ion wavefunctions were reduced considerably by the R-matrix computations and resulted in a good agreement between the calculated and experimental σ_{PI}.

Table 1. The 45 levels of Cl IV that were included in the CC wavefunction expansion for Cl III. The calculated energies in Ry obtained from SUPERSTRUCTURE (SS) are compared with the measured values [25] available in the NIST [26] compilation table.

	Config	SLπ	2J	E (NIST, Ry)	E (SS, Ry)
1	$3s^23p^2$	^3P	0	0.0	0.0
2	$3s^23p^2$	^3P	2	0.004483	0.004553
3	$3s^23p^2$	^3P	4	0.012228	0.012601
4	$3s^23p^2$	^1D	4	0.125460	0.153621
5	$3s^23p^2$	^1S	0	0.296597	0.350797
6	$3s3p^3$	$^5S^o$	4	0.592325	0.587425
7	$3s3p^3$	$^3D^o$	2	0.93635	1.116700
8	$3s3p^3$	$^3D^o$	4	0.93666	0.911418
9	$3s3p^3$	$^3D^o$	6	0.93741	0.912132
10	$3s3p^3$	$^3P^o$	4	1.09585	1.090706
11	$3s3p^3$	$^3P^o$	2	1.09602	1.091088
12	$3s3p^3$	$^3P^o$	0	1.09625	1.091201
13	$3s^23p3d$	$^1D^o$	4	1.09625	1.201007
14	$3s3p^3$	$^3S^o$	2	1.482837	1.509883
15	$3s^23p3d$	$^3F^o$	4		1.482837
16	$3s^23p3d$	$^3F^o$	6		1.487126
17	$3s^23p3d$	$^3F^o$	8		1.492982
18	$3s3p^3$	$^1P^o$	2	1.51946	1.589771
19	$3s^23p3d$	$^3P^o$	4	1.65525	1.775908
20	$3s^23p3d$	$^3P^o$	2	1.65917	1.779791
21	$3s^23p3d$	$^3P^o$	0	1.66124	1.781748
22	$3s^23p3d$	$^3D^o$	2	1.70414	1.814810
23	$3s^23p3d$	$^3D^o$	4	1.70565	1.816602
24	$3s^23p3d$	$^3D^o$	6	1.70722	1.817951
25	$3s3p^3$	$^1D^o$	4		1.806554
26	$3s^23p3d$	$^1F^o$	6		1.980420
27	$3s^23p4s$	$^3P^o$	0	1.959461	1.987296
28	$3s^23p4s$	$^3P^o$	2	1.962772	1.991670
29	$3s^23p4s$	$^3P^o$	4	1.972602	1.999506
30	$3s^23p4s$	$^1P^o$	2	1.99981	1.970271
31	$3s^23p3d$	$^1P^o$	2		2.087061
32	$3s3p^23d$	^5F	2		2.115199
33	$3s3p^23d$	^5F	4		2.116611
34	$3s3p^23d$	^5F	6		2.118763
35	$3s3p^23d$	^5F	8		2.121696
36	$3s3p^23d$	^5F	10		2.125465
37	$3p^4$	^1D	4		2.147704
38	$3s3p^23d$	^5D	0		2.174180
39	$3s3p^23d$	^5D	2		2.174565
40	$3s3p^23d$	^5D	4		2.175339
41	$3s3p^23d$	^5D	6		2.176516
42	$3s3p^23d$	^5D	8		2.178121
43	$3p^4$	^3P	0	2.291576	2.042081
44	$3p^4$	^3P	2	2.293894	2.039031
45	$3p^4$	^3P	4	2.300008	2.032375

Photoionization cross sections were obtained with the inclusion of the radiation damping effects [22]. For precise positions of the resonances, the calculated core ion excitation energies were replaced by the available observed energies and the cross sections were shifted to match the measured ionization energies of the levels. These cross sections were compared directly with the measured cross sections for identification of the resonances of various levels to which they may belong. The cross sections were also convolved with a Gaussian profile of width of 10 meV using the program GUSAVPX [27] for comparison with the observed spectral features in the experiment. However, the width was varied in a few instances to avoid significant dissolution of resonances.

4. Experiment

A photon beam and a Cl III ion beam propagating in opposite directions were merged over a common collinear path. As a consequence of their interaction, ions from the Cl III ion beam are photoionized, yielding Cl IV. The resulting ions were separated from the parent ion beam and counted. Relevant parameters of both beams and their overlap were monitored.

The experiment was implemented in the now decommissioned ion-photon beam end-station at the Advanced Light Source (ALS) at Lawrence Berkeley National Laboratory. This technique was based on the merged-beams technique [28] and was described in more detail in previous measurements of the photoionization of Ne^+ [29] and Se^+ [30]. A short description of the experimental technique with some details pertinent to the present study is presented.

The Cl III ion beam was generated with an ECR ion source. This ion source had an insertion oven in which a small sample of $FeCl_3$ was evaporated and ionized in the chamber by collisions with the ECR electrons. Ions were then extracted from the ion source by a repulsive potential of 6 kV. These ions were mass-to-charge selected by a 90° bending electromagnet. The mass resolution of the electromagnet is sufficient to discern and therefore discard any possible contamination of the Cl III ($m/q = 17.73$) ion beam with O^+ ($m/q = 16$).

Next, a set of electrostatic lenses focused the ion beam into the center of a voltage-biased cylindrical mesh located in a chamber of the experiment where the ions were merged with photons from the ALS (for the ions to reach this chamber, they were first deflected by a set of 90° spherical electrostatic bending plates). In this chamber, Cl III ions were ionized by the counter-propagating photons. In the following stage, a second electromagnet served to separate the Cl IV ions from the parent Cl III ions. Cl IV ions were counted using electrons emitted from an emission-biased metal plate located in front of a channel electron multiplier plate. A typical Cl III ion beam current was 50 nA.

The photon beam was generated by a 10-cm-period undulator installed in the trajectory of a constant current of 0.5 A in the 1.9 GeV synchrotron storage ring, producing a collimated photon beam of width less than 1 mm and with a divergence less than 0.05°. This photon beam was then incident onto a grazing spherical-grating monochromator. The photon energy was scanned by rotating the grating and translating the exit slit of the monochromator while adjusting the undulator gap to maximize the photon beam intensity.

The photon flux was measured with a calibrated silicon photodiode [30] with a 5% uncertainty. A precision current meter provided a normalization signal to the data acquisition system. The photon beam was mechanically pulsed using a controlled chopper-wheel to separate the Cl IV signal ions produced by photoionization from those produced by collisions with the residual gas in the apparatus.

In this experiment, absolute cross sections were not measured. The reported measured intensity is therefore arbitrary. However, the signal was normalized to the ion beam current and to the photon beam current. The photon beam current was corrected by the number of photons or via the known quantum efficiency of the photon silicon diode detector as a function of the photon energy. Due to the presence of higher-order components in the photon beam [31], the error in the intensity can be large. However, this experiment is concerned primarily with the energy resonances, so the corrections to the photon beam intensity will not be discussed here.

The photon energy was nominally established by the rotating-grating control system. Further calibration with an ionization gas-cell increased the accuracy of the measured photon energy. This calibration was accomplished by using the ionization energies for He [32] and Kr [33] as references in the energy interval from 21.218 eV to 63.355 eV. We estimate that the photon energy error, from this technique, has a maximum value of ±10 meV. This same energy calibration technique was employed for the single-photon photoionization spectrum measurement of Cl$^+$ [6], achieving an agreement with the NIST data base [26] out to three decimal places (in eV).

5. Results

We present the photoionization spectrum of Cl III with resonant features (as defined in the introduction) studied both experimentally and theoretically, and benchmarked with each other. The measured spectrum is the combined spectra of the ground and low-lying excited levels of Cl III that can be present in the primary beam. The theoretical spectra are obtained for each individual low-lying level and are compared with the measured one for identification of features. The results are illustrated and discussed below.

The observed spectrum, shown in Figure 1, was measured in energy-intervals of 1 eV. Each individual spectrum was overlapped with its neighbors by 0.5 eV. All the individual scans were then merged into a single spectrum after normalization. This method was employed in the early stage of the experiment and was implemented to reduce the effects of possible mechanical backlash from the monochromator positioning mechanism. The process of integrating the pieces of the spectrum was conducted by grouping the individual bins of spectra when variations among them were very small, so that their normalization factors were close to one. Whenever needed, this process of "gluing" also included small energy shifts to overlap observed resonances rather than merely combining the spectra. The overall photon energy error caused by this procedure was estimated to be no greater than ±8 meV. Therefore, considering a ±10 meV additional uncertainty from the gas-cell energy calibration, the total photon energy uncertainty for this experiment is ±13 meV. The resulting measured ion-yield spectrum in Figure 1 shows several resonant series structures and a slow monotonic rise in the background.

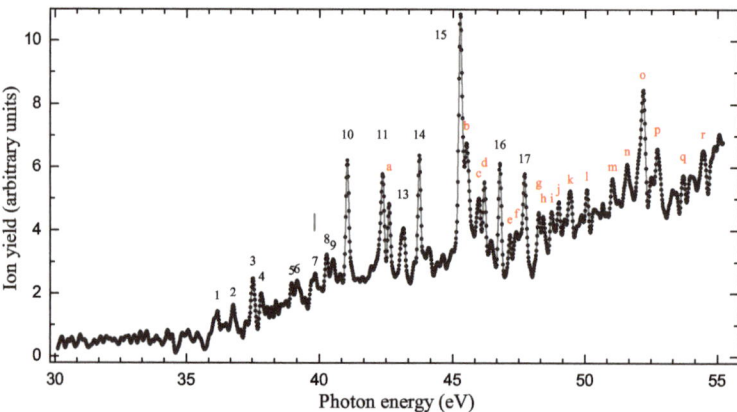

Figure 1. (Color online) Single-photon photoionization ion-yield spectrum of Cl III into Cl IV measured with a photon energy resolution of 10 meV. Measured data are (black) dots joined by a straight line. The vertical line (gray) corresponds to the ionization limit of 39.80 eV. Alpha (red) and numeric labels correspond to some of the most well-defined and prominent peaks listed in Tables 2 and 3. Labeling was arbitrarily chosen to improve clarity.

Figure 2 presents the theoretically-calculated σ_{PI} of the eight lowest levels of Cl III in various panels, ground (b) $3s^23p^3(^4S^o_{3/2})$ and excited (c) $3s^23p^3(^2D^o_{5/2})$, (d) $3s^23p^3(^2D^o_{3/2})$,

(e) $3s^23p^3(^2P^o_{3/2})$, (f) $3s^23p^3(^2P^o_{1/2})$, (g) $3s3p^4(^4P_{5/2})$, (h) $3s3p^4(^4P_{3/2})$, (i) $3s3p^4(^4P_{1/2})$ while (panel a) presents the measured spectrum. The eight levels were selected because they can generate resonances within the energy range of the experiment. Each panel shows complex resonances formed from the Rydberg series of autoionizing levels belonging to the core ion excitations. The complexity arises from overlapping of the series of resonances. However, in each panel, one can identify the origin of resonances and features of the levels that are seen in the observed spectrum. The arrows in the panels point to identifiable resonances in the observed spectrum, as listed in Tables 2 and 3. It can be seen that there are resonances in the observed spectrum that appear in σ_{PI} of multiple levels. The low energy resonances of the 3 levels of $3s3p^4(^4P)$, $j = 1/2, 3/2, 5/2$, are much stronger in the predicated σ_{PI} compared to the peaks in the observed σ_{PI}. This can be explained by the low abundances of these excited levels in the experimental beam in the low energy region.

Table 2. Cl^{2+} single-photon photoionization resonance energies E_0 and widths derived from Gaussian fits to the resonant peaks. Labels correspond to those in Figure 1.

E_0 (eV)	Width (eV)	Label
42.605	0.103	a
45.514	0.182	b
45.963	0.135	c
46.180	0.096	d
47.147	0.102	e
47.391	0.104	f
48.255	0.072	g
48.422	0.096	h
48.740	0.139	i
49.000	0.098	j
49.425	0.256	k
50.060	0.105	l
51.034	0.148	m
51.586	0.101	n
52.177	0.126	o
52.736	0.122	p
53.715	0.237	q
54.450	0.234	r

Table 3. Cl^{2+} single-photon photoionization resonance energies E_0 and widths derived from Gaussian fits to the resonant peaks. Labels correspond to those in Figure 1.

E_0 (eV)	Width (eV)	Label
36.161	0.069	1
36.771	0.112	2
37.521	0.128	3
37.846	0.109	4
38.966	0.129	5
39.165	0.106	6
39.829	0.072	7
40.269	0.128	8
40.506	0.148	9
41.038	0.101	10
42.354	0.118	11
43.119	0.236	13
43.735	0.093	14
45.280	0.113	15
46.764	0.115	16
47.706	0.110	17

To reproduce the observed features, the calculated single-photon photoionization cross sections of each level were convolved with a Gaussian profile of (experimental beam) width of 10 meV. No percentage contributions were considered since the abundances are not known. For each LS state, the convolved σ_{PI} of its component fine structure levels were added with statistical weight factors and then all the cross sections were summed together. The resultant spectra are presented in Figure 3. While (panel a) presents the measured cross sections, (panel b) presents the summed features from the convoled σ_{PI} of the five lowest initial levels of states $^4S°$, $^2D°$, and $^2P°$, and (panel c) corresponds to the summed σ_{PI} of the three ^4P initial levels. The same set of arrows pointing identifiable resonances are also displayed here. The overall agreement for the existences of the resonances between the predicted and observed spectra, except for the heights of the resonances, is illustrated much better in this figure. Almost all observed resonances are accounted for in the predicted cross sections.

The measured cross sections show a monotonic rising background in Figure 1, which appeared to be an artifact of the measurement technique or of the normalization procedure. This trend can be explained to be real. We used the same normalization procedure which is identical to that used for several other measurements in which this tendency was not observed. Of particular relevance is the case of the photoionization of Cl II [6,7] for which the data were measured under the same experimental conditions as of Cl III. The rising trend was not observed in that experiment. The origin in the present spectrum is not obvious from the calculated cross sections of individual levels in Figure 2. However, the background of the summed cross sections convolved with the measured width in Figure 3 predicts such a trend, thus confirming the validity of this feature.

The present experimental photoionization spectra is comprised of the superposition of a non-resonant background signal and resonant structures. Resonant peaks are normally assigned to a Rydberg series. A Rydberg series consists of a progression of resonances at energies $E_\nu = E_c - z^2/\nu^2$ where ν is the effective quantum number (although often designated by the principal quantum number n); z is the charge of the ion, and E_c is the excitation energy of the core ion to which the Rydberg series belongs. The intensities and shapes of resonances belonging to a series are also correlated. Our criteria in assigning a Rydberg series consists of relating at least four resonant peaks by energy and by intensity. In addition, the lower peak principal quantum number n should be $n > 3$ (the ground state configuration of Cl III is $3s^23p^3$). In the present measurements, resonant structures are intermingled with strong superposition of various Rydberg series belonging to closely-lying core ion excitation levels. No particular Rydberg series of resonances were identified under the required criteria.

For the purpose of identification of features in the spectrum, we use the theoretical cross sections in Figure 2 as a guide. As mentioned above, Cl III ions were generated (in this experiment) with an ECR ion source. This ion source produces undetermined populations of excited electronic states in the ion beam. Hence, the relative populations of excited states in the ion beam cannot be established, although their presence can be inferred. For instance, the lower panels of Figure 2 show that the zero offset of the relative ion yield of Figure 1 and the structure below 42 eV can be attributed to excited Cl III in the states $(3s3p^4)^4P_{1/2,3/2,5/2}$; $(3s^23p^3)^2P°_{1/2,3/2}$ and $^2D°_{3/2,5/2}$. By matching the positions of the observed resonances and relevant shapes with the calculated ones in Figure 2, we can identify the initial states originating most of the energy resonances in the spectrum, and thus, benchmark the experiment and theory with each other.

The peak at 41.038 eV (labeled as 10 in Figure 1 and in Table 3) is correlated to the Cl III $^2P°_{1/2,3/2}$ initial state. The most prominent resonance in the experimental spectrum at 45.280 eV (labeled as 15 in Figure 1 and in Table 3) can be correlated to the interference of resonant structures from initial Cl III ions in states $^4S°$ (panel b) and $(3s3p^4)^4P$ (panels g to i). Although the intensity does not appear to correspond, it should be noted that the spectrum of Figure 1 is relative. In addition, the indeterminacy of the relative contributions from all possible states could be an explanation for this disagreement. This is, for example, the case for the lower-energy interval of the spectrum where resonance intensities predicted by the theory do not match the experimental spectrum.

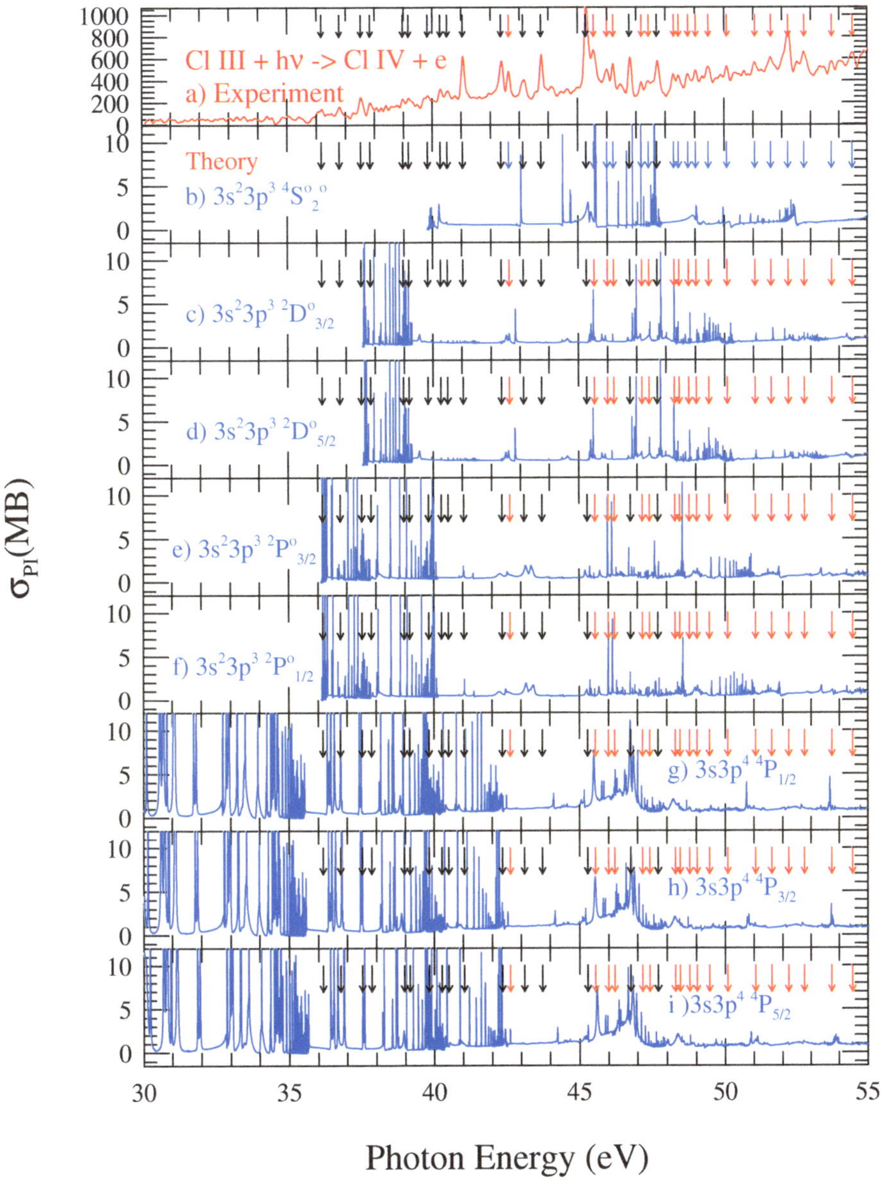

Figure 2. (Color online) Calculated single-photon photoionization ion-yield spectrum of Cl III into Cl IV (blue) in panels (**b–i**). The top panel (red) corresponds to the measured spectrum of Figure 1. Subsequent panels present the σ_{PI} of the eight lowest levels of Cl III, ground (**b**) $3s^23p^3(^4S^o_{3/2})$ and excited (**c**) $3s^23p^3(^2D^o_{5/2})$, (**d**) $3s^23p^3(^2D^o_{3/2})$, (**e**) $3s^23p^3(^2P^o_{3/2})$, (**f**) $3s^23p^3(^2P^o_{1/2})$, (**g**) $3s3p^4(^4P_{5/2})$, (**h**) $3s3p^4(^4P_{3/2})$, (**i**) $3s3p^4(^4P_{1/2})$. The first five levels (**b–f**) show contributions to the observed spectrum by the presence of resonant structures while the three excited levels of $3s3p^4(^4P)$ (**g–i**) show lesser contributions.

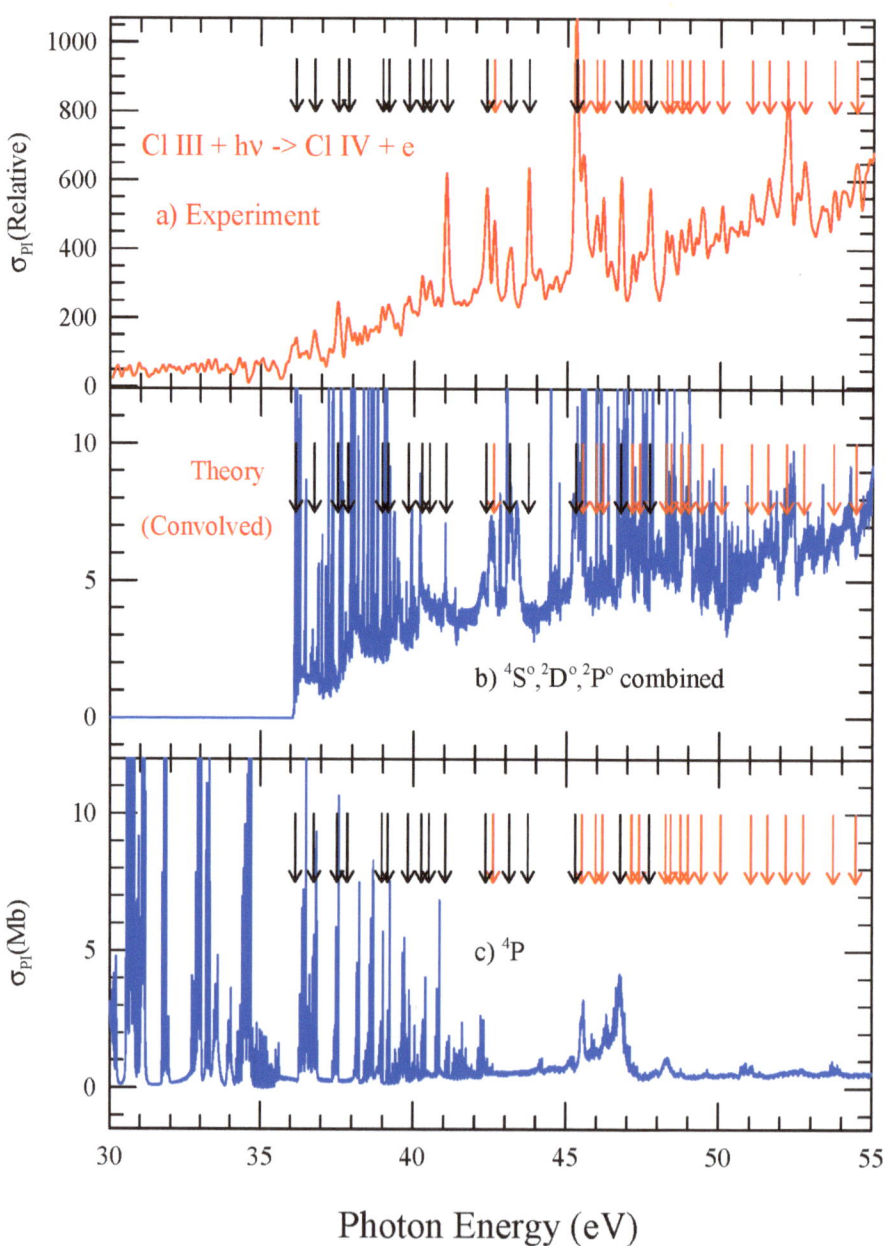

Figure 3. (Color online) Calculated single-photon photoionization (convolved with the Gaussian profile of the experimental beam-width) ion-yield spectrum of Cl III into Cl IV (blue). The top panel corresponds to the measured spectrum of Figure 1 (red). Panel (**b**) presents the summed features from the convolved σ_{PI} of the lowest five initial levels of states $^4S°$, $^2D°$ and $^2P°$ only. Panel (**c**) corresponds to the sum σ_{PI} of the three ^4P initial levels only.

In the present theory, the experimental resolution was simulated by a convolution of the theoretical data with a Gaussian function. Gaussian is an approximate profile. The width of the convolution function was set to 10 meV, but was also reduced in some energy regions to avoid dissolution of some resonances and match the experimental spectrum. These could explain some differences in the resonances intensities. The observed resonance at about 43.8 eV is not found in the predicted spectra of any of the five levels. However, a weak structure can be seen close to the energy in the convolved sum of cross sections for ^4P in Figure 3 (the lowest panel). Apart from these considerations, the origin of this particular resonance is unknown.

Often, the predicted resonant lines are suppressed and smeared by the experiment beam width, which is not of a pure Gaussian shape. We note from the comparison of shapes in the observed (panel a) and predicted (panel b) of Figure 3 that the main contributors to the observed spectra came from the five lower-lying levels, $3s^23p^3(^4S^o_{3/2})$, $3s^23p^3(^2D^o_{5/2})$, $3s^23p^3(^2D^o_{3/2})$, $3s^23p^3(^2P^o_{3/2})$, and a relatively small contribution from the three levels of $3s3p^4(^4P_{1/2,3/2,5/2})$. Good agreement between the observed and the predicted combined spectra in (panel b) of Figure 3 is obtained.

6. Conclusions

A study of the single-photon photoionization of the Cl III ion was carried out with two sophisticated approaches; an experimental approach using the high-resolution ALS synchrotron at LBNL, and a theoretical approach using the powerful R-matrix method. In addition to its fundamental interest, this ion is important because it has been detected in the interstellar medium, and photoionization spectra can therefore provide information on the physical condition and chemical evolution of astrophysical objects. In this experiment, a relative-intensity ion-yield spectrum of Cl IV originating from the photoionization of Cl III was measured with a photon energy resolution of 10 meV. The resulting spectrum was complex, as a monotonically-increasing background signal with several overlapping resonant structures was found. To interpret the spectrum, a relativistic, close-coupling calculation using the Breit–Pauli R-matrix method was implemented. The resulting theoretical spectrum predicts most of the observed resonant structures and the observed increase in the non-resonant background.

Author Contributions: Conceptualization: G.H. and S.N.N.; Methodology: S.N.N. (theory). Software: S.N.N. (R-matrix codes and relevant codes for theoretical analysis). Resources: Ohio Supercomputer Center (computation), Advanced Light Source at Lawrence Berkeley National Laboratory (experiment). Investigation: all authors in the experimental team contributed equally to the experiment execution and analysis and data reduction. All authors have read and agreed to the published version of the manuscript.

Funding: G.H. acknowledges partial funding from UNAM-DGAPA-PAPIIT-IN107420. SNN acknowledges the Ohio Supercomputer Center grant for carrying out all computations. D.H acknowledges financial support from the Swedish Research Council (2016-03650 and 2020-03505).

Data Availability Statement: Photoionization cross section data are available online at NORAD-Atomic-Data website https://norad.astronomy.osu.edu/.

Conflicts of Interest: The authors declare no conflict of interest.

References

1. Feldman, P.; Ake, T.; Berman, A.F.; Moos, H. Detection of Chlorine Ions in the Far Ultraviolet Spectroscopic Explorer Spectrum of the Io Plasma Torus. *Astrophys. J.* **2001**, *554*, L123–L126. [CrossRef]
2. Feldman, P.D. A Spectroscopic Tour of the Solar System with FUSE. In *Astrophysics in the Far Ultraviolet: Five Years of Discovery with FUSE, ASP Conf. Series; Proceedings of the Conference, Victoria, BC, Canada 2–6 August 2004*; Sonneborn, G., Moos, H., Andersson, B.-G., Eds.; NASA/Goddard Space Flight Center: Greenbelt, MD, USA, 2006; Volume 348, p. 307.

3. Schneider, N.M.; Park, A.H.; Kuppers, M.E. Spectroscopic Studies of the Io Torus during Galileo Encounters: Remote Plasma Diagnostics and the Detection of Cl^{++}. In Proceedings of the 32nd Meeting of the Division for Planetary Sciences of the American Astronomical Society, Pasadena, CA, USA, 23–27 October 2000; Volume 32, p. 1047.
4. Sylwester, B.; Phillips, K.J.H.; Sylwester, J.; Kuznetsov, V.D. The Solar Flare Chlorine Abundance from RESIK X-ray Spectra. *Astrophys. J.* **2011**, *738*, 49. [CrossRef]
5. Moomey, D.; Federman, S.R.; Sheffer, Y. Revisiting the Chlorine Abundance in Diffuse Interstellar Clouds from Measurements with the Copernicus Satellite. *Astrophys. J.* **2012**, *744*, 174. [CrossRef]
6. Hernández, E.; Juárez, A.; Kilcoyne, A.; Aguilar, A.; Hernández, L.; Antillón, A.; Macaluso, D.; Morales-Mori, A.; González-Magaña, O.; Hanstorp, D.; et al. Absolute measurements of chlorine Cl^+ cation single photoionization cross section. *J. Quant. Spectosc. Radiat. Transf.* **2015**, *151*, 217–223. [CrossRef]
7. Nahar, S.N. Photoionization features of the ground and excited levels of Cl II and benchmarking with experiment. *New Astron.* **2021**, *82*, 101447. [CrossRef]
8. Seaton, M.J. Atomic data for opacity calculations. I. General description. *J. Phys. B* **1987**, *20*, 6363. [CrossRef]
9. Bizau, J.M.; Champeaux, J.P.; Cubaynes, D.; Wuilleumier, F.J.; Folkmann, F.; Jacobsen, T.S.; Penent, F.; Blancard, C.; Kjeldsen, H. Absolute cross sections for L-shell photoionization of the ions N^{2+}, N^{3+}, O^{3+}, O^{4+}, F^{3+}, F^{4+} and Ne^{4+}. *Astron. Astrophys.* **2005**, *439*, 387–399. [CrossRef]
10. Nahar, S. The IRON Project: Photoionization of Fe ions. *arXiv* **2018**, arXiv:1801.05410.
11. Nahar, S. Database NORAD-Atomic-Data for atomic processes in plasma (Nahar-OSU-Radiative-Atomic-Data). *Atoms* **2020**, *8*, 68. [CrossRef]
12. Schectman, R.M.; Federman, S.R.; Brown, M.; Cheng, S.; Fritts, M.C.; Irving, R.E.; Gibson, N.D. Oscillator Strengths for Ultraviolet Transitions in Cl II and Cl III. *Astrophys. J.* **2005**, *621*, 1159–1162. [CrossRef]
13. Tayal, S.S. Strong term dependence of wavefunctions and series perturbations in singly ionized chlorine. *J. Phys. At. Mol. Opt. Phys.* **2003**, *36*, 3239. [CrossRef]
14. McLaughlin, B.M. Photoionization of Cl^+ from the $3s^2 3p^4$ $^3P_{2,1,0}$ and the $3s^2 3p^4$ $^1D_2$, $^1S_0$ states in the energy range 19–28 eV. *Mon. Not. R. Astron. Soc.* **2016**, *464*, 1990–1999. [CrossRef]
15. Berkowitz, J.; Greene, J.P.; Cho, H.; Goodman, G.L. Photoionisation of atomic phosphorus. *J. Phys. B At. Mol. Phys.* **1987**, *20*, 2647–2656. [CrossRef]
16. Westphal, M.S.; Pradhan, A.K. Monte Carlo simulations of biophysical factors for viability of life in exoplanetary atmospheres. In Proceedings of the Workshop on Astrophysical Opacities, Kalamazoo, Michigan, 1–4 August 2017; ASP Conference Series 2018; Volume 515, pp. 249–252.
17. Pradhan, A.K.; Nahar, S.S.N. *Atomic Astrophysics and Spectroscopy*; Cambridge University Press: Cambridge, MA, USA, 2011.
18. Burke, P.; Robb, W. The R-matrix theory of atomic processes. *Adv. At. Mol. Phys.* **1976**, *11*, 143–214.
19. Berrington, K.A.; Burke, P.G.; Butler, K.; Seaton, M.J.; Storey, P.J.; Taylor, K.T.; Yan, Y. Atomic data for opacity calculations. II. Computational methods. *J. Phys. B* **1987**, *20*, 6379. [CrossRef]
20. Berrington, K.A.; Eissner, W.B.; Norrington, P.H. RMATRX1: Belfast atomic R-matrix codes. *Comput. Phys. Commun.* **1995**, *92*, 290–420. [CrossRef]
21. Nahar, S.N.; Pradhan, A.K. Unified treatment of electron-ion recombination in the close-coupling approximation. *Phys. Rev. A* **1994**, *49*, 1816–1835. [CrossRef] [PubMed]
22. Zhang, H.L.; Nahar, S.N.; Pradhan, A.K. Close-coupling R -matrix calculations for electron-ion recombination cross sections. *J. Phys. B* **1999**, *32*, 1459. [CrossRef]
23. Eissner, W.; Jones, M.; Nussbaumer, H. Techniques for the calculation of atomic structures and radiative data including relativistic corrections. *Comput. Phys. Commun.* **1974**, *8*, 270–306. [CrossRef]
24. Nahar, S.N.; Eissner, W.; Chen, G.X.; Pradhan, A.K. Atomic data from the Iron Project: LIII. Relativistic allowed and forbidden transition probabilities for Fe XVII. *Astron. Astrophys.* **2003**, *408*, 789–801. [CrossRef]
25. Martin, W.C.; Kaufman, V.; Sugar, J.; Musgrove, A. Preliminary Compilation of Wavelengths and Energy-Levels for the Spectra of Chlorine (Unpublished), 1992–1997. Available online: https://physics.nist.gov/asd (accessed on 1 July 2020).
26. Kramida, A.; Ralchenko, Y.; Reader, J.; NIST ASD Team. *NIST Atomic Spectra Database (ver. 5.9)*; National Institute of Standards and Technology: Gaithersburg, MD, USA, 2021. Available online: http://physics.nist.gov/asd (accessed on 1 July 2020). [CrossRef]
27. Nahar, S. Photoionization cross sections of O II, O III, O IV, and O V: Benchmarking R-matrix theory and experiments. *Phys. Rev. A* **2004**, *69*, 042714. [CrossRef]
28. Phaneuf, R.A.; Havener, C.C.; Dunn, G.H.; Muller, A. Merged-beams experiments in atomic and molecular physics. *Rep. Prog. Phys.* **1999**, *62*, 1143. [CrossRef]
29. Covington, A.M.; Aguilar, A.C.; Covington, I.R.; Gharaibeh, M.F.; Hinojosa, G.; Shirley, C.A.; Phaneuf, R.A.; Álvarez, I.; Cisneros, C.; Dominguez-Lopez, I.; et al. Photoionization of Ne^+ using synchrotron radiation. *Phys. Rev. A* **2002**, *66*, 062710. [CrossRef]
30. Esteves, D.A.; Bilodeau, R.C.; Sterling, N.C.; Phaneuf, R.A.; Kilcoyne, A.L.D.; Red, E.C.; Aguilar, A. Absolute high-resolution Se^+ photoionization cross-section measurements with Rydberg-series analysis. *Phys. Rev. A* **2011**, *84*, 013406. [CrossRef]
31. Müller, A.; Schippers, S.; Hellhund, J.; Holste, K.; Kilcoyne, A.L.D.; Phaneuf, R.A.; Ballance, C.P.; McLaughlin, B.M. Single-photon single ionization of W^+ ions: Experiment and theory. *J. Phys. B At. Mol. Opt. Phys.* **2015**, *48*, 235203. [CrossRef]

32. Domke, M.; Schulz, K.; Remmers, G.; Kaindl, G.; Wintgen, D. High-resolution study of $^1P^0$ double-excitation states in helium. *Phys. Rev. A* **1996**, *53*, 1424–1438. [CrossRef]
33. King, G.C.; Tronc, M.; Read, F.H.; Bradford, R.C. An investigation of the structure near the $L_{2,3}$ edges of argon, the $M_{4,5}$ edges of kryton and the $N_{4,5}$ edges of xenon, using electron impact with high resolution. *J. Phys. B* **1977**, *10*, 2479–2495. [CrossRef]

Disclaimer/Publisher's Note: The statements, opinions and data contained in all publications are solely those of the individual author(s) and contributor(s) and not of MDPI and/or the editor(s). MDPI and/or the editor(s) disclaim responsibility for any injury to people or property resulting from any ideas, methods, instructions or products referred to in the content.

Article

L-Shell Photoionization of Magnesium-like Ions with New Results for Cl^{5+}

Jean-Paul Mosnier [1,*], Eugene T. Kennedy [1], Jean-Marc Bizau [2], Denis Cubaynes [2,3], Ségolène Guilbaud [2], Christophe Blancard [4,5], M. Fatih Hasoğlu [6] and Thomas W. Gorczyca [7]

1. School of Physical Sciences and National Centre for Plasma Science and Technology (NCPST), Dublin City University, Dublin 9, Ireland
2. Institut des Sciences Moléculaires d'Orsay, UMR 8214, Rue André Rivière, Bâtiment 520, Université Paris-Saclay, 91405 Orsay, France
3. Synchrotron SOLEIL, L'Orme Des Merisiers, Saint-Aubin, BP 48, CEDEX, 91192 Gif-sur-Yvette, France
4. Commissariat à l'Énergie Atomique, DAM, DIF, 91297 Arpajon, France
5. Commissariat à l'Énergie Atomique, Laboratoire Matière en Conditions Extrêmes, Université Paris-Saclay, 91680 Bruyères le Châtel, France
6. Department of Computer Engineering, Hasan Kalyoncu University, Gaziantep 27010, Turkey
7. Department of Physics, Western Michigan University, Kalamazoo, MI 49008, USA
* Correspondence: jean-paul.mosnier@dcu.ie

Abstract: This study reports on the absolute photoionization cross sections for the magnesium-like Cl^{5+} ion over the 190–370 eV photon energy range, corresponding to the L-shell (2s and 2p subshells) excitation regime. The experiments were performed using the Multi-Analysis Ion Apparatus (MAIA) on the PLéIADES beamline at the SOLEIL synchrotron radiation storage ring facility. Single and double ionization ion yields, produced by photoionization of the 2p subshell of the Cl^{5+} ion from the $2p^63s^2\,^1S_0$ ground state and the $2p^63s3p\,^3P_{0,1,2}$ metastable levels, were observed, as well as 2s excitations. Theoretical calculations of the photoionization cross sections using the Multi-Configuration Dirac-Fock and R-matrix approaches were carried out, and the results were compared with the experimental data. The Cl^{5+} results were examined within the overall evolution of L-shell excitation for the early members of the Mg-like isoelectronic sequence (Mg, Al^+, Si^{2+}, S^{4+}, Cl^{5+}). Characteristic photon energies for P^{3+} were estimated by interpolation.

Keywords: photoionization; atomic data; inner-shell excitation; chlorine ion

1. Introduction

Photon-driven excitation and ionization along the magnesium isoelectronic sequence have long been of interest in experimental and theoretical investigations e.g., [1–18]. This is not surprising, given the nature of magnesium-like systems with their simple closed-subshell ground state configuration $2p^63s^2$. The $^1S_0$ ground state tends to make it a relatively stable ion configuration in many plasma environments; therefore, from a theoretical point of view, photoionization calculations should focus mainly on the effects of excited state configurations. However, this ideal picture largely fails in practice, as electron correlations, e.g., two-electron excitations, turn out to be extremely important in this system, leading to a significant departure from the simple one-electron excitation picture. Many early investigations concentrating on the first ionization threshold energy region highlighted the complexity of such electron interactions, along with consequent demands on appropriate theories, see e.g., [17] and the references therein. As short wavelength photon interactions lead to inner-shell excitations, which couple with multiple excitations of the valence electrons, photoionization calculations become even more challenging due to the increased number of unfilled (sub)shells. Reliable photoionization laboratory investigations of magnesium-like ions in the inner-shell excitation photon energy regime are therefore of continuing interest as they provide robust benchmarking opportunities for different theoretical approaches.

Experimental and theoretical investigations of the photoionization of free atomic ions are becoming ever more important, stimulated by the abundant data obtained from X-ray satellite observatories such as Chandra and XMM-Newton. Future space missions such as XRISM and Athena promise even greater sensitivity and higher spectral resolution. Optimizing the value of astrophysical spectral observations, and the concomitant provision of greater physical insights, requires improved theoretical modelling and simulation of astrophysical plasma environments. The latter rely on fundamental atomic data, such as those provided by ongoing systematic and complementary experimental and theoretical investigations of the interaction of ionizing photons with free atomic ions. This is evidenced by the recent growth in laboratory astrophysics initiatives and associated data banks [19–24].

In the absence of experimental results, much of the atomic photoionization data exploited in modelling ionized environments has to rely on atomic physics calculations. Investigations of isoelectronic magnesium-like ions (twelve bound electrons) provide one important avenue for the calibration of different theoretical approaches. Along the isoelectronic sequence from neutral magnesium through Al^+, Si^{2+}, P^{3+}, S^{4+} to Cl^{5+} ions, the increasing core charge leads to significant changes in the relative effects of the nuclear charge versus electron–electron interactions on the photoionization properties. The sequence provides an ideal case study to examine the ability of theoretical models to predict and interpret these changes, which involve level crossing and plunging configurations [17].

Plasma-based light source experiments were successful in recording resonances and ionization thresholds for early members of the magnesium sequence for both ground and excited states, see, for example [1,5–7,25].

The biggest step, however, in the systematic investigation of the inner-shell photoionization of positive ions in general has resulted from dedicated merged photon-ion beam facilities at synchrotron radiation storage rings, notably, Daresbury (UK), SuperACO and later SOLEIL (France), ASTRID (Denmark), ALS (USA), and PETRA III (Germany), see e.g., [26–30] and the references therein. The great advantage of such merged beam experiments is that they isolate specific ion stages and allow quantitative results to be obtained at high spectral resolution. This often results in absolute cross sections for different ionization channels, thereby providing the highest quality fundamental data for atomic and plasma modelling, as well as for benchmarking the most advanced atomic theories.

Absolute cross sections for L-shell (2s and 2p) photoionization along the magnesium series have been previously obtained through synchrotron radiation-based merged-beam experiments for Al^+ [9,31], Si^{2+} [10], S^{4+} [18], and Fe^{14+} [32], with the latter produced in an electron beam ion trap. These experiments have been complemented by various theoretical investigations. These investigations have interpreted the evolution of the observed resonances in the early part of the sequence, showing that considerable changes in the relative intensities and positions of resonances take place as the core charge increases along the sequence. The L-shell single and double photoionization cross sections of magnesium itself have been studied extensively using synchrotron radiation in high resolution experiments, but only on a relative scale, see [16] and the references therein.

The next member of the sequence is Cl^{5+}. Chlorine is an extremely reactive element, appearing in many gaseous, liquid, and solid molecular forms of considerable scientific and industrial significance. Similarly to all halogens, it is strongly oxidizing as it is one electron short of the neighboring inert gas configuration. Chlorine ions play key roles in many terrestrial and astrophysical environments; therefore, understanding their interaction with radiation is important [33–38].

In this article, we report the first study of L-shell photoionization of magnesium-like Cl^{5+} for photon energies between 190 and 370 eV, straddling the 2s and 2p inner-shell excitation regimes. It is interesting to compare the success of the Multi-Configuration Dirac Fock (MCDF) and the R-matrix theoretical treatments of Cl^{5+} with the success of the previous analogous treatments of S^{4+} [18]. Our results provide experimental absolute photoionization cross sections for Cl^{5+}, a comparison with the theoretical predictions of both the MCDF and the R-matrix approaches, as well as insights into the evolution along the isoelectronic sequence from Al^+ to Cl^{5+}. To the best of our knowledge, no such photoionization data for P^{3+} is available in the open literature. From the observed trends along the sequence, we estimated the values of some relevant characteristic energies for P^{3+}.

2. Experimental Details

The experimental results for Cl^{5+} reported in this study were obtained using the dedicated merged photon-ion apparatus, MAIA (Multi-Analysis Ion Apparatus), on the ultra-high resolution soft X-ray (10 eV to 1 keV photons) PLÉIADES beamline at SOLEIL. Well-characterized synchrotron photon and ion beams overlap, and the resulting photoionization ions are selectively measured, providing cross section information for the different ionization channels. An advantage of working with ions over neutral species is that the number density in the overlap region can be determined. This allows the absolute cross sections to be measured. A comprehensive description of MAIA, and how absolute photoionization cross section results may be determined, is provided in [26]. Here we provide a shorter description to include the specific experimental parameters used for the Cl^{5+} investigations.

HCl gas was introduced into an Electron Cyclotron Resonance (ECR) ion source in order to generate chlorine ions. The Cl^{5+} ions were extracted and accelerated by an applied potential difference of −4 kV. They were then guided by a 90^0 bending magnet into the overlap region to meet the counter-propagating synchrotron radiation beam. An input power of 32 W at 12.36 GHz was used to optimize the production of Cl^{5+} ions. The magnetic filter enabled the selection of the most abundant $^{35}Cl^{5+}$ isotope. The ion current, measured in a Faraday cup placed after the magnet exit, was of the order of 10 µA. This beam was then focused and shaped to match the size of the counter-propagating photon beam. The remaining current of ions interacting with the photons was typically of the order of 300 nA. The length of the interaction region was determined by a 57 cm long tube placed in the path of the two beams and polarized at a voltage of −2kV. Three transverse profilers, located at the center and at both ends of this pipe, respectively, allowed measurement of the overlap of the photon beam and the ion beam; a Form Factor [26] of 32,000 m^{-1} was reached. The primary ion beam current was measured after the interaction region using a Faraday cup. The photoionized ions, either singly ionized Cl^{6+} or doubly ionized Cl^{7+}, were separated from the primary Cl^{5+} beam by a second dipole magnet, selected in speed by an electrostatic analyzer, and measured with a microchannel-plate detector coupled to a counter. A photon chopper was used to subtract the contribution of ions produced by collisions in the residual gas (background pressure of 1.5×10^{-9} mbar) to the Cl^{6+} or Cl^{7+} signal. The photon beam flux was monitored by a calibrated photodiode, and a typical current of ~100 µA at 195 eV photon energy and 150 meV bandwidth was measured. Knowing the photodiode current, the Cl^{5+} ion current, the form factor characterizing the overlap of the two beams, the length of the interaction region, the ion speed, and the efficiencies of the photodiode and channel-plate detectors, and recording the Cl^{6+} and the Cl^{7+} signals as functions of the photon energy, allowed the determination of the single and double absolute photoionization cross sections of the Cl^{5+} ions. During the Cl^{5+} experiments, SOLEIL was operated with a current of 450 mA. Circular left polarization, which delivers the highest flux in the photon range of interest, was used. The photon energy, corrected for the Doppler shift due to ion velocity, was calibrated using a gas cell containing argon, where the $2p_{3/2} \rightarrow 4s$ transition in argon at 244.39 eV [39] was measured. The energy uncertainties were of the order of 20 meV, but varied depending on the resonance (see tables). The total uncertainty of the measured cross sections was estimated to be not greater than 15%, and was mostly due to the combined effects of the inaccuracy of the determination of the beam overlaps (the form factor), the efficiency of the detector, and the photon flux.

The heating microwaves interacted directly with the electrons at the ion source through the ECR mechanism, and a complicated range of processes involving energetic electrons was responsible for producing the ions. The detailed electron energy distribution was not known and the ECR ion source plasma was not in equilibrium [40]. A well-recognized result is that in the merged beam technique, the ions in the interaction region can often exist in excited states as well as the ground state [26–30]. The population of excited levels depends on their lifetimes and whether they survive the journey from the source into the interaction region. In general, this can be the case for metastable levels, some of which can have relatively long lifetimes. We will see below how the issue of excited states for the specific case of magnesium-like ions can be satisfactorily resolved.

3. Theoretical Details

For the recent analogous investigations of photoionization of S^{4+}, we used two quite disparate theoretical approaches—MCDF and R-matrix theory [18]. We used the same methods to predict and interpret the new Cl^{5+} results presented here. While long-established and used to calculate many photoionization cross sections for a wide variety of ion species, both methods are being continuously reviewed and improved. In the MCDF approach, the twelve-electron magnesium-like ion is tackled directly. The quality of the results critically depends on including the most relevant electron configurations, although this is subject to calculational limitations. The R-matrix approach treats the magnesium-like challenge as a problem of an incident electron scattering off the eleven-electron sodium-like ion. Therefore, the quality of the final overall results depends on the accuracy of the intermediate target description. In many cases, some of the thresholds of the target system are already known, and this information can be used to help in the R-matrix approach. This was indeed the case for S^{4+} [18], but not for Cl^{5+}. It is therefore interesting to see how the R-matrix code works for the latter when compared with the former. Both length and velocity gauge calculations were carried out using both approaches, and very satisfactory agreement was shown each time. Consequently, for the sake of brevity, only the length gauge results are presented in this paper.

Because of the similarity of the computations for the Cl^{5+} system with those previously carried out for the S^{4+} ion [18], we provide only brief descriptions of the R-matrix and MCDF calculations. In order to focus on the problem at hand, viz., the L-shell photoionization of Cl^{5+} and its behavior along the beginning of the magnesium sequence, we do not provide lengthy mathematical descriptions to support the fundamental aspects of the theoretical descriptions. Only the most important details are provided, i.e., those specific to the case of Cl^{5+}. For additional information on the more fundamental and mathematical aspects of the MCDF and R-matrix theories, see the following recent references that will provide didactic descriptors as well as extensive lists of anterior references, namely, [41–43] for MCDF and [44–47] for R-Matrix.

3.1. R-Matrix Calculations

The calculational convenience of transitioning from Mg-like S^{4+} to Cl^{5+} is that the electronic orbital and configurational descriptions are the same once proper Z-scaling is accounted for. Previous R-matrix calculations for S^{4+} have been described in detail [18,48,49]; the changes for the new calculations of Cl^{5+} simply require adding one more proton to the nucleus, while keeping the same number of electrons. Indeed, from simple Z-scaling of the radial coordinate $\rho = Zr$, the radial orbital $P(\rho) \to Z^{1/2} P(r)$, and the cross section $\sigma(Z) \sim \sigma(1)/Z^2$, the plots of the corresponding radial orbitals $P(r)$ (shown in Figure 1) and the resultant photoabsorption cross sections are both remarkably similar. The Z-scaling also naturally leads to a reduction in the electron correlation effect, manifested via the electron–electron repulsion $1/\left|\vec{r}_1 - \vec{r}_2\right|$ that scales as $1/Z$ and becomes less significant at higher Z. Therefore, Cl^{5+} is expected to show a more hydrogenic structure than S^{4+}. This behavior is seen in Figure 1, where both sets of orbitals exhibit very similar patterns, with a slightly greater localization of the orbitals near the nuclear core for Cl^{5+}.

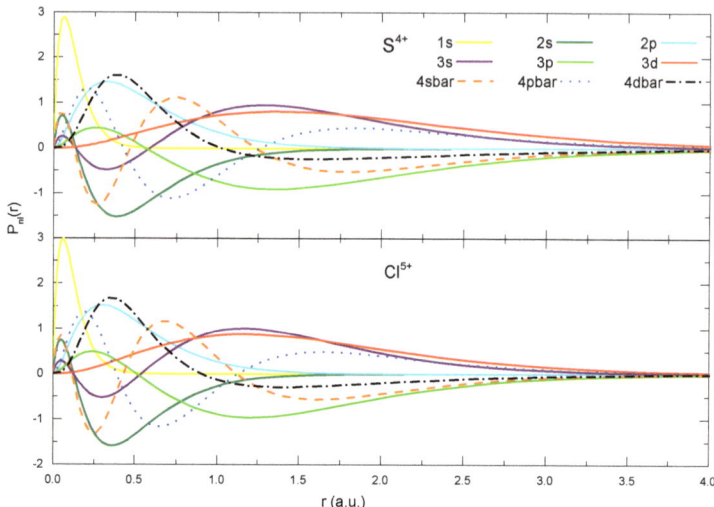

Figure 1. Plots of radial orbitals $P_{nl}(r)$ as a function of the radial coordinate r (in atomic units) used in the present R-matrix calculations. The same orbitals are plotted using the same colour code for both S^{4+} and Cl^{5+}.

For the present Cl^{5+} work, the main series considered within the R-matrix formulation are the $2s^22p^63l$ channels of Cl^{6+} and the $2s^22p^53s3l$ and $2s^22p^53s^2$ inner-shell channels. Note that again, as for S^{4+}, the $2s2p^63s3p$ channels are omitted to keep the computation tractable. This means that certain weak resonances, which are seen in the MCDF results, will be absent. Nevertheless, the main $2p \rightarrow nd$ Rydberg resonance series are well featured in the computed end results, and the weaker interspersed $2p \rightarrow (n+1)s$ series and $2s \rightarrow np$ series at higher photon energy are also included.

It can never be overstated that the primary source of uncertainty in any R-matrix calculation is due to a lack of convergence in the initial and/or final state energies, given that the calculation is based on a variational principle. This leads, in turn, to a corresponding non-convergence of the photon transition energies. In order to simplify the comparison, we present ab initio results for Cl^{5+} rather than using certain empirical energy information [50], knowing in advance that certain energy shifts must be aligned with the experimental results. Additionally, the R-matrix results are preconvolved with a resonance width of $\Gamma_{spectator} = 2.5 \times 10^{-3}$ Ryd for the spectator Auger decay process. This is to ensure the full resolution of the various infinite Rydberg series of otherwise narrowing resonances ($\Gamma_{participator} \sim 1/n^3$). $\Gamma_{participator}$ refers to the resonance width of the participator Auger decay in which the initially photon-excited electron engages in the Auger process. While in contrast, this does not occur in spectator Auger decay.

As a final note, the R-matrix method does include the Fano interference [51] between direct photoionization and indirect photoexcitation–autoionization. Therefore, asymmetric profiles can be revealed in general. This is in contrast to perturbative methods, such as the MCDF method, where the resonance is treated as a single final state, giving a Breit–Wigner (symmetric Lorentzian) resonance profile added incoherently to the (uncoupled) smooth background photoionization cross section. Hence, the asymmetrical profile of the experimental results of the $2s \rightarrow np$ Rydberg series, in particular, can only be reproduced by the R-matrix calculations here.

3.2. MCDF Calculations

An updated version of the MCDF code developed by Bruneau [42] has been used to compute the relevant photoexcitation and photoionization cross sections. Calculations were performed using a full intermediate coupling scheme in a jj basis set. The calculations were restricted to electric dipole transitions using both the Coulomb and the Babushkin gauges, which correspond to the velocity and the length form of the electric dipole operator in the non-relativistic limit, respectively [42]. Preliminary calculations were performed to compute a set of one-electron wave functions resulting from the energy minimization of the Slater transition state [52] using the following configuration set: [Ne]3snl, [Ne]3s, and [F]3s^2, and [He]2s2p^{6}3s^2, where n = 3,...7 and l = s, p, d, and where [He], [F], and [Ne] mean 1s^2, 1s^{2}2s^{2}2p^5, and 1s^{2}2s^{2}2p^6, respectively. This one-electron wave function set (OWFS) was used to calculate the 2s and 2p photoexcitation cross sections from the $^1S_0$ level of the [Ne]3s^2 ground configuration and from the $^3P_{0,1,2}$ metastable levels of the [Ne]3s3p configuration. With regard to the photoexcitation cross sections from the ground level, the following photo-excited configurations were retained: [F]3s^{2}3d, [F]3s3p^2, [F]3s^{2}n'l', and [He]2s2p^{6}3s^{2}nd, where n' = 4,..., 7 and l' = s, d. For the photoexcitation cross sections from the [Ne]3s3p $^3P_{0,1,2}$ metastable levels, the photo-excited configurations retained were: [F]3s^{2}3p, [F]3s3p3d, [F]3s3pn'l', [He]2s2p^{6}3s3p^2, and [He]2s2p^{6}3s3pn'p. The initial OWFS was also used to compute the autoionization rates for [F]3s^{2}3d and [F]3s^{2}3p/3s3p3d levels photoexcited from the [Ne]3s^2 $^1S_0$ ground level and the [Ne]3s3p $^3P_{0,1,2}$ metastable levels, respectively. The largest calculated autoionization rates were found for the [F]3s^{2}3d $^1P_1$ and $^3P_1$ excited levels, with corresponding Auger widths equal to 86 and 92 meV, respectively. The Auger widths for the [F]3s^{2}3p excited levels were all lower than 3.6 meV. The direct 2p and 3s photoionization cross sections were also calculated using the initial OWFS for the [Ne]3s^2 $^1S_0$ and [Ne]3s3p $^3P_{0,1,2}$ initial levels. The 3p photoionization cross sections were also computed for the [Ne]3s3p $^3P_{0,1,2}$ levels.

4. Results

4.1. L-Shell Photoionization of Cl^{5+}

This section presents the results for the single and double ionization of Cl^{5+} ions over the photon energy range corresponding to the L-shell excitations (both 2s and 2p). Figure 2 shows a schematic energy level diagram indicating some of the main ionization pathways due to photon absorption in the L-shell excitation regime. In particular, it shows excitation from the ground 2p^{6}3s^2 ^{1}S, and the metastable 2p^{6}3s3p ^{3}P levels which lie just over 12 eV above the ground state (the vertical upward lines indicate photoabsorption). The final inner-shell excited states shown lying above ~211 eV then give rise to Auger decay resonances in the photoionization cross section. These can interact with the underlying direct photoionization process, and the downward dotted arrows illustrate the ensuing production of either Cl^{6+} or Cl^{7+} ions due such non-radiative decay process(es) of the original 2p vacancy.

Figure 3 shows the measured photoionization cross section over the photon energy region from 190 to 370 eV. The 285–320 eV photon region was not scanned as resonances were not expected in this interval. All cross sections between 190 and 285 eV were measured in the very dominant single ionization (SI) channel, while all the cross sections above 326 eV were measured in the double ionization (DI) channel (which becomes the dominant channel above the 2p^{-1} ionization limit). We also show the results of the ab initio MCDF and R-matrix calculations where the theoretically predicted resonances are folded with the experimentally determined bandpasses for the different spectral regions. Because of the weakness of the resonances near 200 eV and those above 320 eV, their cross sections (both experimental and theoretical) were multiplied by a factor of ten, which accounts for the apparent vertical displacements of the cross sections on the right-hand side of Figure 3.

We first discuss how we estimated the contribution of metastable levels to the overall photoionization cross section. We then analyse the experimental cross section in comparison with the MCDF and R-matrix predictions in separate contiguous energy regions corresponding to changing photon interaction processes and taking into account the contribution due to metastable levels.

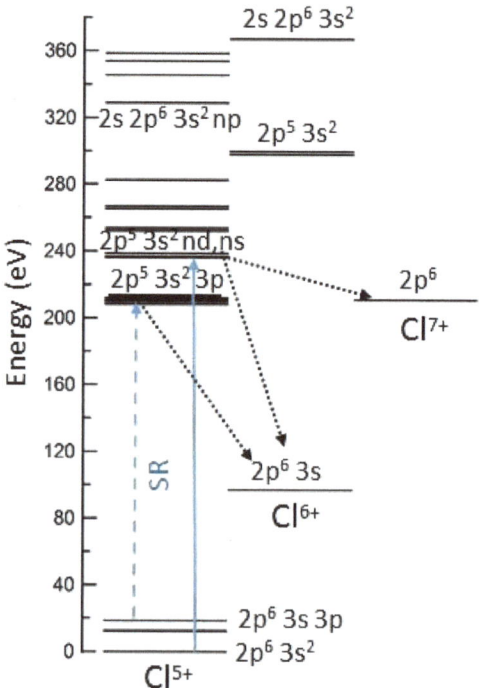

Figure 2. Schematic energy level diagram for Cl^{5+} showing excitation energies and ionization thresholds. The vertical solid and dashed lines indicate absorption pathways for synchrotron radiation (SR) photons, starting from the ground level $3s^2$ ^1S and the metastable $3s3p$ ^3P levels, respectively. The dotted lines show how the inner-shell excitations lead to single ionization (Cl^{6+}) or double ionization (Cl^{7+}) channels.

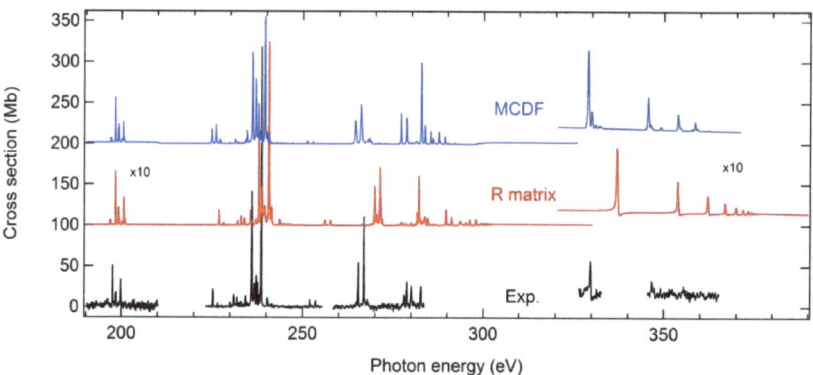

Figure 3. Photoionization cross section data, measured (black trace) and calculated using the R-matrix and MCDF theories (red and blue traces, respectively) for Cl^{5+} for photon energies lying between 190 and 370 eV. The experimental data (labelled Exp.) are the single ionization below the photon energy of 300 eV and the double ionization above the photon energy of 300 eV. The region around 200 eV corresponds with 2p excitations from the metastable $2p^63s3p$ ^3P levels. The resonances lying to higher photon energies arise predominantly from 2p (and 2s) excitations from the ground state $2p^63s^2$ $^1S_0$ level. The cross section values in both the 200 eV and above 320 eV regions are ×10 in order to enhance visibility.

4.1.1. Metastable Resonance Region

The important contribution of metastable levels to photoabsorption of plasmas containing magnesium-like ions has been recognized in some early dual laser produced plasma experiments [6]. These experiments showed that for Mg, Al$^+$, and Si^{2+}, the resonances arising from the $2p^63s3p\ ^3P$ levels appear in a separate photon energy region to the resonances arising from the ground state [5,7]. This feature allows the estimation of the metastable states population fractions by comparing the theoretically predicted cross sections from these states with the measured ones at those photon energies. Similar advantageous behavior was observed for S^{4+} [18], and was again the case for the current Cl^{5+} results.

In Figure 3, the photoionization resonances arising from the excitation from the ^3P metastable levels ($2p^63s3p\ ^3P \to 2p^53s^23p\ ^3S,\ ^3P,\ ^3D$) can be seen in the 195–201 eV photon energy region, well separated from the ground state resonances lying above 220 eV photon energy. We can estimate the relative populations of the metastable levels by comparing the experimental data with the MCDF and R-matrix predicted cross sections. Figure 4 shows details of the measured cross section due to the metastable levels and the comparative cross sections from both the MCDF (blue line) and the R-matrix (red line) calculations. The detailed comparison of the relative strengths of the individual resonances is determined by the relative populations of the individual $^3P_{0,1,2}$ J-levels. The summation of the cross section over all the resonances depends on the relative metastable to ground state ratio. The best fit of the theoretical with experimental data implies relative populations of 77% $^1S_0$ + 4% $^3P_0$ + 1% $^3P_1$ + 18% $^3P_2$. As for S^{4+}, the very low contribution of the $2p^63s3p\ ^3P_1$ state is readily explained by its E1 radiative decay lifetime of ~2.6 µs [52], which contributes to significantly repopulating the ground state by the time the fast (3.23×10^5 ms^{-1}) Cl^{5+} sample ions reach the interaction zone. These relative population factors are used in all future figures where we compare the experimental data with the theoretical predictions.

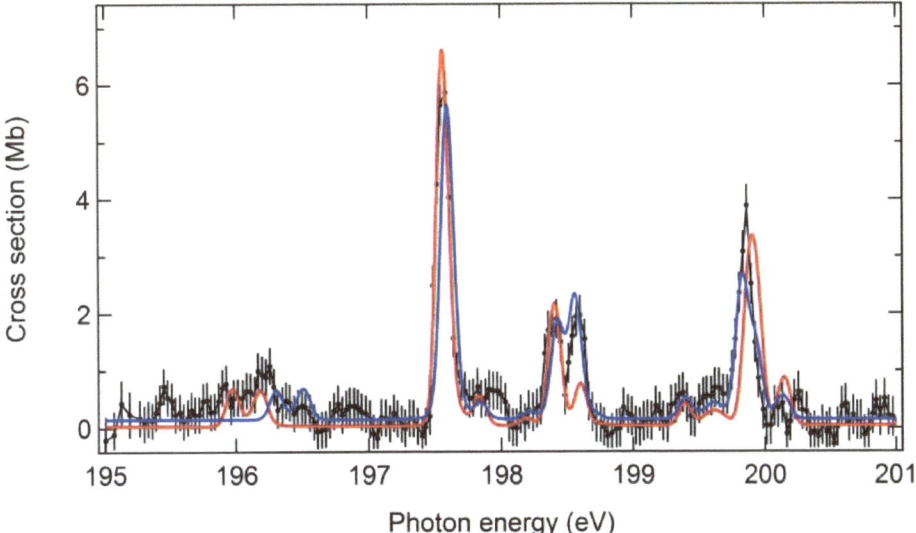

Figure 4. Experimental (single ionization) and scaled theoretical photoionization cross sections of Cl^{5+} in the 195–201 eV photon region. The observed resonance structure corresponds to 2p excitations from the metastable $2p^63s3p\ ^3P$ to $2p^53s^23p\ ^3S,\ ^3P,\ ^3D$ levels. The relative ground and $2p^63s3p\ ^3P$ metastable level populations are derived from fitting the theoretical MCDF (blue) and R-matrix (red) cross sections, shown here after convolution with a Gaussian profile of 110 meV FWHM and shifted by −0.54 eV meV and −0.64 meV, respectively, to the experimental data (black line with superimposed statistical error bars).

It should be noted that in comparing the experimental data of Figure 4 with theory, the ab initio MCDF predictions were shifted by -0.54 eV, while the ab initio R-matrix results were shifted by -0.64 eV. The integrated cross sections over the metastable resonances, taking into account the relative populations, were: experimental data (2.9 ± 0.4 Mb.eV), MCDF (2.36 Mb.eV), and R-matrix (2.91 Mb.eV). In Table 1, we show experimental and theoretical (both R-matrix and MCDF) resonance energies, together with resonance line strengths (cross section integrated over a single resonance profile) and assignments for the structures in the 195 eV to 200 eV photon energy region arising from inner-shell excitations of the valence-excited metastable 3P levels. The assignments are based on best intensity and energy position matchings of the observed resonances and the MCDF predictions.

Table 1. Experimental energies, theoretical energies, line strengths, and assignments in *LSJ* notation (based on MCDF results) for Cl^{5+} resonances in the 195 eV–200 eV photon energy range, arising from $2p \rightarrow 3s$ inner-shell excitations from the valence-excited metastable $2p^6 3s 3p$ $^3P_{0,1,2}$ states.

	Energy (eV)/Strength (Mb.eV)			
Experimental *	R-Matrix	MCDF		Assignment
195.66(4)/0.09(2)	196.61/0.083	196.84/0.064		$^3P_2$-$^3S_1$
195.95(3)/0.15(3)	196.83/0.074	197.05/0.064		$^3P_0$-$^3S_1$
197.29(2)/0.85(13)	198.20/0.864	198.13/0.700		$^3P_2$-$^3D_3$
197.54(4)/0.09(2)	198.47/0.118	198.38/0.036		$^3P_1$-$^3D_2$
198.12(3)/0.27(4)	199.04/0.268	198.97/0.211		$^3P_0$-$^3D_1$
198.31(4)/0.28(5)	199.24/0.076	199.10/0.273		$^3P_2$-$^3P_2$
199.09(4)/0.07(3)	200.04/0.054	199.94/0.048		$^3P_2$-$^3D_1$
199.33(4)/0.09(2)	200.18/0.013	200.08/0.012		$^3P_1$-$^3D_1$
199.49(4)/0.11(2)	200.25/0.028	200.15/0.024		$^3P_0$-$^3D_1$
199.59(2)/0.47(8)	200.51/0.478	200.36/0.306		$^3P_2$-$^3D_2$
	200.57/0.252	200.47/0.156		$^3P_2$-$^3P_1$
199.97(5)/0.05(2)	200.79/0.067	200.68/0.50		$^3P_0$-$^3P_1$

* The number in brackets is the experimental uncertainty on the last digit, i.e., 195.66(4) eV is the same as (195.66 $\pm$ 0.04) eV. The same convention is used for all the data presented in this column.

4.1.2. $2p \rightarrow 3d$ Excitation Region

The dominant resonance structures observed between 235 and 240 eV in Figure 3 are due to the expected 2p–3d excitations, accompanied by considerably weaker 2p–ns resonances. These are shown in detail in Figure 5. The relative strengths of the individual resonances within this group vary considerably as one progresses from neutral magnesium toward Cl^{5+}. The detailed spectroscopic assignments of the corresponding resonances for the early members of the sequence have given rise to considerable early discussions [1,2,5,16], with the labelling to some extent depending on the particular theoretical approach. This is not surprising because, as previously noted in Section 3., considerable challenge arise in the theoretical calculations due to the complexity associated with multiply excited outer electron configurations combining with the inner-shell excited 2p (or 2s) hole. Further complexity arises from the nominally closed shell ground state of $3s^2$ $^1S_0$ mixing with configurations such as $3p^2$ [5,9,10]. The theoretical calculations indicate that some of the weaker resonances in Figure 5 can be attributed to such electron correlation effects between energetically close-lying, multiply excited states.

As can be seen in Figure 5, the R-matrix results mimic the experimental data in this region somewhat better than the corresponding MCDF predictions, however, only by a small amount. Both theoretical approaches require systematic energy shifts in order to bring the strongest resonances into approximate agreement with the experimental data (MCDF and R-matrix shifts of -0.46 eV and -1.86 eV, respectively). When compared with the experimental data, the R-matrix relative strengths of the main resonances are somewhat underestimated, while the energy separation of the strongest resonances is slightly overestimated (typically by an amount of less than 0.5 eV) in the MCDF calculations. For the latter approach, the use of an even more extended set of ground state and photoexcited configurations would likely improve the value of this energy separation, via both improved OWFS and description of ground and excited state correlations. This relatively minor improvement would require significant additional computational efforts, which were not available to the authors. The cross sections integrated over the photon energies of the spectral region of Figure 5 are: experimental data (129 ± 19 Mb.eV), MCDF (111.2 Mb.eV), and R-matrix (118.9 Mb.eV), respectively.

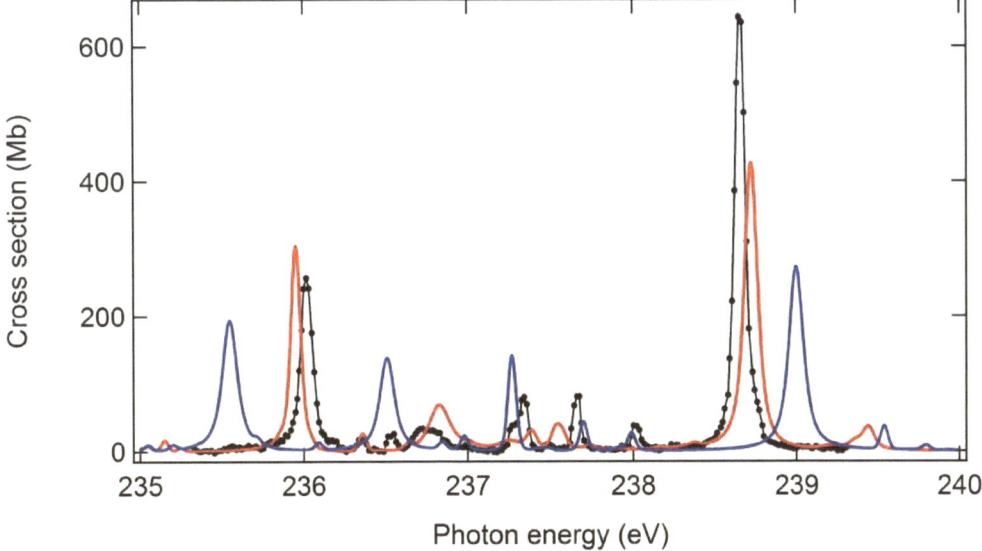

Figure 5. Photoionization cross sections for Cl^{5+} in the photon energy region corresponding to 2p → 3d excitations: experimental data (single ionization, black trace), MCDF (blue), and R-matrix (red) theoretical predictions, respectively. The theoretical predictions are folded with the experimental bandpass (Gaussian profile of 50 meV FWHM) and weighted according to semi-empirically determined population percentages (see text). The MCDF and R-matrix theoretical predictions are shifted by −0.46 eV and −1.86 eV, respectively.

4.1.3. Region of 2p → nd Excitations

Figure 6 shows the corresponding comparisons for the experimental and ab initio theoretical predictions for the resonances in the 2p → nd (n > 3) photon energy region. The systematic energy shifts required to bring the theoretical predictions into approximate agreement with the experimental data were +0.97 eV for the MCDF and a larger shift of −4.36 eV for the R-matrix calculations. The integrated cross sections over the spectral region covered by Figure 6 were in good agreement, and were: experimental data (87 ± 13 Mb.eV), MCDF (71.8 Mb.eV), and R-matrix (85.4 Mb.eV), respectively. Both the R-matrix and MCDF results suggest that the strong resonance feature observed at 282.66 eV likely corresponds with the first member (n = 3) of the 2s → np $^1P_1$ series. Numerical fitting of the Rydberg series for the 2p → nd series using the basic $E_n = I_p - \frac{13.6 \times 6^2}{(n-\delta)^2}$ hydrogenic quantum defect formula (in eV) led to values of 299.6(2) eV and 0.22(1) for the (2s^{2}2p^{5}3s^2) ^{2}P ionization limit I_p and quantum defect δ, respectively. In the experimental conditions, it was not possible to reliably access the energy values of the two fine-structure limits $^2P_{1/2,3/2}$. The comments in Section 4.1.2 about the use of more extended OFWS and configuration sets are also applicable here.

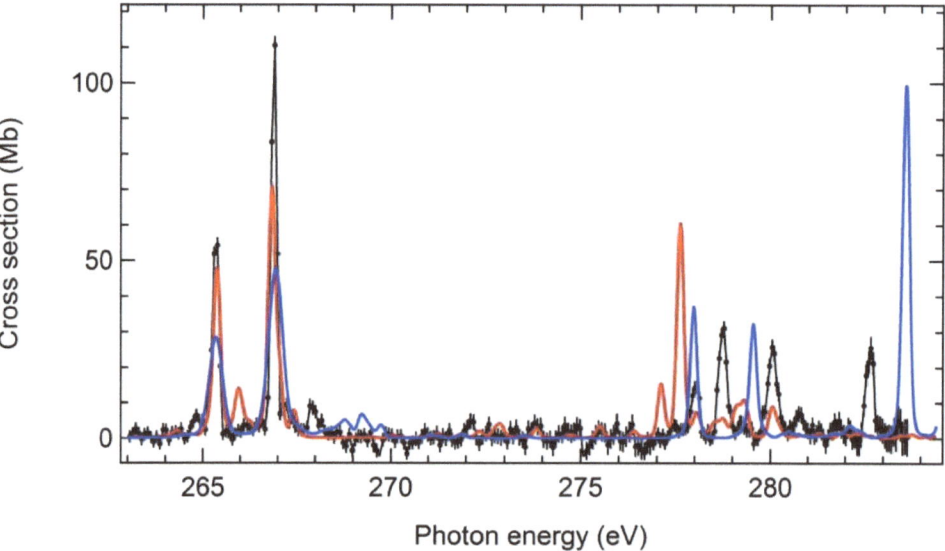

Figure 6. Photoionization cross sections for Cl^{5+} in the photon energy region corresponding to 2p → nd excitations: experimental data (single ionization, black trace), MCDF (blue), and R-matrix (red) theoretical predictions, respectively. The theoretical predictions are folded with the experimental bandpass (Gaussian profile of 190 meV FWHM) and weighted according to semi-empirically determined population percentages (see text). The MCDF and R-matrix theoretical predictions are shifted by +0.97 eV and −4.36 eV, respectively.

4.1.4. Region of 2s → np Excitations

Above the 2p^{5}3s^2 ^{2}P limit, we observed a regular series of relatively weak resonances, analogous in shape to those previously observed [18] for S^{4+}. As experimental absolute cross sections were not available here, the data were instead normalized to the theoretical cross sections, as shown in Figure 7. The strongest resonance lay at ∼330 eV, identified as the 2s → 4p member of the series. The most striking feature of the resonances in this region was the strongly asymmetrical Fano profiles due to the strong autoionization interaction with the underlying continua. As noted in Section 3.1, the R-matrix calculation includes the Fano interference between the resonance and the underlying continua, and mimics rather well the observed asymmetric profiles. Hydrogenic extrapolation (see previous paragraph) of the experimental 2s → np series led to the best fit values of 373.2(6) eV and 0.70(1) for the 2s $^2S_{1/2}$ ionization threshold energy and quantum defect of the 2snp $^1P_1$ levels, respectively.

In Table 2, we gather together the experimental measurements and the corresponding theoretical predictions for all resonances observed between 225 and 370 eV, most of which have been shown in detail in Figures 5–7. Table 2 shows that the vast majority of the observed resonances are reasonably accounted for by both theories in terms of resonance energy, strength, and width. In addition, the supernumerary low-intensity peaks seen in Figure 5 are readily interpreted as transitions from the 2p^{6}3s3p $^3P_{0,2}$ metastable states to ^3S,^3D states of the multiply excited 2p^{5}3s3p3d configuration. This is consistent with our interpretation of the 195 eV–201 eV low-energy photon region (see Section 4.1.1).

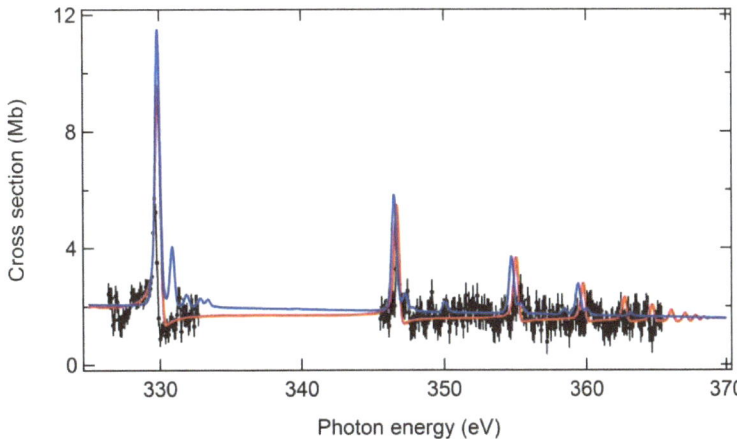

Figure 7. Photoionization cross sections for Cl^{5+} in the photon energy region corresponding to excitations from the 2s inner-shell: MCDF (blue) and R-matrix (red) theoretical predictions, and experimental data (double ionization, black trace) normalized to the theoretical curves. The theoretical predictions are folded with the experimental bandpass (Gaussian profile of 340 meV FWHM) and weighted according to semi-empirically determined population percentages (see text). The MCDF and R-matrix theoretical predictions are shifted by +1.0 eV and −7.0 eV, respectively.

Table 2. Experimental and theoretical (MCDF and R-matrix) energies, as measured line strengths and, where available, Auger widths for the Cl^{5+} resonances due to 2p → nd and 2s → np inner-shell excitations from the $2p^63s^2$ $^1S_0$ ground state. Where identifiable from the output of the MCDF theory, the dominant electron configuration and leading LSJ character of the final state of the resonance is given in the outermost right column.

Energy (eV)/Strength (Mb eV)/Auger Width (meV) * Experimental **	Energy (eV)/Strength (Mb eV)/Auger Width (meV) * R-matrix	Energy (eV)/Strength (Mb eV)/Auger Width (meV) * MCDF	Assignment (MCDF)
225.28(3)/4.0(7)/x	226.77/3.94/x	224.78/5.06/x	$2p^53s3p^2$ $^3D_1$
226.56(7)/0.7(4)/x	228.08/0.63/x	225.93/6.57/x	$2p^53s3p^2$ $^3P_1$
230.03(7)/0.9(4)/x	231.91/1.07/x		
231.01(3)/2.9(6)/x	232.88/2.75/x	231.25/1.32/x	$2p^53s3p^2$ $^3P_1$
231.93(4)/2.2(5)/x	233.80/2.12/x		
232.49(7)/0.8(4)/x			
233.05(5)/1.2(4)/x	235.18/0.19/x		
233.64(10)/0.5(4)/x			
234.26(4)/2.4(5)/x		234.48/2.83/x	$2p^53s3p3d$ $^3D_3$
234.73(6)/1.0(4)/x			
235.98(2)/26(4)/51(4)	237.81/29/37	236.01/39.8/45	$2p^53s^23d$ $^3D_1$
236.35(3)/1.1(4)/x			
236.54(3)/1.5(4)/x			
236.71(3)/9(2)/158(65)	238.69/20.3/134	236.97/28/92	$2p^53s^23d$ $^3P_1$
237.27(3)/2.2(3)/x			
237.34(3)/4.7(8)/x		237.73/5.5/x	$2p^53s3p3d$ $^3D_2$***
237.49(4)/0.5(4)/x			

Table 2. Cont.

Energy (eV)/Strength (Mb eV)/Auger Width (meV) * Experimental **	Energy (eV)/Strength (Mb eV)/Auger Width (meV) * R-matrix	Energy (eV)/Strength (Mb eV)/Auger Width (meV) * MCDF	Assignment (MCDF)
237.66(3)/4.8(8)/x		238.16/2.7/x	$2p^53s3p3d\ ^3D_3$***
238.01(3)/2.3(5)/x		238.45/1.72/x	$2p^53s3p3d\ ^3D_1$***
238.61(2)/78(12)/46(2)	240.58/58.1/69	239.46/42.6/86	$2p^53s^23d\ ^1P_1$
240.26(4)/2.0(5)/x		240/2.24/x	$2p^53s3p3d\ ^3S_1$***
240.47(7)/9(4)/x			
251.98(2)/1.5(2)/x	255.97/1.9/87	251.10/1.0/x	$2p^53s^24s\ ^3P_1$
253.55(2)/1.4(2)/x	257.51/1.8/87	252.71/0.8/x	$2p^53s^24s\ ^3P_1$
265.33(3)/13(2)/x	269.73/12.2/59	264.36/14.0/x	$2p^53s^24d\ ^3D_1$
266.88(3)/21(3)/x	271.18/17.9/53	265.96/21.8/x	$2p^53s^24d\ ^3P_1$
278.01(3)/4.1(7)/x			
278.75/8.5(1.3)/x		276.98/9.09/x	$2p^53s^25d\ ^3P_1$
280.05(3)/8.4(1.3)/x		278.56/7.57/x	$2p^53s^25d\ ^3P_1$
282.66(3)/6.5(1.0)/x		282.60/21.9/x	$2s2p^63s^23p\ ^1P_1$
329.6(1)/x/x	336.88/x/x	328.85/4.27/x	$2s2p^63s^24p\ ^1P_1$
346.7(1)/x/x	353.38/x/x	345.47/1.77/x	$2s2p^63s^25p\ ^1P_1$
355.3(1)/x/x	362.03/x/x	353.69/0.88/x	$2s2p^63s^26p\ ^1P_1$
360.4(1)/x/x	366.67/x/x	358.38/0.49/x	$2s2p^63s^27p\ ^1P_1$

* The symbol x indicates partial absence of data for that entry. No entry at all means that there are no data available. ** The number in brackets is the experimental uncertainty on the last digit of the data, i.e., 235.98(2)/26(4)/51(4) is the same as (235.98 ± 0.02) eV/(26 ± 4) Mb.eV/(51 ± 4) meV. The same notation convention is used for all the data presented in this column. *** These resonances are due to 2p → 3d excitations from the initial $2p^63s3p\ ^3P_{0,2}$ metastable states.

4.2. Evolution along the Mg-like Sequence

Studies of isoelectronic sequences of atomic energy levels [53] and absorption oscillator strengths (f-values) [54], line strengths or cross sections are of fundamental interest. New atomic data can be inferred from isoelectronic sequences via the interpolation/extrapolation to neighboring atoms and ions of sets of f-values/energies exhibiting smooth and systematic variations. Further, trends along isoelectronic sequences are also of great general interest because a smooth trend can be modeled using Z-dependent perturbation theory. Irregular, non-monotonic trends are also known to occur, and are commonly related to the crossing of levels of correlation–mixed states from interacting electronic configurations or spin–orbit mixed states [55]. The magnesium spectrum is a case in point for correlation mixing due to the strongly perturbing effects of the $3p^2$ configuration on the $3s3d$ configuration as well as over the entire $3dns$ series indeed. As pointed out in the introduction, these effects are found to persist in the presence of a $2p$ vacancy. Therefore, it is interesting to view the Cl^{5+} results within the overall context of the magnesium isoelectronic sequence.

Figure 8 provides a full view of both the single and double experimental photoionization cross sections, measured in synchrotron radiation-based, merged-beam experiments, including Al$^+$ [9] (ASTRID), Si^{2+} [10] (SuperACO), S^{4+} [18] (SOLEIL), and Cl^{5+} (SOLEIL, this study). Although it is not possible to make detailed and/or quantitative intercomparisons between the four traces of Figure 8 due to their very different photon energy ranges with associated different spectral dispersions (determined by the beamline instrumentation), as well as the choice of experimental monochromator bandpass(es), the following qualitative remarks can be made. As expected, the resonance structures shift markedly to higher energies as the nuclear charge increases. The corresponding trends are explored in greater detail below for selected transitions. Progressing along the series, the progressive discretization of the SI resonance structure (i.e., mostly the $2p^53s^2nd$ resonances) into a seemingly hydrogenic Rydberg series for Cl^{5+}, as well as the concomitant redistribution of the line strengths into the first $2p^53s^23d$ series member, are apparent.

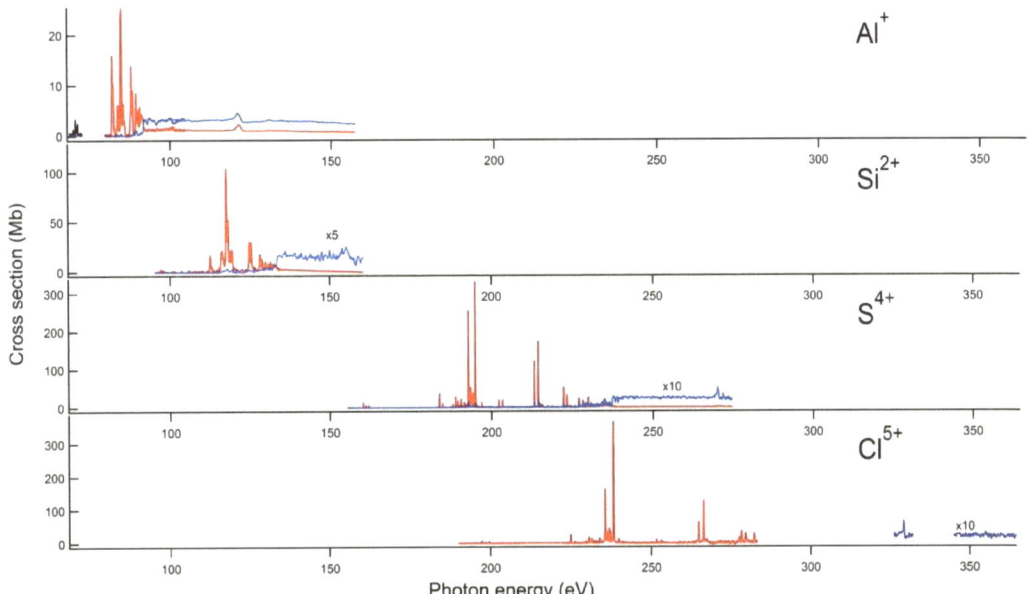

Figure 8. Single (red) and double (blue) photoionization cross sections for the magnesium-like ions Al^+ [9], Si^{2+} [10], S^{4+} [18], and Cl^{5+} [this study] in the photon energy region corresponding to L-shell excitations.

In order to better visualize and characterize the main features of isoelectronic evolution, Figure 9 shows the total cross sections for the ions Al^+ to Cl^{5+}; however, these are displayed on modified photon energy scales so that the resulting spectra line up for better intercomparison. To accomplish this, the energy scales are multiplied by appropriate numerical factors and suitably shifted so that the main $2p \to 3d$ and $2p \to 4d$ resonance structures line up (i.e., the same $2p \to 3d$, $2p \to 4d$ energy difference is maintained for the four ions) below each other. It is clear from Figure 9 that for S^{4+} and Cl^{5+}, the overall resonance structures are very similar. This confirms that by S^{4+}, the increased core charge has become the dominating factor, simplifying the spectra as they become more hydrogenic in character. For the early members of the sequence, the $2p \to 3d$ structures show considerable changes as we move from Al^+ to S^{4+}: the supernumerary resonances to the lower energy side of the $2p \to 3d$ group, in particular, are strong for Al^+, weaker for Si^{2+}, and much weaker for S^{4+} and Cl^{5+}. This shows the rapidly diminishing effects of the main $2p^53s(3p^2)$ series perturber (see above).

Table 3 shows the main energy level and strength data (resonance energies and inner ionization thresholds) for all the early members of the sequence including magnesium. Where necessary, explanations regarding the origin of some data entries are provided in the footnote of Table 3. We plotted this data in two different ways in order to establish isoelectronic trends and interpolate with the missing P^{3+} ion.

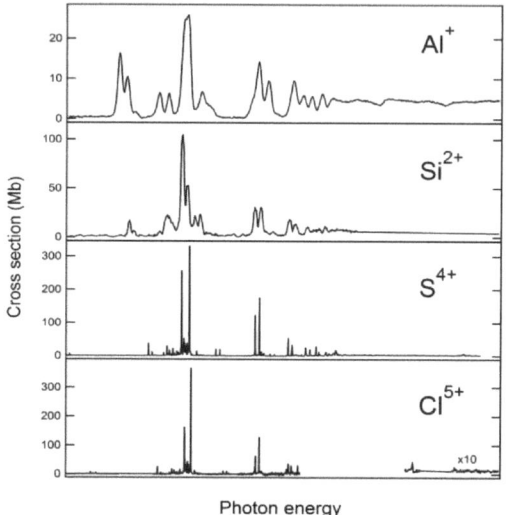

Figure 9. Photoionization cross sections for the magnesium-like ions Al$^+$ [9], Si^{2+} [10], S^{4+} [18] (sum of the single and double ionization channels) and Cl^{5+} (this study: single ionization below or double ionization above the photon energy of 300 eV) in the photon energy region corresponding to L-shell excitations, with modified photon energy scales (see text).

Table 3. Top line: Experimental energy (eV), middle line: Non radiative (NR) width (meV), and bottom line: Strength (Mb eV) for the strongest 2p → 3d $^1P_1$, $^3D_1$, $^3P_1$ resonances, energy of the 2p limit (eV), and experimental energy of the 2s → 3p $^1P_1$ resonance along the magnesium-like sequence. Experimental data and their references are provided for all members of the sequence up to Cl^{5+}, apart from P^{3+}, for which interpolated values are deduced. The symbol x indicates that data are not available. The number in brackets is the experimental uncertainty on the last digit of the data, i.e., 55.492(1) is the same as (55.492 ± 0.001) etc. The same convention is used for all the data presented in this table where error values are available.

Atom/Ion	2p→3d $^1P_1$	2p→3d $^3D_1$	2p→3d $^3P_1$	2p Limit	2s→3p $^1P_1$
Mg I	55.492(1) 4.6(9) x	55.677(1) 5.0(1) x	55.838(1) 8.2(4) x	57.658(2) [e]	x
Al II	84.99(2) x 5.6 [b]	85.36(2) x 0.54 [b]	85.57(2) x 1.5 [b]	91.75(15) [c]	121.5(5)
Si III	117.6(1) x 37(7) [b]	118.1(1) x 17(3) [b]	118.8(1) x 5(1) [b]	133.5(1) [c]	155.7(4)
P IV (present work) [a]	154.4 x x	153.9 x x	154.7 x x	180.5	193.6
S V	194.88 51 46(7) [b]	192.74 47 32(5) [b]	193.42 149 13(2) [b]	237.35 [c]	235.4
Cl VI (present work)	238.61(2) 46(2) 68(10) [b]	235.98(2) 51(4) 26(4) [b]	236.71(3) 158(65) 9(2) [b]	299.6 [d]	282.66

[a] For P IV, the entries are obtained by interpolation of the isoelectronic plots of Figure 10 and italicized for clarity. [b] Not corrected for possible contributions of metastable states. [c] Values estimated from the experimental photon energy of the onset of double ionization. [d] From hydrogenic fit (see text). [e] Statistically averaged over the $^2P_{1/2}$ and $^2P_{3/2}$ components.

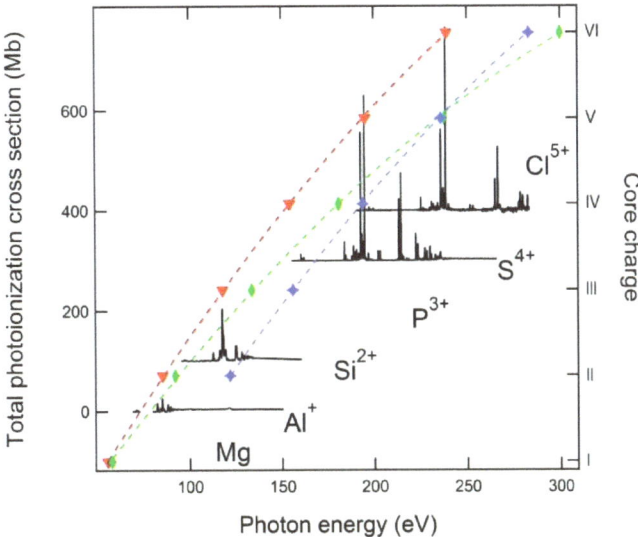

Figure 10. Photoionization cross sections for the magnesium-like sequence Al$^+$ [9], Si^{2+} [10], S^{4+} [18] (sum of the single and double ionization channels) and Cl^{5+} (this study: single ionization below, or double ionization above, the photon energy of 300 eV) in the photon energy region corresponding to L-shell excitations. No experimental data as yet exist for P^{3+}. The scale ticks on the right-hand side indicate the net core charge of the ion. The energies of the strongest 2p → 3d resonance (red triangles), the 2s → 3p resonance (blue stars), and the 2p^{-1} threshold (green diamonds) are shown for each ion. The data are accurately represented by 3rd-order polynomial fits (red-, blue-, and green-colored broken lines, respectively), yielding the corresponding photon energies for P^{3+} by interpolation.

Firstly, as shown in Figure 10, the cross section data (represented to scale for all the ions where electronic data are available) are displayed as a function of photon energy, and parametrically, as a function of the net core charge ζ seen by a Rydberg electron ($\zeta = 1$ for Mg, $\zeta = 2$ for Al$^+$, etc...). The spectra are shifted vertically from one to the next by an arbitrary amount of 100 Mb for clarity. The three curves in Figure 10 show the movement of the strongest 2p → 3d (the energy differences between the three possible final states of Table 3 are too small to be seen in Figure 9) and 2s → 3p resonances, and the 2p ionization threshold as the net core charge increases from one to six. Noting the very smooth evolution of the curves along the sequence, the data are very accurately represented by 3rd-order polynomial fits (red-, blue- and green-colored broken lines, respectively), it seems justified to derive results for the missing member P^{3+} by interpolation, we show the corresponding data in Table 3.

We note that for the first four members, the 2s → 3p resonance is found at energies above the 2p threshold. For a value of the net core charge close to five (vicinity of S^{4+}), this situation is reversed. This is discussed in greater detail below.

Secondly, we examine the isoelectronic trends of characteristic (resonance and threshold) energies and absorption oscillator strength, as shown in Figure 11a,b, respectively. Figure 11a plots the $E_{2p \rightarrow 3d}$ and $E_{2s \rightarrow 3p}$ transition energies divided by the net core charge, referred to the energy value of the first inner limit 2p^{5}3s^2 ^2P (E_{2p}), divided by the net core charge (ζ), i.e., $\left(E_{2p \rightarrow 3d} - E_{2p}\right)/\zeta$ and $\left(E_{2s \rightarrow 3p} - E_{2p}\right)/\zeta$, respectively. The energy differences between the three possible final states $^1P_1, ^3D_1, ^3P_1$ are too small to be seen in Figure 11a. Further relevant data in the form of the theoretical 2s^{-1} threshold energies taken from [56], referred to the 2p^{-1} threshold in the same manner, are also included in Figure 11a.

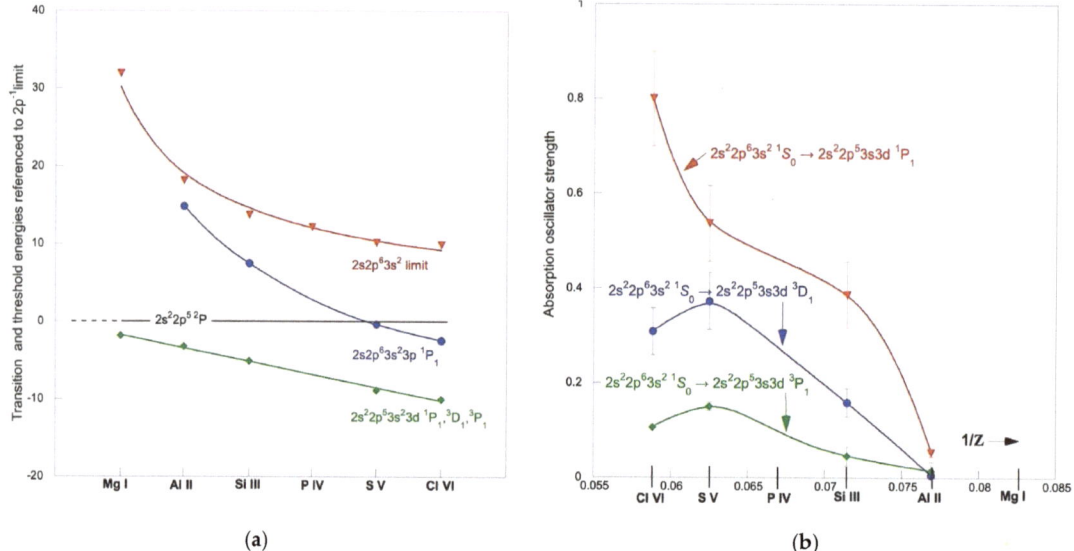

Figure 11. Isoelectronic plots for selected L-shell transitions and thresholds along the magnesium electronic sequence. (**a**) Transition and threshold energies (in eV) relative to the 2p limit, divided by the net core charge (spectral roman number) for the first six members of the sequence (the experimental data for P IV is unknown). (**b**) Absorption oscillator strength plotted as a function of $1/Z$ for the 2p → 3d $^1P_1$, $^3D_1$, $^3P_1$ transitions. Further details for both plots are given in the text. The solid lines joining the points in both plots are cubic-spline numerical fits to guide the eye more easily through the data.

From Figure 11a, we observe a regular (linear) series behavior for the $2p^53s^23d$ average term energy, while for both the $2s2p^63s^23p\ ^1P_1$ and the $2s^{-1}$ threshold energies, the series behaviors are clearly non-linear, although smooth. Notably, the energy position of the first member of the asymmetric profile $2s \to 3p\ ^1P_1$ excitation moves closer to the $2p^{-1}$ threshold as the core charge increases, and drops just below and well below the $2p^{-1}$ threshold for S^{4+} [18] and Cl^{5+}, respectively. This crossing of the $2p^{-1}$ threshold by the $2s2p^63s^23p\ ^1P_1$ state is accompanied by a noticeable change from a Fano to a Lorentzian profile and a prominence of the resonance in the single ionization channel only. As one moves further up the sequence, this downward movement toward the $2p^{-1}$ threshold (and finally below) is expected to be replicated for the higher members of the series (n > 3). We note the similarity of this behavior with the previously observed behavior for the $2s \to np$ transitions in the neon-like sequence [57]. This suggests a lesser role played by the aforementioned $(3s^2 + 3p^2 + 3d^2)$ correlations on the $2s \to 3p$ transition compared with the $2p \to 3d$ for the early members of the Mg-like sequence. If we interpolate the data for P^{3+} from Figure 11a, we obtain the values 154 eV (2p-3d) and 195 eV (2s-3p), which compare favorably with the corresponding predictions from Figure 10, as shown in Table 3. The value of the $2p^{-1}$ threshold for P^{3+}, which is needed to obtain the data just quoted, was estimated from [58] to be (183 ± 1) eV. This is in reasonable agreement with the interpolated value shown in Table 3.

In Figure 11b, we plot the oscillator strengths $f_{i \to j}$ for the 2p → 3d $^1P_1$, $^3D_1$, $^3P_1$ transitions obtained from the integrated cross sections (in Mb eV) shown in Table 3, using the standard equation $\int_{line} \sigma(E) dE = 109.8 f_{i \to j} \int_{line} \Phi_{ij}(E) dE$, where Φ_{ij} is the normalized line profile as a function of $1/Z$, where Z is the atomic number. Correction factors of 0.95, 0.975, 0.78, and 0.77 were applied to the Al$^+$, Si^{2+}, S^{4+}, and Cl^{5+} strength values, respectively, in Table 3 to take into account the different initial ground state populations. The curves for the 2p → 3d $^3D_1$, $^3P_1$ transitions are seen to present a maximum (near S^{4+}), which is characteristic of series where configuration interaction effects are known to be important [55]. This is fully compatible with our previous discussions regarding the effects of the $(3s^2 3d + 3s3p^2)$ mixing along the magnesium sequence. As the nuclear charge increases, the oscillator strength transfers into the fully LS allowed $2p^1 S_0 \rightarrow 3d\ ^1P_1$ transition, in tandem with the weakening of the configuration interaction effects discussed above.

5. Conclusions

The development of the storage ring-based, merged-beam technique has allowed the systematic study of inner-shell photoionization for extended isoelectronic sequences. For magnesium-like ions, the current results extend the sequence up to Cl^{5+}. The similarity of the S^{4+} and Cl^{5+} ion yields implies that the photoionization behavior settles down when compared with earlier members of the sequence. The new experimental results for Cl^{5+} are compared with ab initio MCDF and R-matrix calculations. Differences between the experimental data and the theoretical predictions underscore the importance of the ongoing benchmarking of theory. While the relative energies and strengths of resonances are reasonably predicted by theory, it is clear that significant systematic energy shifts are required to bring the theoretically predicted resonance structures into reasonable coincidence with experimental data. Interpolation of the results along the sequence provides estimates for the energies of the 2p–3d and 2s–3p resonances and the $2p^{-1}$ and $2s^{-1}$ thresholds for the P^{3+} ion, which have not yet been experimentally studied.

Author Contributions: D.C., J.-M.B., E.T.K., J.-P.M. and S.G. conducted experiments and data collection. The theoretical predictions were contributed by C.B., M.F.H. and T.W.G. The figures and tables were prepared by J.-M.B., J.-P.M. and D.C. All the authors contributed to the analyses and conclusions. E.T.K., J.-P.M., T.W.G. and J.-M.B. prepared the original draft. All the authors reviewed subsequent draft versions. All authors have read and agreed to the published version of the manuscript.

Funding: This research received no external funding.

Data Availability Statement: Data supporting the results reported here can be obtained by directly contacting the authors.

Acknowledgments: T.W.G. was supported in part by NASA. The authors would like to thank the SOLEIL and PLÉIADES beamline staff, J. Bozek, C. Nicolas, and A. Milosavljevic for their support throughout the experiment. ETK and JPM thank SOLEIL for partial support.

Conflicts of Interest: The authors declare no conflict of interest.

References

1. Esteva, J.M.; Mehlman, G. Autoionization Spectra of Magnesium (Mg I, Mg II and Mg III) in the 50- to 110-eV Energy Range. *Astrophys. J.* **1974**, *193*, 747–754. [CrossRef]
2. Mehlman, G.; Weiss, A.W.; Esteva, J.M. Revised Classification of Mg II Levels between 59 and 63 EV. *Astrophys. J.* **1976**, *209*, 640–641. [CrossRef]
3. Shorer, P.; Lin, C.D.; Johnson, W.R. Oscillator Strengths for the Magnesium Isoelectronic Sequence. *Phys. Rev. A* **1977**, *16*, 1109–1116. [CrossRef]
4. Butler, K.; Mendoza, C.; Zeippen, C.J. Oscillator Strengths and Photoionization Cross-Sections for Positive Ions in the Magnesium Isoelectronic Sequence. *Mon. Not. R. Astron. Soc.* **1984**, *209*, 343–351. [CrossRef]
5. Costello, J.T.; Evans, D.; Hopkins, R.B.; Kennedy, E.T.; Kiernan, L.; Mansfield, M.W.D.; Mosnier, J.-P.; Sayyad, M.H.; Sonntag, B.F. The 2p-Subshell Photoabsorption Spectrum of Al^+ in a Laser-Produced Plasma. *J. Phys. B At. Mol. Opt. Phys.* **1992**, *25*, 5055–5068. [CrossRef]
6. Mosnier, J.P.; Costello, J.T.; Kennedy, E.T.; Kiernan, L.; Sayyad, M.H. Even-Parity Autoionizing States in the Extreme-Ultraviolet Photoabsorption Spectra of Mg, Al^+, and Si^{2+}. *Phys. Rev. A* **1994**, *49*, 755–761. [CrossRef]
7. Sayyad, M.H.; Kennedy, E.T.; Kiernan, L.; Mosnier, J.-P.; Costello, J.T. 2p-Subshell Photoabsorption by Si^{2+} Ions in a Laser-Produced Plasma. *J. Phys. B At. Mol. Opt. Phys.* **1995**, *28*, 1715–1722. [CrossRef]
8. Fang, T.K.; Nam, B.I.; Kim, Y.S.; Chang, T.N. Resonant Structures of Overlapping Doubly Excited Autoionization Series in Photoionization of Mg-like Al^+ and Si^{2+} Ions. *Phys. Rev. A* **1997**, *55*, 433–439. [CrossRef]
9. West, J.B.; Andersen, T.; Brooks, R.L.; Folkmann, F.; Kjeldsen, H.; Knudsen, H. Photoionization of Singly and Doubly Charged Aluminum Ions in the Extreme Ultraviolet Region: Absolute Cross Sections and Resonance Structures. *Phys. Rev. A* **2001**, *63*, 052719. [CrossRef]

10. Mosnier, J.-P.; Sayyad, M.H.; Kennedy, E.T.; Bizau, J.-M.; Cubaynes, D.; Wuilleumier, F.J.; Champeaux, J.-P.; Blancard, C.; Varma, R.H.; Banerjee, T.; et al. Absolute Photoionization Cross Sections and Resonance Structure of Doubly Ionized Silicon in the Region of the $2p^{-1}$ Threshold: Experiment and Theory. *Phys. Rev. A* **2003**, *68*, 052712. [CrossRef]
11. Ho, H.C.; Johnson, W.R.; Blundell, S.A.; Safronova, M.S. Third-Order Many-Body Perturbation Theory Calculations for the Beryllium and Magnesium Isoelectronic Sequences. *Phys. Rev. A* **2006**, *74*, 022510. [CrossRef]
12. Kim, D.-S.; Kim, Y.S. Theoretical Photoionization Spectra in the UV Photon Energy Range for a Mg-like Al^+ Ion. *J. Phys. B At. Mol. Opt. Phys.* **2008**, *41*, 165002. [CrossRef]
13. Pradhan, G.B.; Jose, J.; Deshmukh, P.C.; Radojević, V.; Manson, S.T. Photoionization of Mg and Ar Isonuclear Sequences. *Phys. Rev. A* **2009**, *80*, 053416. [CrossRef]
14. Kim, D.-S.; Kwon, D.-H. Theoretical Photoionization Spectra for Mg-Isoelectronic Cl^{5+} and Ar^{6+} Ions. *J. Phys. B At. Mol. Opt. Phys.* **2015**, *48*, 105004. [CrossRef]
15. Khatri, I.; Goyal, A.; Dioulde Ba, M.; Faye, M.; Sow, M.; Sakho, I.; Singh, A.K.; Mohan, M.; Wagué, A. Screening Constant by Unit Nuclear Charge Calculations of Resonance Energies and Widths of the $3pns\ ^{1,3}P°$ and $3pnd\ ^1P°$ Rydberg Series of Mg-like (Z = 13–26) Ions. *Radiat. Phys. Chem.* **2017**, *130*, 208–215. [CrossRef]
16. Wehlitz, R.; Juranić, P.N. Relative Single- and Double-Photoionization Cross Sections of Mg around the $2p \to nl$ Resonances. *Phys. Rev. A* **2009**, *79*, 013410. [CrossRef]
17. Safronova, U.I.; Johnson, W.R.; Berry, H.G. Excitation Energies and Transition Rates in Magnesiumlike Ions. *Phys. Rev. A* **2000**, *61*, 052503. [CrossRef]
18. Mosnier, J.-P.; Kennedy, E.T.; Cubaynes, D.; Bizau, J.-M.; Guilbaud, S.; Hasoglu, M.F.; Blancard, C.; Gorczyca, T.W. L-Shell Photoionization of Mg-like S^{4+} in Ground and Metastable States: Experiment and Theory. *Phys. Rev. A* **2022**, *106*, 033113. [CrossRef]
19. Kallman, T.R.; Palmeri, P. Atomic Data for X-Ray Astrophysics. *Rev. Mod. Phys.* **2007**, *79*, 79–133. [CrossRef]
20. Foster, A.R.; Smith, R.K.; Brickhouse, N.S.; Kallman, T.R.; Witthoeft, M.C. The Challenges of Plasma Modeling: Current Status and Future Plans. *Space Sci. Rev.* **2010**, *157*, 135–154. [CrossRef]
21. Kallman, T.R. Modeling of Photoionized Plasmas. *Space Sci. Rev.* **2010**, *157*, 177–191. [CrossRef]
22. Savin, D.W.; Brickhouse, N.S.; Cowan, J.J.; Drake, R.P.; Federman, S.R.; Ferland, G.J.; Frank, A.; Gudipati, M.S.; Haxton, W.C.; Herbst, E.; et al. The Impact of Recent Advances in Laboratory Astrophysics on Our Understanding of the Cosmos. *Rep. Prog. Phys.* **2012**, *75*, 036901. [CrossRef] [PubMed]
23. Mendoza, C.; Bautista, M.A.; Deprince, J.; García, J.A.; Gatuzz, E.; Gorczyca, T.W.; Kallman, T.R.; Palmeri, P.; Quinet, P.; Witthoeft, M.C. The XSTAR Atomic Database. *Atoms* **2021**, *9*, 12. [CrossRef]
24. Nahar, S. Database NORAD-Atomic-Data for Atomic Processes in Plasma. *Atoms* **2020**, *8*, 68. [CrossRef]
25. Kennedy, E.T.; Costello, J.T.; Mosnier, J.-P.; Cafolla, A.A.; Collins, M.; Kiernan, L.; Koeble, U.; Sayyad, M.H.; Shaw, M.; Sonntag, B.F.; et al. Extreme-Ultraviolet Studies with Laser-Produced Plasmas. *Opt. Eng.* **1994**, *33*, 3984–3992. [CrossRef]
26. Bizau, J.M.; Cubaynes, D.; Guilbaud, S.; El Eassan, N.; Al Shorman, M.M.; Bouisset, E.; Guigand, J.; Moustier, O.; Marié, A.; Nadal, E.; et al. A Merged-Beam Setup at SOLEIL Dedicated to Photoelectron–Photoion Coincidence Studies on Ionic Species. *J. Electron. Spectrosc. Relat. Phenom.* **2016**, *210*, 5–12. [CrossRef]
27. Kjeldsen, H. Photoionization Cross Sections of Atomic Ions from Merged-Beam Experiments. *J. Phys. B At. Mol. Opt. Phys.* **2006**, *39*, R325–R377. [CrossRef]
28. Phaneuf, R.A.; Kilcoyne, A.L.D.; Müller, A.; Schippers, S.; Aryal, N.; Baral, K.; Hellhund, J.; Aguilar, A.; Esteves-Macaluso, D.A.; Lomsadze, R. *Cross-Section Measurements with Interacting Beams*; American Institute of Physics: Gaithersburg, MD, USA, 2013; pp. 72–78. [CrossRef]
29. Schippers, S.; Kilcoyne, A.L.D.; Phaneuf, R.A.; Muller, A. Photoionization of ions with synchrotron radiation: From ions in space to atoms in cages. *Contemp. Phys.* **2016**, *57*, 215–229. [CrossRef]
30. Schippers, S.; Buhr, T.; Borovik, A., Jr.; Holste, K.; Perry-Sassmannshausen, A.; Mertens, K.; Reinwardt, S.; Martins, M.; Klumpp, S.; Schubert, K.; et al. The Photon-Ion Merged-Beams Experiment PIPE at PETRAIII—The First Five Years. *X-ray Spectrom.* **2020**, *49*, 11–20. [CrossRef]
31. Hudson, C.E.; West, J.B.; Bell, K.L.; Aguilar, A.; Phaneuf, R.A.; Folkmann, F.; Kjeldsen, H.; Bozek, J.; Schlachter, A.S.; Cisneros, C. A Theoretical and Experimental Study of the Photoionization of AlII. *J. Phys. B At. Mol. Opt. Phys.* **2005**, *38*, 2911–2932. [CrossRef]
32. Simon, M.C.; Crespo López-Urrutia, J.R.; Beilmann, C.; Schwarz, M.; Harman, Z.; Epp, S.W.; Schmitt, B.L.; Baumann, T.M.; Behar, E.; Bernitt, S.; et al. Resonant and Near-Threshold Photoionization Cross Sections of Fe 14 +. *Phys. Rev. Lett.* **2010**, *105*, 183001. [CrossRef]
33. Blake, G.A.; Anicich, V.G.; Huntress, W.T., Jr. Chemistry of Chlorine in Dense Interstellar Clouds. *Astrophys. J.* **1986**, *300*, 415. [CrossRef]
34. Feldman, P.D.; Ake, T.B.; Berman, A.F.; Moos, H.W.; Sahnow, D.J.; Strobel, D.D.; Weaver, H.H. Detection of Chlorine Ions in the Far Ultraviolet Spectroscopic Explorer Spectrum of the IO Plasma Torus. *Astrophys. J.* **2001**, *554*, L123–L126. [CrossRef]
35. Kounaves, S.P.; Carrier, B.L.; O'Neil, G.D.; Stroble, S.T.; Claire, M.W. Evidence of Martian Perchlorate, Chlorate, and Nitrate in Mars Meteorite EETA79001: Implications for Oxidants and Organics. *Icarus* **2014**, *229*, 206–213. [CrossRef]

36. Neufeld, D.A.; Wiesemeyer, H.; Wolfire, M.J.; Jacob, A.M.; Buchbender, C.; Gerin, M.; Gupta, H.; Güsten, R.; Schilke, P. The Chemistry of Chlorine-Bearing Species in the Diffuse Interstellar Medium, and New SOFIA/GREAT*Observations of HCl$^+$. *Astrophys. J.* **2021**, *917*, 104. [CrossRef]
37. Maas, Z.G.; Pilachowski, C.A.; Hinkle, K. Chlorine Abundances in Cool Stars. *Astron. J.* **2016**, *152*, 196. [CrossRef]
38. Wallström, S.H.J.; Muller, S.; Roueff, E.; Le Gal, R.; Black, J.H.; Gérin, M. Chlorine-Bearing Molecules in Molecular Absorbers at Intermediate Redshifts. *Astron. Astrophys.* **2019**, *629*, A128. [CrossRef]
39. Ren, L.-M.; Wang, Y.-Y.; Li, D.-D.; Yuan, Z.-S.; Zhu, L.-F. Inner-Shell Excitations of 2p Electrons of Argon Investigated by Fast Electron Impact with High Resolution. *Chin. Phys. Lett.* **2011**, *28*, 053401. [CrossRef]
40. Thuillier, T.; Benitez, J.; Biri, S.; Rácz, R. X-ray Diagnostics of ECR Ion Sources—Techniques, Results, and Challenges. *Rev. Sci. Instrum.* **2022**, *93*, 021102. [CrossRef]
41. Bieron, J.; Froese-Fischer, C.; Jonsson, P. Special Issue "The General Relativistic Atomic Structure Package—GRASP". 2022. Available online: https://www.mdpi.com/journal/atoms/special_issues/the_grasp#info (accessed on 14 March 2023).
42. Bruneau, J. Correlation and Relaxation Effects in Ns2-Nsnp Transitions. *J. Phys. B At. Mol. Phys.* **1984**, *17*, 3009. [CrossRef]
43. Grant, I.P. Gauge Invariance and Relativistic Radiative Transitions. *J. Phys. B At. Mol. Phys.* **1974**, *7*, 1458. [CrossRef]
44. Savukov, I.M. Special Issue "Atomic Structure Calculations of Complex Atoms". 2021. Available online: https://www.mdpi.com/journal/atoms/special_issues/AtomicStructureCalculations_ComplexAtoms (accessed on 14 March 2023).
45. Schneider, B.I.; Hamilton, K.R.; Bartschat, K. Generalizations of the R-Matrix Method to the Treatment of the Interaction of Short-Pulse Electromagnetic Radiation with Atoms. *Atoms* **2022**, *10*, 26. [CrossRef]
46. Sardar, S.; Xu, X.; Xu, L.-Q.; Zhu, L.-F. Relativistic R-Matrix Calculations for Photoionization Cross-Sections of C IV: Implications for Photorecombination of C V. *Mon. Not. R. Astron. Soc.* **2018**, *474*, 1752–1761. [CrossRef]
47. Delahaye, F.; Ballance, C.P.; Smyth, R.T.; Badnell, N.R. Quantitative Comparison of Opacities Calculated Using the R-Matrix and Distorted-Wave Methods: Fe XVII. *Mon. Not. R. Astron. Soc.* **2021**, *508*, 421–432. [CrossRef]
48. Burke, P.G. *R-Matrix Theory of Atomic Collisions*; Springer: New York, NY, USA, 2011.
49. Berrington, K.A.; Eissner, W.B.; Norrington, P.H. RMATRX1: Belfast Atomic R-Matrix Codes. *Comput. Phys. Commun.* **1995**, *92*, 290–420. [CrossRef]
50. Kramida, A.; Ralchenko, Y.; Reader, J.; NIST ASD Team. *NIST Atomic Spectra Database*; Version 5.10; NIST: Gaithersburg, MD, USA, 2022. [CrossRef]
51. Fano, U. Effects of Configuration Interaction on Intensities and Phase Shifts. *Phys. Rev.* **1961**, *124*, 1866–1878. [CrossRef]
52. Froese Fischer, C.; Tachiev, G.; Irimia, A. Relativistic Energy Levels, Lifetimes, and Transition Probabilities for the Sodium-like to Argon-like Sequences. *At. Data Nucl. Data Tables* **2006**, *92*, 607–812. [CrossRef]
53. Curtis, L.J. Bengt Edlén's Handbuch Der Physik Article—26 Years Later. *Phys. Scr.* **1987**, *35*, 805–810. [CrossRef]
54. Wiese, W. Regularities of Atomic Oscillator Strengths in Isoelectronic Sequences. In *Beam-Foil Spectroscopy*; Springer: Berlin/Heidelberg, Germany, 1976; pp. 145–178.
55. Hasoğlu, M.F.; Nikolić, D.; Gorczyca, T.W.; Manson, S.T.; Chen, M.H.; Badnell, N.R. Nonmonotonic Behavior as a Function of Nuclear Charge of the K -Shell Auger and Radiative Rates and Fluorescence Yields along the $1s2s^22p^3$ Isoelectronic Sequence. *Phys. Rev. A* **2008**, *78*, 032509. [CrossRef]
56. Verner, D.A.; Yakovlev, D.G.; Band, I.M.; Trzhaskovskaya, M.B. Subshell Photoionization Cross Sections and Ionization Energies of Atoms and Ions from He to Zn. *At. Data Nucl. Data Tables* **1993**, *55*, 233–280. [CrossRef]
57. Chakraborty, H.S.; Gray, A.; Costello, J.T.; Deshmukh, P.C.; Haque, G.N.; Kennedy, E.T.; Manson, S.T.; Mosnier, J.-P. Anomalous Behavior of the Near-Threshold Photoionization Cross Section of the Neon Isoelectronic Sequence: A Combined Experimental and Theoretical Study. *Phys. Rev. Lett.* **1999**, *83*, 2151–2154. [CrossRef]
58. Brilly, J.; Kennedy, E.T.; Mosnier, J.P. The 2p-Subshell Absorption Spectrum of Al III. *J. Phys. B At. Mol. Opt. Phys.* **1988**, *21*, 3685–3693. [CrossRef]

Disclaimer/Publisher's Note: The statements, opinions and data contained in all publications are solely those of the individual author(s) and contributor(s) and not of MDPI and/or the editor(s). MDPI and/or the editor(s) disclaim responsibility for any injury to people or property resulting from any ideas, methods, instructions or products referred to in the content.

Review

Measurement of Photoionization Cross-Section for the Excited States of Atoms: A Review

Muhammad Aslam Baig

National Center for Physics (NCP), Quaid-i-Azam University, Islamabad 45320, Pakistan; baig@qau.edu.pk or aslam.baig@ncp.edu.pk or baig77@gmail.com

Abstract: A review of experimental studies of the measurement of the photoionization cross-section for the excited states of the alkali atoms, alkaline earth atoms, and rare-gas atoms is presented, with emphasis on using multi-step laser excitation, ionization, and the saturation technique. The dependence of the photoionization cross-section from different intermediate states populated in the first step and ionized in the second step are discussed, including results on the photoionization cross-sections measured above the first ionization threshold. Results based on different polarizations of the exciting and the ionizing dye lasers are also discussed. Examples are provided, illustrating the photoionization cross-sections measured using thermionic diode ion detector, atomic beam apparatus in conjunction with a time-of-flight mass spectrometer and DC/RF glow discharge cell as an optogalvanic detection.

Keywords: photo-absorption; photoionization cross-sections; autoionizing resonances; saturation technique; optogalvanic; atomic beam; time-of-flight mass spectrometry

1. Introduction

Photoionization of atoms is one of the most fundamental processes in the interaction of radiation with matter that plays an important role in different fields of research. In particular, photoionization from the excited states has numerous applications in the stellar atmospheres, controlled thermonuclear research plasma, radiation protection, laser design, and radiative recombination. Therefore, accurate measurement of photoionization cross-sections from the excited states of atoms is a much more stimulating mission. Photo-excitation, photo-ionization, and recombination processes are occurring naturally all the time due to the radiation matter interaction, in the presence of the radiation emitted by the Sun. Since all the elements listed in the Periodic Table possess specific electronic configurations, electrons occupy different shells, and therefore the atoms in the ground state can be promoted to the excited states due to the photo-absorption process. However, if high energy radiation of (>10 eV) is available, then the outermost bound electron can be detached, resulting in a decomposition of atoms into ions and free electrons; a process termed photoionization.

$$Atom + h\nu \rightarrow ion + e^-.$$

Hydrogen is the simplest one-electron system, the electron from its ground state (1s $^2S_{1/2}$) can be promoted to the first excited state (2p $^2P_{1/2,3/2}$) via a dipole transition in the presence of 10.2 eV (121.57 nm) radiation. The interaction of radiation of energy 13.606 eV (911.36 nm) with the hydrogen atoms yields photoionization, depending on the energy density of the available radiation source, since one photon can only excite/ionize one atom. Since the dipole selection rules for excitation from the ground state limit the parity of the excited levels, therefore, only selected processes can be investigated. The advent of tune-able dye lasers has opened up vast possibilities to populate an excited state with a well-defined quantum number and then to excite or ionize atoms via multistep or multiphoton excitations. Thus, with monochromatic radiation of much higher intensity, ionization can be achieved through an intermediate state:

$$\text{Atom (ground state)} + h\nu \rightarrow \text{atom (excited sate)} + h\nu \rightarrow \text{Ionization} + e^-$$

Thus, hydrogen atoms from the ground state (1s $^2S_{1/2}$) can be promoted to the (2s $^2S_{1/2}$) state via a two-photon excitation in the presence of laser radiation at 243.14 nm. The situation is different in the case of the two-electron system. The helium atoms in the ground state (1s^2 $^1S_0$) can be promoted to the first excited state (1s2p $^1P_1$) in the presence of the 58.43 nm radiation and photo-ionized by the 24.212 eV (51.21 nm) radiation; He$^+$ (1s $^2S_{1/2}$) plus a free electron in the ϵp continuum. The ground state of all the alkali atoms is the same as that of hydrogen, whereas that of alkaline earth is similar to helium. Therefore, the excited states as well the ionization fragments will be identical. As the atoms are excited to the higher principal quantum number levels, the successive energy difference between the Rydberg levels decreases ($\sim 2R/n^3$), where R is the Rydberg constant (13.606 eV) and n is the principal quantum number. The intensities of the Rydberg series also decrease as $(1/n^3)$ due to varying spatial overlap of the wave functions between the lower and the excited levels. Consequently, there is a natural limit up to which the highest member of a Rydberg series ($n \sim 740$) can be experimentally measured, ($n \sim \sqrt[3]{\left(\frac{4\pi R \tau}{h}\right)}$), limitations due to the decreasing successive energy difference between the highly excited states and the lifetime or the widths of the spectral lines ($\Delta E \, \Delta t = \frac{h}{2\pi}$). Near the ionization threshold, numerous unresolved Rydberg levels exist that give rise to a sudden jump in the photo-absorption spectrum, which then monotonically decreases as the energies of the photons surpass the first ionization potential. Thus, one can measure the photoionization cross-section at the ionization threshold of any atom provided a suitable ultraviolet radiation continuum source is available. The intensity of an emission line due to an electric dipole transition from an upper level at energy E_u to a level at energy E_l is represented by Cowan [1], Axner et al. [2], Demtroeder [3], and Nahar and Paradhan [4]:

$$I^{em} = \left(\frac{h\nu}{4\pi}\right) A_{ul} N_u ,$$

where N_u is the total number of atoms in the upper level, $h\nu$ is the energy of the emitted photon, and A_{ul} is the Einstein coefficient, which is related to the lifetime of the upper level. Thus, the higher the population of the upper level, the higher the intensity of the spectral line will be. The situation is slightly different in the case of absorption from a lower level to an upper level, as a source of radiation is required to induce the absorption process. The intensity I^{abs} transmitted through a vapor column of length L is represented as:

$$I^{abs} = I_0[1 - \exp(-\sigma_a \, n \, L] \cong I_0 \, \sigma_a \, n \, L,$$

where I_0 is the intensity of the incident radiation beam, n is the number density in the lower level, L is the length of the uniformly distributed sample and σ_a is the absorption cross-section. The difference between photo-excitation and photo-ionization is that in the former case the excitation is to a discrete state, whereas in the latter case the final state is a continuum channel. The transition probability is governed by the dipole transition moment. The probability of photoionization of atoms to ionic states is proportional to the photoionization cross-section, governed by a dipole transition between the initial and final state. The photo-ionization cross-section is proportional to the sum of the squares over all the available final states. In the dipole transitions, the angular momentum and parity selection rules are strictly followed; ($\ell_f = \ell_i \pm 1$), and even to odd or odd to even). Ionization from a specific subshell to various continuum channels are; $ns \rightarrow \epsilon p$, $np \rightarrow \epsilon s, \epsilon d$ and $nd \rightarrow \epsilon p, \epsilon f$. Furthermore, the transitions following the ($\ell_f \rightarrow \ell_i + 1$) selection rule are more probable, possessing higher intensities. The photoionization cross-section is usually much higher near the ionization threshold, and it decreases monotonically by increasing the photon energy above the first ionization threshold. The situation is even more interesting when the experiments are performed at photon energies much higher than the ionization potential. In hydrogen, the photoionization cross-section is maximum at the ionization threshold (13.606 nm) and it decreases monotonically at higher photon energies. A similar trend in photo-ionization prevails for all the alkali atoms, maximum at the ionization threshold and decreasing at higher photon energies. However, the situation gets complex in the case of alkaline earth elements or inert gases as numerous broad and asymmetric autoionizing resonances appear just above the first ionization threshold that is attributed either to the simultaneous excitation of both the valence electrons or to inner shell excitation.

Another characteristic of measuring the photo-ionization cross-section for the excited states of atoms is by populating the first excited state in the first step and then promoting the atoms from this excited state to the ionization threshold. The advantage of such experimental arrangements is that one needs much lower energy photons to photo-ionize atoms as compared with that from the ground state. In this review, we will present different experimental techniques to measure the photo-ionization cross-sections from the excited states of atoms concentrating on alkali atoms, alkaline earth atoms, and inert gases. Several groups have contributed to the photo-absorption measurement of atoms using conventional light sources, Beutler at Berlin, GV Marr at Reading University, WRS Garton at Imperial College, London, and ML Ginter at the Maryland University USA. The work was further extended in the VUV and XUV region using synchrotron radiation, Madden and Codling at NBS, JP Connerade at Bonn University, Sontag at Hamburg, Heinzmann at BESSY, and Ueda in Japan. Now, state-of-the-art storage rings coupled with insertion devices and free electron lasers are being extensively used to measure the photoionization cross-sections of atoms, molecules, and ions. There are several excellent books describing the basics of photoionization measuring techniques and data compilation: Samson [5], Berkowitz [6,7], and Schmidt [8]. The photoabsorption/photoionization cross-sections of atoms from the ground state of the alkali metal atoms have been widely studied by many researchers, such as Rothe [9,10], Marr and Creek [11], Weisheit [12], Ambartzumian et al. [13], and Duong et al. [14]. The spectra of alkaline earth atoms have also been extensively studied, Dichburn and Hudson [15], Garton and Codling [16–18], Hudson et al. [19], Carter et al. [20], Brown et al. [21–23], Wynne and Herman [24], Ueda et al. [25–28], Baig and Connerade [29], Griesmann et al. [30–33], Yie et al. [34], Chu et al. [35], and Maeda et al. [36,37] whereas the spectra of inert gases have been studied by Huffman et al. [38], Rundel [39], Ito et al. [40], Yoshino et al. [41–43], Bonin et al. [44], Baig and Connerade [45], and Ito et al. [46]. There are several reviews addressing the measurement of photoionization of atoms using synchrotron radiation, lasers, and a combination of synchrotron radiation with lasers for multistep excitation and ionization: The experimental data on the photoabsorption spectra of atoms due to inner-shell excitations using synchrotron radiation by Connerade and Baig in the "*Handbook on Synchrotron Radiation*" Marr [47], photoionization and collisional ionization of excited atoms using synchrotron radiation and laser radiation by Wuilleumier et al. [48], the XUV spectroscopy of metal atoms Sontag and Zimmermann [49], the pump-probe experiments in atoms involving laser and synchrotron radiation by Wuilleumier and Mayer [50], a combination of lasers and synchrotron radiation in studies of atomic photoionization by Mayer [51], photoionization cross sections of atomic ions from merged-beam experiments by Kjeldsen [52], the photo-dynamics of excited Ne, Ar, Kr and Xe atoms near the threshold by Sukhorukov et al. [53], experiments at FLASH by Bostedt et al. [54], photoionization of ion with synchrotron radiation: from ions in space to atoms in cages by Schippers et al. [55], roadmap of ultrafast X-ray atomic and molecular physics by Young et al. [56], roadmap on photonic, electronic, and atomic collision physics: light-matter interaction by Ueda et al. [57], and photoionization of astrophysically relevant atomic ions at PIPE by Schippers and Muller [58]. Numerous theoretical models have been developed for the calculation of the photoionization cross-sections. Cooper [59] used as a model the light absorption by a single electron moving in potential similar to the Hartree–Fock potential [H.F] appropriate to the outer subshell of each atom and reported for the rare gases He, Ne, Ar, and Kr, for Na, and the closed-shell ions Cu+ and Ag+. The sum rules were used to analyze the oscillator strength spectral distribution at higher energies. Manson and Cooper [60] used a one-electron model with Herman–Skillman central potential and calculated the photoionization in the soft x-ray range and explained the combined Z and energy dependence of the photoionization cross-sections for different subshells. Kennedy and Manson [61] calculated the photoionization in the noble gases using Hartree–Fock wave functions. Burke and Taylor [62] used the R-matrix method to calculate the electron-atom and ion collision cross-sections atomic polarizabilities to study the atomic photoionization processes in neon and argon. Aymar et al. [63] calculated the photoionization cross-sections for the s, p, and d Rydberg states of lithium, sodium, and potassium in the framework of the single-electron model (non-relativistic) using a parametric central potential. Jonson et al. [64] developed the relativistic random-phase approximation (RRPA) from the linearized time-dependent Hartree–Fock theory (H.F) and determined the excitation energies and oscillator strengths along with the helium, beryllium, magnesium, zinc, and neon isoelectronic sequences. Savukov [65] calculated the photoionization cross-section for the alkali-metal atoms in the framework of relativistic many-body perturbation theory (RMBPT) using quasi-continuum B-spline orbitals. It was inferred that the agreement with the experiment is improved compared to random phase approximation (RPA) and Dirac–Hartree–Fock approximation. The photoionization of potassium atoms from the ground and 4p, 5s–7s, and 3d–5d excited states have been calculated by Zatsarinny and Tayal [66] using the Dirac-based B-spline R-matrix method. The effect of the core polarization by the outer electron was included through the polarized pseudo-states. There was excellent agreement with the experiment for the cross-sections of the 4s photoionization and accurate description of the near-threshold Cooper–Seaton minimum. Kim and Tayal [67] used the non-iterative variational R-matrix method combined with multichannel quantum defect theory at the R-matrix surface to calculate the photoionization of the ground state of magnesium atom in the energy region between 3s and 4p thresholds. Johnson et al. [68] analyzed the Beutler–Fano autoionizing resonances in the rare gas atoms using the relativistic multichannel quantum defect theory. The configuration interaction Pauli–Fock including the core polarization (CIPFCP) method has been applied by Petrov et al. [69,70] to calculate the total and partial cross-sections for the photoionization of excited noble gases. The photoionization cross-sections for the highly excited state and ions have been calculated for several atoms and ions by Nahar's group [71–73] and references therein. Recently, the photoionization and electron-ion recombination of $n = 1$ to very high n-values of hydrogenic ions have been studied by Nahar [74] who also made available a FORTRAN program to compute photoionization cross-sections, recombination cross-sections, and rate coefficients for any principal quantum number and orbital angular momentum shell.

In the next section, the basics of the saturation technique to measure the photoionization for the excited states of atoms using multi-step laser excitation and ionization technique are presented.

2. Saturation Technique

Let us consider the simplest case of a two-step excitation and ionization process to measure the photoionization cross-section for the excited states of atoms. In the first step, the ground state atoms are resonantly excited to the first excited state by the excitation laser pulse and in the second step, the excited atoms are promoted to the continuum by the ionizing laser pulse. When both the laser beams are linearly polarized, the polarization vectors are parallel and only transitions between the magnetic sublevels $\Delta m = 0$ are allowed. The rate equations are used to develop a relationship between the produced photoions and the photoionization cross-section for that excited state. Assuming that collisions and spontaneous emission are discounted in a pure two-step photoionization process, the rate equations are written as Letokove [75], Burkhart et al. [76], He et al. [77], and Saleem et al. [78]:

$$\frac{dN_0}{dt} = -\sigma_a \phi_{ex}(N_0 - N_{ex}), \tag{1}$$

$$\frac{dN_{ex}}{dt} = \sigma_a \phi_{ex}(N_0 - N_{ex}) - \sigma(\lambda_{io})\phi_{io} N_{ex}, \tag{2}$$

$$\frac{dN_I}{dt} = \sigma(\lambda_{io})\phi_{io} N_{ex}, \tag{3}$$

Here, σ_a is the photoabsorption cross-section, ϕ_{ex} is the photon flux of the exciting laser, $\sigma(\lambda_{io})$ is the photoionization cross-section of the excited state at the ionizer laser wavelength λ_{io}, ϕ_{io} is the photon flux of the ionizing laser, N_0 is the population of the ground state, N_{ex} is the population of the excited state, and N_I is the number of ions produced as a result of two-step ionization. The first requirement for the saturation technique is that the intensity of the exciting laser pulse must be sufficiently high enough to saturate the excited state so that both the populations stay in equilibrium i.e., $N_0 \cong N_{ex}$. The total number of atoms is, $N_T = N_0 + N_{ex} = 2N_{ex}$. Adding Equations (1) and (2):

$$\begin{aligned}\frac{d(N_0+N_{ex})}{dt} &= \frac{d(N_T)}{dt} = -\sigma(\lambda_{io})\phi_{io} N_{ex} \\ \Rightarrow \frac{dN_{ex}}{dt} &= -\frac{1}{2}\sigma(\lambda_{io})\phi_{io} N_{ex} \\ \Rightarrow N_{ex} &= \frac{N_0}{2}\exp\left(-\frac{1}{2}\int_{-\infty}^{t}\sigma(\lambda_{io})\phi_{io}(t')dt'\right).\end{aligned} \tag{4}$$

Here, N_0 is the number density in the ground state before the arrival of the exciting laser pulse, and the exponential term containing t' determines the decreasing number of excited atoms at any time t after the arrival of the ionizer laser pulse. Integrating Equation (3) and inserting the value of N_{ex} from Equation (4);

$$N_I = \int_{-\infty}^{\infty} \sigma(\lambda_{io})\phi_{io}(t)\frac{N_0}{2}\exp\left(-\frac{1}{2}\int_{-\infty}^{t}\sigma(\lambda_{io})\phi_{io}(t')dt'\right)dt. \tag{5}$$

The solution of the equation yields:

$$N_I = N_0\left[1 - \exp\left(-\frac{1}{2}\int_{-\infty}^{\infty}\sigma(\lambda_{io})\phi_{io}(t)dt\right)\right].$$

At saturation, $N_0 \cong N_{ex}$, therefore;

$$N_I = N_{ex}\left[1 - \exp\left(-\frac{1}{2}\sigma(\lambda_{io})\Phi_{io}\right)\right]. \tag{6}$$

Here Φ_{io} is the fluence (energy/area) of the ionizing laser as seen by the atoms at the beam center: $\Phi_{io} = \int_{-\infty}^{\infty} \phi_{io}(t)dt$. However, the fluence is related to the energy E (J) of the ionizing laser pulse as: $\Phi_{io} = \frac{\lambda_{io}E}{hcA}$. By substituting it in Equation (6):

$$N_I = N_{ex}\left[1 - \exp\left(-\frac{1}{2}\sigma(\lambda_{io})\frac{\lambda_{io}E}{hcA}\right)\right]. \tag{7}$$

However, the number of produced ions N_I is related to the total charge Q that is produced by the ions in the ionizing volume V(cm^3) as: $N_I = Q/eV$. While performing experiments, photoions produced as a result of two-step photoionization are registered as a voltage signal, which is related to the charge as $Q = \left(\frac{Voltage signal}{R}\right) \times \Delta t$, where R is the terminating resistance (ohm) at the oscilloscope and Δt (seconds) is the pulse width (FWHM) of the photoion signal peak. A final equation turns out to be:

$$N_I = N_{ex} V \left[1 - \exp\left(-\frac{1}{2}\sigma(\lambda_{io})\frac{\lambda_{io} E}{hcA}\right)\right]. \quad (8)$$

The absolute value of the photoionization cross-section $\sigma(\lambda_{io})$ for a particular excited state at a specific ionizing laser wavelength λ_{io} is extracted from a least-squares fit to the experimentally measured ionization data (N_I) as a function of the energy (J) of the ionizing laser beam (E). This relation shows that the photoionization signal approaches saturation as the energy of the ionizing laser is increased to a much higher value. However, every detection system encounters certain limitations. This equation is applicable under the following assumptions:

- The first step transition remains saturated (i.e., $N_0 \approx N_{ex}$), during the exciting laser pulse duration (10 ns);
- The laser intensity of the exciting laser is kept much higher than that required to saturate the transition. Consequently, the Rabi frequency is high, and the spontaneous emission may be ignored during the laser pulse;
- The intensity of the ionizing laser is sufficiently higher than that required to completely ionize atoms in the excited state.

The photoionization cross-section σ is then extracted from a least-squares fit to the experimental photoion data registered as a function of the energy density (Energy/area) of the ionizing laser. The error associated with this method is (i) in determining the cross-sectional area in the interaction region, (ii) in measuring the pulse energy with an energy meter, (iii) in the transmission of the optical windows, and (iv) in the fitting process. Thus, to extract an accurate absolute value of the photoionization cross-section, it is important to accurately measure the correctional area (A) of the laser beam and the Energy per pulse (E) of the ionizing laser. The energy density/intensity of a laser beam in the interaction region depends on its spot size, which can be calculated under the assumption of a Gaussian laser beam:

$$diameter\ of\ the\ focused\ beam = \left(\frac{4}{\pi}\right)\lambda\left(\frac{f}{D}\right). \quad (9)$$

Here, d is the diameter of the focused laser beam, λ is the laser wavelength, f is the focal length of the focusing lens, and D is the diameter of the laser beam falling on the focusing lens. However, the depth of focus for the laser beam can be calculated as:

$$depth\ of\ focus = \left(\frac{8}{\pi}\right)\lambda\left(\frac{f}{D}\right)^2. \quad (10)$$

The cross-sectional area is also calculated under the diffraction limitations: Demtröder [3].

$$Cross\ sectional\ area = \pi\,\omega_0^2 \left[1 + \left(\frac{z\,\lambda_{io}}{\pi\,\omega_0^2}\right)^2\right]. \quad (11)$$

Here, $\omega_0 = \left(\frac{\lambda f}{\pi \omega_s}\right)$ is a diffraction-limited radius at $Z = 0$, ω_s is the radius of the spot size of the ionizing laser beam on the focusing lens, λ_{io} is the wavelength of the ionizing laser, and z is the distance on the beam propagation axis from the focus. Thus, an accurate value of the cross-sectional area depends on the accuracy of Z and ω_s. A rough estimation of the cross-sectional area can also be determined either by using a photographic method by measuring the burn spot or using a thermopile in combination with a moveable knife-edge intersecting the laser beam. Based on the experimental uncertainties, the cross-section can be determined with $\pm 15\%$ accuracy. Photoionization cross-section is conventionally reported in Mb units, where 1 Mb = 10^{-18} cm^2.

The two-step excitation and ionization experiments are normally conducted using narrow-bandwidth tunable dye lasers, pumped by a high-power laser such as Ruby, Nitrogen, Excimer, or Nd:YAG laser. A 600 nm dye laser beam with a 5 mm diameter, and focused with a lens of 100 mm focal length, the diameter of the focused dye laser beam will be about 15 µ and the depth of the focused beam will be about 610 µ. Once the diameter of the ionizing laser beam is measured, the laser beam's cross-sectional area and the energy density of the dye laser can be determined. Thus, one measures the ion signal as a function of the intensity of the ionizing laser beam. Initially, the ionization signal keeps on increasing as the energy of the ionizing laser is increased. This is because one photon can only ionize one atom. As the intensity of the ionizing laser is increased, the number of ions also increases up to a point where the ionization signal stops increasing even with any further increase in the intensity of the ionizing laser beam. At this point, saturation in the ionization signal sets in, which is why this technique is termed "saturation technique". The experimental data points are used to fit Equation (8), which yields the absolute value of the photoionization cross-section of the excited state and the ground state density of the atom. The determination of N_{ex} is independent of the photoionization cross-section, which is determined from the asymptotic value, while σ is associated with the shape of the experimental data of ionization signal curve against the ionizing laser intensity. The accurate determination of N_{ex} requires that both the transitions must be saturated, and that the ionizing volume is measured accurately. The ionizing volume is defined as the interaction volume of the overlap region of the exciting and the ionizing laser beams in the effective collection region.

At the Atomic and Molecular Physics Laboratory, Quaid-i-Azam University, Islamabad, Pakistan, we have developed different experimental arrangements to measure the photoionization cross-section for the excited states of atoms: a thermionic diode ionic detector based on the heat pipe design by Niemax [79], Baig et al. [80,81], and Yaseen et al. [82], an optogalvanic effect-based detection system by Barbieri et al. [83] and Babin et al. [84], using a DC discharge system, as in Piracha et al. [85], Stockhausen et al. [86], and Hanif et al. [87], a RF discharge system, as in Zia et al. [88,89], and an atomic beam apparatus coupled with a time-of-flight mass spectrometer based on a linear TOF design as in Wiley and McLaren [90] and Saleem et al. [91]. Several other techniques for the measurement of photoionization cross-sections are also being used such as the modulated fluorescence technique by Gilbert et al. [92], a magneto-optical trap (MOT) system by Wippel et al. [93], Madsen and Thomsen [94], and Dinneen et al. [95], the isotope-selective photoionization for calcium ion trapping by Lucas et al. [96], and finally by Marago et al. [97], who measured the photoionization cross-section for the 6p $^2P_{3/2}$ excited laser-cooled cesium atoms. The photoionization cross-sections of the excited states of titanium, cobalt, and nickel were reported by Yang et al. [98]. Cong et al. [99] used resonance-enhanced multiphoton ionization coupled with a time-of-flight mass spectrometer. Zheng et al. [100] reported the measurements of photoionization cross-sections of the excited states of cobalt using a two-color, two-step resonance ionization technique in conjunction with a molecular beam time-of-flight (TOF) mass spectrometer at the threshold and near-threshold regions (0–1.2 eV). A comparison and working principles of different experimental techniques being used for the measurements of the photoionization cross-section from the ground state or excited states of atoms were presented by Saleem et al. [101]. In the next sections, we present details about the measurements of the photoionization cross-sections for the excited states of alkali atoms.

3. Alkali Atoms

The ground state configurations for the alkali atoms are similar to that of hydrogen, having a single electron in the ground state, ms $^2S_{1/2}$ (m = 2, 3, 4, 5, and 6 for Li, Na K, Rb, and Cs, respectively). The atoms from the ground state are excited by the Rydberg states: np $^2P_{1/2,3/2}$, and photoionization activates as the energies of the interacting photons approach the value of the first ionization potential of an atom. The process of photoionization from the ground state yields an ion plus a free electron and the continuum above the first ionization threshold is represented as εp ($\ell = 1$; $J = 1/2, 3/2$) channels.

$$ms\ ^2S_{1/2}\ (\text{excitation}) \rightarrow np\ ^2P_{1/2,3/2}\ (\text{ionization}) \rightarrow \varepsilon p\ \text{continuum}\ (J = 1/2, 3/2).$$

All the alkali atoms show monotonically decreasing photoionization cross-sections above the first ionization threshold, just like the photoionization cross-section in the hydrogen atom. The pioneering work on the measurement of the photoionization cross-section from the excited states of lithium and sodium was reported by Rothe [9,10] using the recombination radiation method. The photoionization cross-section for the 2p excited state at the threshold is 19.7 ± 3.0 Mb, assuming the oscillator strength of the $5s\ ^2S_{1/2} \rightarrow 2p\ ^2P_{1/2}$ transition as 0.00417. The photoionization of the alkali metals atoms, sodium, potassium rubidium, and cesium was measured by Marr [11] and Weisheit [12]. The advent of lasers and dye lasers enabled the measurement of photoionization cross-sections for the excited state of atoms at and above the first ionization threshold. The first excited state $mp\ ^2P_{1/2,3/2}$ in alkali atoms is populated in the first step by tuning a dye laser at an appropriate wavelength, a second dye laser is scanned up to the first ionization threshold, and the $ns\ ^2S_{1/2}$ and $nd\ ^2D_{3/2,5/2}$ Rydberg series are observed. Above the first ionization threshold, the continuum is represented as εs ($\ell = 0$; $J = 1/2$), εd ($\ell = 2$; $J = 3/2, 5/2$). Thus, by selecting excited levels of different ℓ-values in the first step, different $\varepsilon \ell$ ($\ell = 0, 1, 2, 3 \ldots$) channels in the continuum can be explored. The total photoionization cross-section is the sum of the partial cross-sections:

$$\sigma_{j0}(\Delta E)(Mb) = \frac{4}{3}\pi^2 \alpha a_0^2 \Delta E \frac{1}{2J_0+1} \sum_{\ell j} |< n_0 \ell_0 J_0 ||D(r)|| \varepsilon \ell J >|^2, \qquad (12)$$

where $\varepsilon \ell$ are the continuum channels; εs ($J = 1/2$) or εd ($J = 3/2, 5/2$), J_0 is the J-value of the intermediate level ($J = 1/2, 3/2$), $\ell_0 = 1$ for p, ΔE is the energy difference between the ionizing photon energy and the ionization potential, a_0 is the Bohr radius, and α is the fine structure constant.

In Table 1, the laser wavelength to access the resonance levels in the first step and the ionizing laser wavelengths in the second step for all the alkali atoms are enlisted. The first step laser wavelengths are mostly in the red region, whereas the ionizing laser wavelength is in the green region, which can easily be achieved by a dye laser pumped by a Nd: YAG laser (2nd or 3rd harmonics), excimer laser, or nitrogen laser.

Table 1. Two-step excitation and ionization laser wavelengths for the alkali atoms.

Atom	First Step (Excitation) $ms\ ^2S_{1/2} \rightarrow mp\ ^2P_{1/2,3/2}$	Second Step (Excitation) $mp\ ^2P_{1/2,3/2} \rightarrow ns\ ^2S_{1/2}$ $\rightarrow nd\ ^2D_{3/2,5/2}$	Ionization Threshold εs, ($J = 1/2$) εd, ($J = 3/2, 5/2$)
Li I	2p $^2P_{1/2}$ 670.793 nm $^2P_{3/2}$ 670.778 nm	349.8 nm 349.8 nm	43,487.150 cm^{-1}
Na I	3p $^2P_{1/2}$ 589.592 nm $^2P_{3/2}$ 588.995 nm	408.2 nm 408.5 nm	41,449.451 cm^{-1}
K I	4p $^2P_{1/2}$ 769.896 nm $^2P_{3/2}$ 766.489 nm	454.0 nm 455.2 nm	35,009.814 cm^{-1}
Rb I	5p $^2P_{1/2}$ 794.760 nm $^2P_{3/2}$ 780.027 nm	473.6 nm 479.0 nm	33,690.810 cm^{-1}
Cs I	6p $^2P_{1/2}$ 894.347 nm $^2P_{3/2}$ 852.113 nm	494.3 nm 508.2 nm	31,406.467 cm^{-1}

A schematic diagram for the two-step laser excitation and ionization for the alkali atoms is presented in Figure 1.

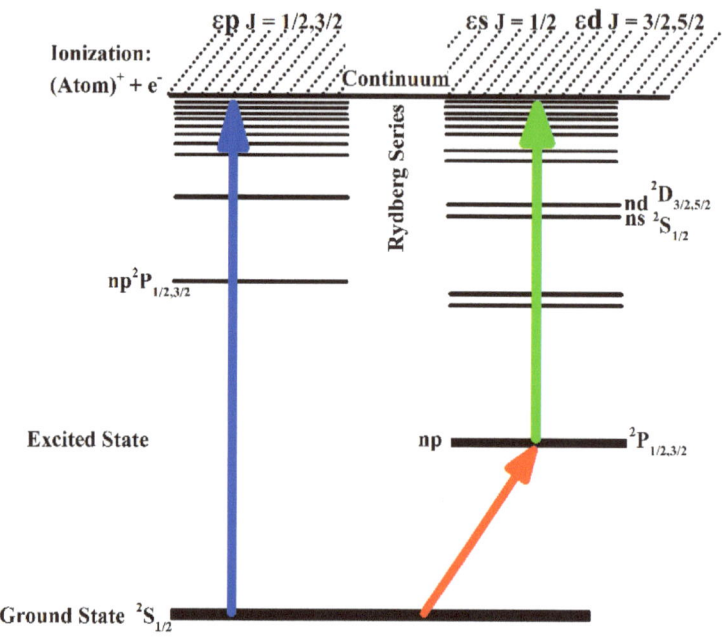

Figure 1. Schematic diagram of the excitation from the ground state and two-step laser excitation and ionization scheme for alkali atoms.

The pioneering work to measure the photoionization cross-section for the excited state of the rubidium atoms using the saturation technique was performed by Ambartzumian et al. [13], who used a Ruby laser pumped dye laser system coupled with a heat pipe containing rubidium vapor and measured the photoionization cross-section for the 6p $^2P_{1/2}$ and 6p $^2P_{3/2}$ excited states. The exciting laser was focused by a lens onto a glass cell containing rubidium vapor and the laser radiation was tuned in resonance to the 5s $^2S_{1/2} \rightarrow$ 6p $^2P_{1/2,\,3/2}$ transitions for selective excitation at 421.56 nm and 420.18 nm, respectively. For producing ionization, a part of the fundamental (694.30 nm) or its second harmonic (347.15 nm) was directed into the cell from the opposite side. The irradiated volume was limited by inserting diaphragms on both ends of the cell and the dependence of the ionization signal on the intensity of the ionizing radiation at constant cell temperature was recorded. The intensity of the ionizing laser was varied, and the corresponding ionization signal was measured. At the low intensity of the ionizing laser, the dependence of the ionization was linear, and with an increase in the ionizing laser intensity, it deviated from linearity and tended to saturation, which means that at high-intensity, the total ionization of the excited transpires. The values of the cross-sections are $(1.7 \pm 0.4) \times 10^{-17}$ cm^2 for the 6p $^2P_{3/2}$ state and $(1.5 \pm 0.4) \times 10^{-17}$ cm^2 for the 6p $^2P_{1/2}$ at the 694.3 nm ionizing laser wavelength. The value of the photoionization cross-section for the 6p $^2P_{3/2}$ state was determined as $(1.9 \pm 0.5) \times 10^{-17}$ cm^2 at the 347.15 nm ionizing laser wavelength. Subsequently, Heinzmann et al. [102] measured the photoionization cross-section for the 7p $^2P_{1/2}$ excited state of cesium at 459.3 nm ionizing laser wavelength as $(6.2 \pm 0.5) \times 10^{-18}$ cm^2, and for the 7p $^2P_{3/2}$ excited states at 455.5 nm ionizing laser wavelength as $(8.8 \pm 1.6) \times 10^{-18}$ cm^2. The photoionization of the 6p $^2P_{1/2,\,3/2}$ fine-structure levels of cesium was measured by Nygaard et al. [103] using a triple-crossed beam experiment covering the wavelength region from 500 nm to 250 nm. The photoionization cross-section for the 6p $^2P_{3/2}$ and $^2P_{1/2}$ excited states was measured at 508.3 nm and 494.4 nm ionizing wavelengths at the first ionization threshold. The absolute photoionization cross-sections for the 5s and 4d excited states of sodium were measured by Smith et al. [104] using two stabilized single-mode CW dye lasers intersecting with the sodium atomic beam to stepwise excite atoms to the desired excited state, and the fundamental wavelength of the Nd:YAG laser at 1064 nm was used as an ionizing laser. The cross-section for the 5s state was determined as $(1.49 \pm 0.13) \times 10^{-18}$ cm^2 and that for the 4d state as $(15.2 \pm 1.70) \times 10^{-18}$ cm^2. The photoionization cross-section for the 7p $^2P_{3/2}$ and 6d $^2D_{3/2}$ excited states of cesium was reported by Gerwert and Kollath [105] and that for the 7 s state of cesium at 540 nm ionizing laser wavelength was measured by Gilbert et al. [92] as $1.14\ (10) \times 10^{-19}$ cm^2 using the modulated fluorescence technique. Bonin et al. [106] measured the absolute photoionization cross-section of the 7d $^2D_{3/2}$ excited state of cesium at different ionizing laser wavelengths using the fluorescence reduction technique.

The absolute value of the photoionization cross-section of the excited 7p state of potassium was measured by Baohua and Zuren [107] at two ionizing laser wavelengths, 321.8 nm, and 643.6 nm, as (0.9 ± 0.6) Mb and (1.9 ± 0.9) Mb, respectively. Maeda and Ambe [108] measured the photoionization cross-sections for the 7d and 8d states of cesium using the technique of measuring the fluorescence signal from the excited states of atoms due to photoionization. The photoionization cross-sections were reported as $(3.5 \pm 1.4) \times 10^{-18}$ cm^2 and $(3.1 \pm 1.2) \times 10^{-18}$ cm^2 for the 7d $^2D_{3/2}$ and 7d $^2D_{5/2}$ states and $(2.4 \pm 1.0) \times 10^{-18}$ cm^2 and $(1.6 \pm 0.6) \times 10^{-18}$ cm^2 for the 8d $^2D_{3/2}$ and 8d $^2D_{5/2}$ states, respectively. The photoionization cross-section for the 6p $^2P_{3/2}$ excited laser-cooled cesium atoms was measured by Marago et al. [97]. Subsequently, Patterson et al. [109] measured the photoionization cross-section of the 6p $^2P_{3/2}$ excited state of cesium confined in a magneto-optical trap. The photoionization rate was measured by monitoring the decay of the trap fluorescence during exposure to ionizing laser radiation, using several lines of Ar-ion laser in the wavelength range 457.9 to 501.7 nm and the photoionization cross-section at 496.5 nm ionizing laser wavelength was reported as $(1.86 \pm 0.15) \times 10^{-17}$ cm^2. Petrov et al. [110] presented the effect of the polarization of the atomic core by the outer electron on the near-threshold photoionization of excited alkali atoms (Na–Cs). Partial and total cross-sections for photo-ionization of the np-electron were computed utilizing the configuration interaction technique with Pauli–Fock atomic orbitals (CIPF) and including the long-range core polarization potential (CP). The variational principle was applied to calculate the core polarization potential. Comparison with previous theoretical results and with available experimental data was presented for the total cross-section σ, for the electron angular distribution parameter β, for the ratio ν = |Dd/Ds| of the reduced electric dipole matrix elements, and the phase shift difference $\Delta = \delta_d - \delta_s$, associated with the d-wave and s-wave continua, respectively. A magneto-optical trap (MOT) system was used by Wippel et al. [93] to measure the photoionization cross-sections of the first excited states of sodium and lithium. A two-element magneto-optical trap (MOT) for Na and Li^6 or Li^6 was used to cool and trap each atom separately. A fraction of the cold atoms was maintained in the first $^2P_{3/2}$ excited state by the cooling laser and the excited state atoms were ionized by the laser light in the near ultra-violet region. Duncan et al. [111] also used the magneto-optical trap (MOT) system to measure the photoionization cross-section of the 5 d $^2D_{5/2}$ excited state of rubidium at the ionizing laser wavelengths ranging from 1064 to 532 nm. The lifetimes and photoionization cross-sections at 10.6 m of the nd Rydberg states of Rb measured in a magneto-optical trap were assessed by Gabinini [112].

The work on the measurement of photoionization cross-sections for the excited states of atoms was initiated by our group in 2002 and we measured the photoionization cross-section for the 3p $^2P_{1/2}$ and $^2P_{3/2}$ excited levels of sodium as 2.16 (43) Mb and 3.74 (74) Mb, respectively, using two-step laser excitation in conjunction with a thermionic diode working in the space charge limited mode and employing the saturation technique (Amin et al. [113]). The photoionization cross-section for the 4d $^2D_{5/2}$ level was measured as 12.2 (2.4) Mb by populating this level via two-photon excitation from the ground (Amin et al. [114]). The cross-section for the 4d $^2D_{3/2}$ level via the 3p $^2P_{1/2}$ level intermediate level and for the 4d $^2D_{3/2,5/2}$ levels via the 3p $^2P_{3/2}$ level was reported as (9.6 ± 1.9) Mb, and (12.8 ± 2.5) Mb, respectively. This work was extended to measure the cross-section for the excited levels of lithium for the 2p $^2P_{1/2,3/2}$, 3d $^2D_{3/2,5/2}$, and 3s $^2S_{1/2}$ excited states at different ionizing laser wavelengths, above the first ionization threshold, as stated by Amin et al. [115]. By changing the ionization photon energy, the smooth frequency dependence of the cross-sections was observed for the 2p and 3d excited states of lithium. The photoionization cross-section for the 4p $^2P_{3/2}$ and $^2P_{1/2}$ states at 355 nm was measured as (7.2 ± 1.1) Mb and (5.6 ± 0.8) Mb and for the 5d $^2D_{5/2,3/2}$ state as (28.9 ± 4.3) Mb (Amin et al. [116]. The cross-section for the 5d $^2D_{3/2}$ state was populated via the 4p $^2P_{1/2}$ intermediate state and for the 5d $^2D_{5/2,3/2}$ states via the 4p $^2P_{3/2}$ intermediate state as (25.1 ± 3.8) Mb and (30.2 ± 4.5) Mb. The photoionization cross-sections for the 3p level of sodium at the first ionization threshold were reported as 7.63 (90) Mb by Rothe [10], and Aymar et al. [63] calculated its value as 7.38 Mb. Preses et al. [117] determined the photoionization cross-section from the 3p $^2P_{3/2}$ state of Na up to the first ionization threshold, with the help of two antiparallel, interpenetrating pulsed laser beams pumped by the frequency-doubled (532 nm) and tripled (355 nm) outputs of a Nd:YAG laser. They determined the value of the cross-section as 8.5 Mb with an estimated uncertainty of about 25%. Petrov et al. [110] calculated the photoionization cross-section of the 3p $^2P_{3/2}$ state at about ≈ 8 Mb. Wippel et al. [93], using the trapping technique, trapped a fraction of Na atoms in the 3p $^2P_{3/2}$ excited state and then ionized them with a laser adjusted at ~407.8 nm. They determined the value of the cross-section at the threshold as 6.9(1) Mb. Miculis and Meyer [118] calculated the photoionization cross-section from 3p $^2P_{3/2}$ at and above the threshold as 6.9 (1) Mb, which is in excellent agreement with the recently reported values 3p $^2P_{1/2}$ as 7.9 (1.3) Mb and 3p $^2P_{3/2}$ as 6.7 (1.1) Mb by Baig et al. [119]. The photoionization cross-section for the 4s $^2S_{1/2}$ state of sodium at the first ionization threshold was measured by Rafiq et al. [120] as (0.65 ± 0.10) Mb using a thermionic diode ion detector. The photoionization cross-sections from the 6p $^2P_{3/2}$ and 7p $^2P_{3/2}$ excited states of potassium have been measured at different ionizing laser wavelengths using an atomic beam apparatus coupled with a time-of-flight mass spectrometer by Yar et al. [121,122]. Haq and Nadeem [123] measured the photoionization cross-section for the 6p $^2P_{3/2}$ state of cesium at the ionization threshold, which was measured at 25 ± 4 Mb as well as at different ionizing lasers wavelengths. The cross-section for the 5p $^2P_{3/2}$ state of rubidium was reported as 18.8 ± 3 Mb by Nadeem and Haq [124], whereas Shahzada et al. [125] reported the photoionization cross-section for the 3p $^2P_{1/2,3/2}$ levels of lithium at the first ionization threshold as (30.0 ± 4.8) Mb, and also determined the oscillator strengths of the 3p $^2P \rightarrow nd$ ^2D Rydberg transitions. Saleem et al. [126] used a two-step selective excitation and ionization technique coupled with an atomic beam apparatus and a time-of-flight (TOF) mass spectrometer and measured the photoionization cross-sections of the lithium isotopes Li^6 and Li^6 for the 2p excited state as (15 ± 2.5) Mb and (18 ± 2.5) Mb, respectively. The excitation scheme is shown in Figure 2.

A technique for the isotopic enrichment of lithium isotope was demonstrated by Saleem et al. [127], who employed the two-step photoionization technique along with a narrow-band dye laser in conjunction with a time-of-flight mass spectrometeLi^{6}r, which yielded a high degree of selectivity by tuning the dye laser at the resonance levels of Li^6 and Li^6. It was inferred that the concentration of the natural abundance of the Li^6 isotope becomes enhanced up to over 47% as the exciter dye laser was tuned to 2p $^2P_{1/2}$ of Li^6, even if the linewidth of the exciter laser was not sufficiently narrow to excite the isotopic level. Although the linewidth of the exciting dye laser was not narrow enough to selectively excite the 2p $^2P_{1/2, 3/2}$ levels of Li^6, TOF-MS separated both the lithium isotopes on the time axis. Consequently, the photoionization signals of the resolved fine structure components as a function of the intensity of the ionizing laser were measured simultaneously. The exciting laser was tuned at 670.8 nm to populate the 2p-excited state via a single-photon excitation. The exciting laser was tuned to the $^2P_{1/2}$ excited state of Li^6, between the $^2P_{3/2}$ of Li^6 and $^2P_{1/2}$ of Li^7 and $^2P_{3/2}$ of Li^7 while the ionizing laser wavelength was fixed for the measurement of the photoionization cross-section for both the lithium isotopes. The photoionization cross-section from the 2p $^2P_{1/2}$ excited state of Li^6

and for the 2p $^2P_{3/2}$ excited states of Li^7 was also measured by tuning the exciter laser to the corresponding excited state. The energy of the exciter laser at any frequency was kept fixed and the energy of the ionizer laser was varied using neutral density filters to achieve the complete ionization of the excited isotopic atoms. The resulting photoion signals from the 2p $^2P_{1/2}$ excited state of Li^7 versus the ionizing laser energy density at 335.4 nm are shown in Figure 3.

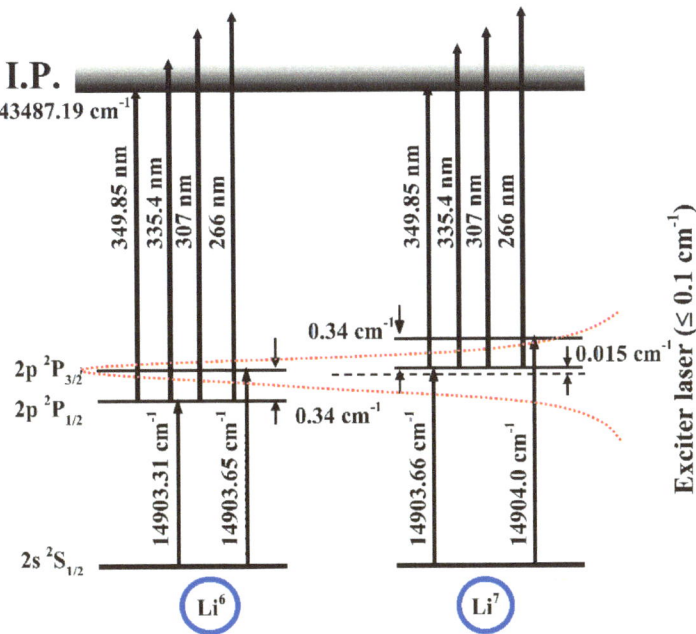

Figure 2. The schematic diagram for two-step photoionization of lithium isotopes.

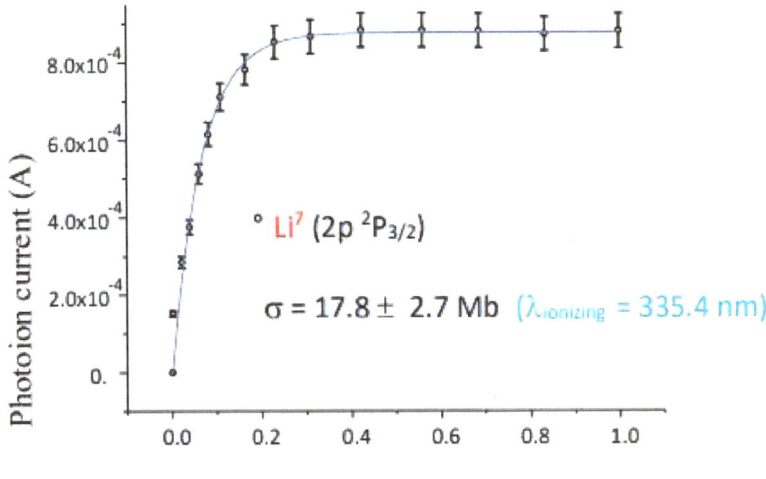

Figure 3. Photoion current signals of Li^7 from the 2p $^2P_{3/2}$ excited states versus the energy density of the ionizer laser at 335.4 nm ionizing laser wavelength.

The photoion signal increases with an increase in the energy density of the ionizer laser up to a certain value, then stops increasing further, and finally saturation sets in. The solid line that passes through the experimental data points is the least square fit to Equation (7), which yields the absolute photoionization cross-section of the 2p fine structure excited states at 335.4 nm ionizer laser wavelength. The measurements of the photoionization cross-section from the 2p $^2P_{1/2, 3/2}$ excited states of lithium using different ionizing laser wavelengths at and above the first ionization threshold are listed in Table 2. The reason for using different ionizing lasers is that by varying the frequency of the ionizing laser, the electrons of different kinetic energies are produced and the behavior of the photoionization in different regions of the continuum can be investigated. The value of the photoionization cross-section decreases as the ionizing laser wavelength is decreased.

Table 2. Experimentally determined photoionization cross-sections from the 2p $^2P_{1/2, 3/2}$ excited states of both isotopes of lithium at different ionizing laser wavelengths.

Lithium Isotope	Excited-State	Ionizing Laser Wavelength	Cross-Section σ (Mb)	Reference
Li^6	2p $^2P_{1/2, 3/2}$	334.4 nm	15 ± 15	Wipple et al. [93]
		335.8 nm	6 (−5, +20)	Wipple et al. [93]
		335.4 nm	15.0 ± 2.3	Saleem et al. [126]
	2p $^2P_{3/2}$	349.85 nm	15.4 ± 2.3	Saleem et. al. [126]
		335.4 nm	14.8 ± 2.2	
		307 nm	10.5 ± 1.6	
		266 nm	6.8 ± 1.0	
	2p $^2P_{1/2}$	349.85 nm	13.8 ± 2.0	
		335.4 nm	13.2 ± 1.9	
		307 nm	9.5 ± 1.4	
		266 nm	5.9 ± 0.9	
Li^7	2p $^2P_{1/2, 3/2}$	334.4 nm	16.2 ± 2.5	Wipple et al. [93]
		335.8 nm	18.3 ± 2.8	Wipple et al. [93]
		335.4 nm	18.0 ± 2.7	Saleem et al. [126]
	2p $^2P_{3/2}$	349.85 nm	19.2 ± 2.9	Saleem et al. [126]
		335.4 nm	17.8 ± 2.7	
		307 nm	12.5 ± 1.9	
		266 nm	8.0 ± 1.2	
	2p $^2P_{1/2}$	349.85 nm	16.5 ± 2.5	
		335.4 nm	15.5 ± 2.3	
		307 nm	10.8 ± 1.6	
		266 nm	6.8 ± 1.0	

The photoionization cross-sections for the 2p, 3s, 3p, 3d, and 4s. The 4p and 4d excited states of both isotopes of lithium using two-step laser excitation and ionization coupled with a time-of-flight mass spectrometer (TOF-MS) have been extensively studied by Hussain et al. [128] and Saleem et al. [129]. The measured values of the photoionization cross-sections, above the first ionization threshold, for the same principal quantum number $n = 3$ but at different orbital angular momentum states $\ell = 0, 1, 2$ (3s, 3p, 3d), are collectively shown in Figure 4.

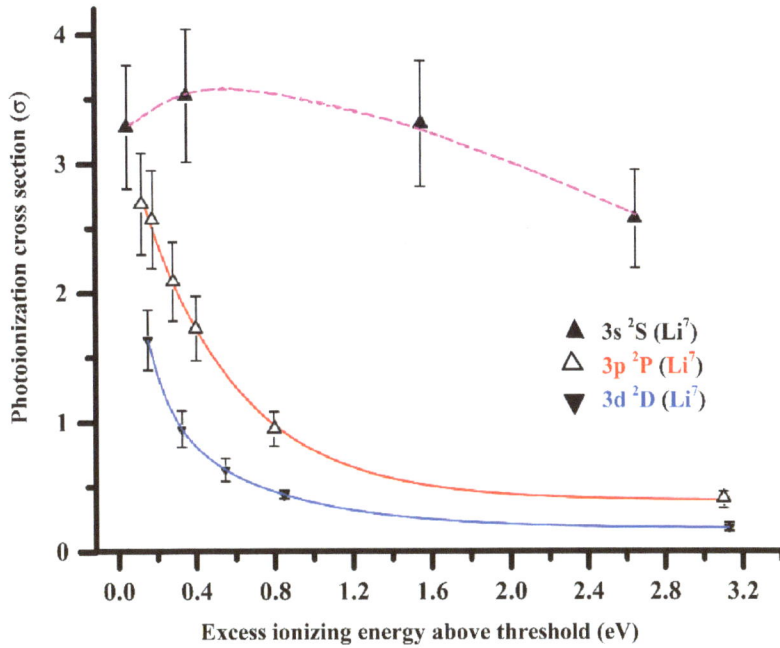

Figure 4. A comparison of the energy-dependent behavior of the measured values of the photoionization cross-section from the 3s ^{2}S, 3p ^{2}P, and 3d ^{2}D.

The values of the cross-section from the 3s and 4s excited states were multiplied by a factor of 25 just to compare the trends of photoionization cross-sections with other states. The dotted lines passing through the values of the cross-section of 3s excited states at different ionizing wavelengths are not the fitted curves but are simply drawn for the comparison with the 3p, and 3d excited states. The solid lines are the exponential decay fit to the experimental data points for the photoionization cross-section from the 3p and 3d excited states. The behavior of the photoionization cross-section for the excited states is correlated with the difference between the initial state quantum defect and the final continuum threshold phase shift. The quantum defects of s and p states of lithium are 0.40 and 0.05, respectively, and effectively zero for $\ell \geq 2$. Thus, only transitions involving s or p states are non-hydrogenic. The oscillator strength distribution in the discrete and continuum regions of the spectrum of lithium was explored by Hussain et al. [130], who determined the photoionization cross-section for the 3s ^{2}S$_{1/2}$ level at the first ionization threshold and then extracted the f-values of the 3s ^{2}S–np ^{2}P Rydberg series. Felfli and Manson [131] remarked that a difference of about 0.5 is necessary to have a Cooper minimum in the continuum. There is only one channel εp through which the electrons from the ns excited states can be promoted to the ionization continuum. The difference between the ns quantum defect and the threshold εp phase shift is 0.37, which is less than 0.5, and therefore a minimum is expected in the discrete region near the threshold. The presence of these minima causes the ns cross-section to be anomalously small at the threshold (Lahiri and Manson [132,133] and the references therein). The measured values of the cross-section for the 3s and 4s excited states at the threshold by Saleem et al. [129] are in good agreement with the calculations by Lahiri and Manson [132] in comparison with Aymar et al. [63]. Saha et al. [134] performed the multi-configuration Hartree–Fock calculation of the photoionization of the excited Na 4d state. Recently, Qi et al. [135] calculated the photo-ionization cross-sections for several excited states of lithium and their calculations are in excellent agreement with that reported by our group.

Interestingly, the photoionization cross-sections for the 3s ^{2}S and 4s ^{2}S excited states of lithium first increase to a maximum value and then decreases with the increase in the ionizing laser photon energy, also deviating from the hydrogenic behavior. A smooth decrease in the photoionization cross-section above the ionization threshold is attributed to the fact that when the ionizing electron gains high kinetic energy, its electron wave becomes more oscillatory and, as a result, the vacated orbital and the photoelectron wave are no more in the same spatial region and consequently the photoionization cross-section for that ionizing photon decreases (Green and Decleva [136]). The behavior of the photoionization cross-section from the np-excited states also differs from that of the ns or nd excited states. The excited electrons from the p-states can be promoted to the ionization continuum through two different channels np → εs and np → εd. The cross-sections decrease monotonically with the increase in the ionizing laser wavelength above the first ionization threshold, but this falloff of the cross-section is not hydrogenic due to the smaller quantum defects for the p-states. Furthermore, near the threshold, the cross-section increases with n and then falls off more rapidly with the excess photon energy. The rapid decay of the cross-section near the threshold with an increase in the principal quantum number n indicates the decreasing contribution of the non-hydrogenic region of the potential. For the photoionization of the atoms from the nd-excited states, there are two possible ionization channels, εp and εf, through which the excited electrons from these states can be uplifted to the ionization region. The nd → εp cross-section is slightly different from hydrogenic, owing to the small p-wave phase shift, but the nd → εf is completely hydrogenic because for $\ell \geq 2$, the quantum defects or phase shifts are effectively zero. The fitted curves through the measured data points for the 3d ^{2}D excited state decrease more sharply with the increase in the ionizing laser photon energy. This decreasing behavior of the cross-section from the nd excited states is hydrogenic as compared to that of the ns or np excited states.

The ionization from the ground state of alkali atoms possesses two partial waves, εp, $J = 1/2$ and $J = 3/2$, due to the spin-orbit effect. They have slightly different phases, and their amplitudes pass through zero at different energies. Thus, photoionization from the ground state passes through a minimum in the region of maximum overlap of the εp wave functions with the ground state wave function as a function of the photon energy. Sandner et al. [137] measured the photoionization of potassium in the vicinity of the minimum in the cross-section using a time-of-flight technique. The photoionization cross-sections for the alkali metal atoms were calculated in the framework of the relativistic many-body perturbation theory (RMBPT) using B-spline orbitals by Savukov [65]. Zatsarinny and Tayal [66] used the Dirac-based B-spline R-matrix method to investigate the photoionization of atomic potassium from the 4s ground and 4p, 5s-7s, 3d-5d excited states. The effect of the core polarization by the outer electron was included through the polarized pseudo-states. Besides the dipole core polarization, they found a noticeable influence of the quadrupole core polarization. Excellent agreement with the experiment for cross-sections of the 4s photoionization, including an accurate description of the near-threshold Cooper–Seaton minimum was observed. A close agreement with the experiment for the 4p photoionization was also noticed, but there were unexpectedly large discrepancies with the available experimental data for photoionization of the 5d and 7s excited states. The existence of a Cooper minimum in the experimentally measured photoionization cross-sections for the 7s ^{2}S$_{1/2}$ excited state of potassium using a two-step laser excitation technique in conjunction with a time-of-flight mass spectrometer was reported by Yar et al. [138]. The photoionization cross-sections for the 7s ^{2}S$_{1/2}$ state were determined as $(9.78 \pm 0.01) \times 10^{-2}$, $(3.30 \pm 0.49) \times 10^{-4}$, $(5.47 \pm 0.82) \times 10^{-3}$, and $(2.06 \pm 0.30) \times 10^{-2}$ Mb at the ionizing laser wavelengths 1064, 700, 532, and 355 nm, respectively. In Figure 5, the measured photoionization cross-section for the 7s excited state of potassium at and above the first ionization threshold is presented along with the theoretical calculations of the cross-section by Zatsarinny and Tayal [66]. A minimum in the cross-section, termed the

Cooper minimum, is evident at ~1.25 eV above the first ionization threshold, which is in agreement with the theoretical prediction.

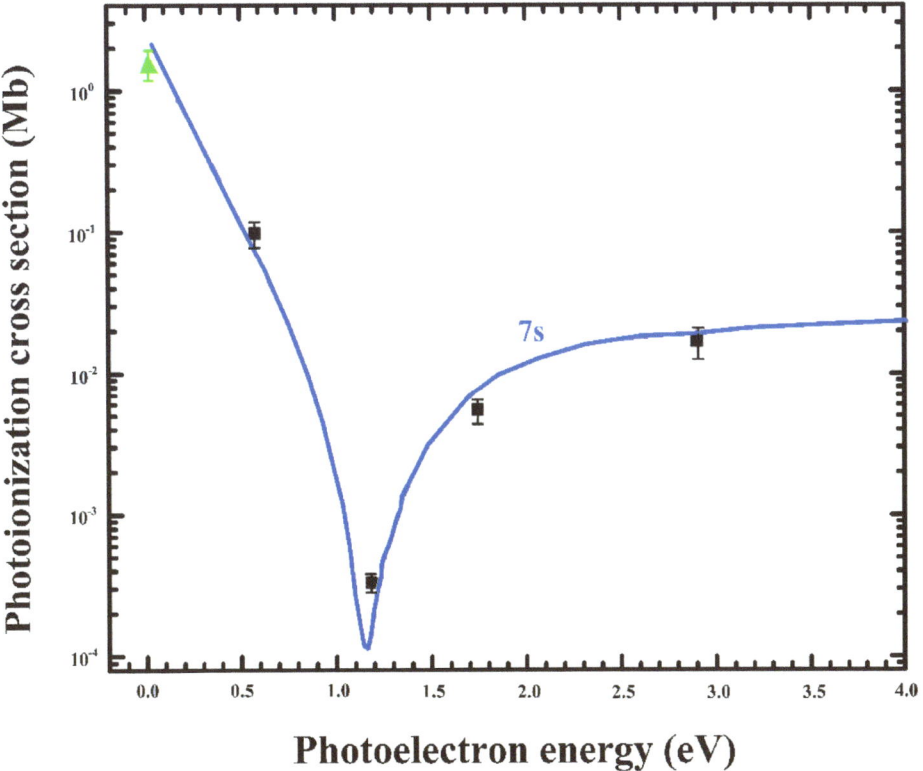

Figure 5. The photoionization cross-section was measured from the 7s excited state of potassium using five ionizing laser wavelengths at and above the first ionization threshold. The continuous curve is the theoretical results of Zatsarinny and Tayal [66], which follow the experimental values.

Once the photoionization cross-section for the excited state at the first ionization threshold is determined, the f-values of the Rydberg transitions attached to the intermediate state can be calculated using a simple relation (Mende et al. [139]).

$$f_n = \frac{mc}{\pi}\left(\frac{4\pi\varepsilon_0}{e^2}\right)\left(\frac{S^n}{S^+}\right)\left(\frac{\lambda^+}{\lambda_n}\right)\sigma^+. \tag{13}$$

Here, f_n is the oscillator strength for the nth transition of a Rydberg series, σ^+ is the photoionization cross-section measured at the ionization threshold wavelength λ^+, S^+ is the ion signal at the ionization threshold, and S^n is the integrated ion signal intensity for the nth transition. Using the absolute value of the cross-section at the ionization threshold, the oscillator strengths of all the discrete Rydberg transitions for which the ionization probability is unity ($p_n \cong 1$) were determined. However, the experimental conditions have to be optimized, such as the oven temperature and the buffer gas pressure, so that the ionization probability for the Rydberg states approaches one. Moreover, the energy difference between the first ionization threshold and $n = 20$ for potassium is ≈ 0.035 eV, whereas the value of $K_B T$ (thermal energy) is ≈ 0.046 eV at T = 540 K. As the energy gap between $n = 20$ and the first ionization threshold is less than the thermal energy, it therefore enhances the ionization efficiency of the Rydberg states for $n \geq 20$. Kalyar et al. [140]

measured the absolute values of the cross-sections from the 4p $^2P_{3/2}$ and $^2P_{1/2}$ excited levels at the ionization threshold as (6.3 ± 0.9) Mb and (5.4 ± 0.8) Mb, respectively, and the oscillator strengths for the 4p $^2P_{1/2} \to nd\ ^2D_{3/2}$ and 4p $^2P_{3/2} \to nd\ ^2D_{3/2,5/2}$ Rydberg transitions were also deduced by using the measured cross-sections of the 4p $^2P_{1/2}$ and $^2P_{3/2}$ levels at the ionization threshold.

The Rydberg series excited from the 4p $^2P_{1/2}$ excited state of potassium up to the first ionization threshold is presented in Figure 6. The optical oscillator strengths for the 4p $^2P_{3/2} \to nd\ ^2D_{3/2,5/2}$ (20 ≤ n ≤ 70) Rydberg transitions were calibrated using the measured absolute value of the photoionization cross-section (6.3 ± 0.9) Mb for the 4p $^2P_{3/2}$ level at the ionization threshold, whereas the values of S^n, λ^n and S^+ were extracted from the recorded spectrum, as marked in Figure 6. The photoionization cross-sections for the 7p $^2P_{3/2}$ excited state of potassium near the first ionization threshold using a two-step laser excitation/ionization technique in conjunction with a time-of-flight (TOF) mass spectrometer were reported as (3.52 ± 0.52) Mb, (2.21 ± 0.33) Mb, (1.35 ± 0.22) Mb, and (0.54 ± 0.08) Mb at four ionizing laser wavelengths; 1064 nm, 643.61 nm, 532 nm, and 355 nm, respectively, by Yar et al. [138]. Recently, Collister et al. [141] measured the non-resonant photoionization cross-section of the 7P $^2P_{3/2}$ state of francium for 442 nm light as (20.8 ± 7.1) Mb using a magneto-optical trap system. The photoionization rate was deduced from the change in the trap lifetime. The results are consistent with a simple extrapolation of known cross-sections for other alkali atoms. In Table 3, we enlist the laser wavelengths for the first-step excitation, laser wavelengths of the ionizing laser, and measured photoionization cross-sections for the resonance lines of lithium, potassium, sodium, rubidium, and cesium at the first ionization threshold.

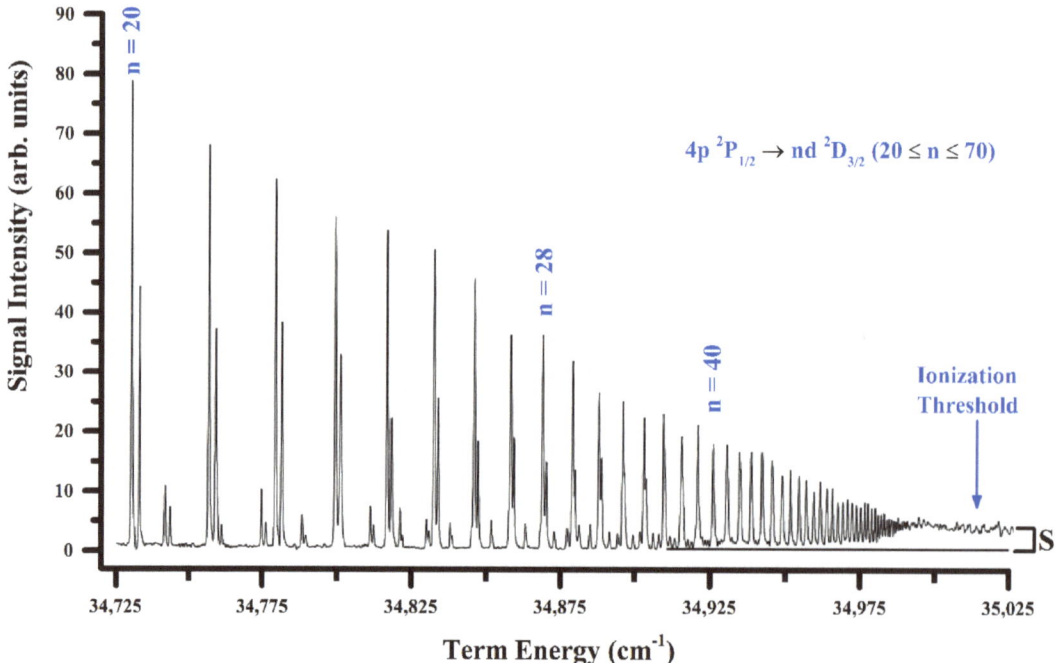

Figure 6. Rydberg series excited from the 4p $^2P_{1/2}$ excited state of potassium up to the first ionization threshold.

Table 3. Experimentally measured values of the photoionization cross-section for the resonance transitions of alkali atoms.

Atom	First Step (Excitation) $ms\ ^2S_{1/2} \to mp\ ^2P_{1/2,3/2}$	Second Step (Ionization) $mp\ ^2P_{1/2,3/2} \to \varepsilon s\ J = 1/2$ $\to \varepsilon d\ J = 3/2, 5/2$	Photoionization Cross-Section at Threshold Mb (1 Mb = 10^{-18} cm^2)
Li I	2p $^2P_{1/2}$ 670.793 nm $^2P_{3/2}$ 670.778 nm	349.8 nm 349.8 nm	16.5 ± 2.5 19.2 ± 2.9
Na I	3p $^2P_{1/2}$ 589.592 nm $^2P_{3/2}$ 588.995 nm	408.2 nm 408.5 nm	7.9 ± 1.3 6.7 ± 1.1
K I	4p $^2P_{1/2}$ 769.896 nm $^2P_{3/2}$ 766.489 nm	454.0 nm 455.2 nm	5.4 ± 0.8 6.3 ± 0.9
Rb I	5p $^2P_{1/2}$ 794.760 nm $^2P_{3/2}$ 780.027 nm	473.6 nm 479.0 nm	18.8 ± 3.0
Cs I	6p $^2P_{1/2}$ 894.347 nm $^2P_{3/2}$ 852.113 nm	494.3 nm 508.2 nm	25 ± 4

In the next section, the photoionization cross-section for the excited states of the alkaline earth atoms is presented.

4. Alkaline Earth Atoms

The ground state configurations for the alkaline earth atoms possess a filled s-subshell, $ms^2\ ^1S_0$ (m = 3, 4, 5, and 6 for Mg, Ca, Sr, and Ba, respectively). The atoms from the ground state are excited to the Rydberg states forming singlet or triplet states; $msnp\ ^1P_1,\ ^3P_{0,1,2}$. Photoionization sets in as the energy of the interacting photons is equal to the first ionization potential of the atom. Photoionization from the ground state is represented as ms ($^2S_{1/2}$) εp ($\ell = 1; J = 0, 1, 2$) continuum channels. The alkaline earth atoms show some broad and asymmetric line profiles above the first ionization threshold due to the simultaneous excitation of both the valence electrons. The absorption spectra of alkaline earth atoms from the ground state to the ionization threshold and above the threshold have been extensively studied by several groups around the globe.

A multistep-laser excitation technique has also been employed to measure the photoionization cross-section for the excited states of these atoms. The data on the highly excited states of atoms and particularly alkaline earth atoms have been studied by many researchers around the world, and has been compiled by Gallagher [142] and Connerade [143]. In Table 4, we summarize the laser wavelengths for the first step excitation and the second step excitation/ionization to measure the photoionization cross-sections experimentally. Except for Mg, where the first step laser is in the UV region, all the other required laser wavelengths are in the visible region, and the experiments can be performed with ease.

The first absolute measurement of the cross-section for photoionization from the selectively excited 3s3p $^1P_1$ atomic state of magnesium at the 3p^2 $^1S_0$ autoionizing resonance was reported by Bradley et al. [144] using two tunable pulsed lasers. The peak of the cross-section was reported as $(8 \pm 4) \times 10^{-16}$ cm^2 at 300.0 nm and the half-width of resonance was measured as 2.5 nm. Besides, the ratio of the cross-sections to the 3p^2 $^1S_0$ and $^1D_2$ states was measured as 14:1. The effect of polarization of the lasers was exploited to identify the angular momentum states of the autoionizing levels. Madsen and Thomsen [94] measured the resonant photo-ionization of the 3s3p $^1P_1$ level of magnesium at the near-resonant light at 285 nm using a magneto-optical trap. The extracted absolute photo-ionization cross-section was $(8.1 \pm 2.3) \times 10^{-17}$ cm^2, which is in good agreement with the theoretical calculations. Rafiq et al. [145] measured the photoionization cross-section from the 3s3p $^1P_1$ excited state of magnesium in the energy region from the first ionization threshold up to 1.4 eV excess energy using two-step photoionization and the saturation ionization technique in conjunction with an atomic beam source and a time-of-flight mass spectrometer. The absolute value of the photoionization cross-sections from the 3s3p $^1P_1$ excited state near the 3s ionization threshold was measured as (90 ± 16) Mb (at 354.5 nm ionizing wavelength) for the dominating isotope (^{24}Mg), whereas the value at the peak of the $3p^2\ ^1S_0$ auto-ionizing resonance was determined as (785 ± 141) Mb. This measured value is in close agreement with that reported by Bradley [144].

Table 4. Two-step excitation and ionization laser wavelengths for the Alkaline Earth Atoms.

Atom	First Step (Excitation) $ms^2\ ^1S_0 \to msnp\ ^1P_1, ^3P_1$	Second Step (Excitation) $msmp\ ^1P_1 \to ms\ np\ ^1S_0, ^3S_1$ $^3P_1 \to ms\ nd\ ^1D_2, ^3D_{1,2,3}$	Ionization Potential $\varepsilon s\ (J = 0, 1)$ $\varepsilon d\ (J = 0,1,2,3)$
Mg I	3s3p $^1P_1$ 285.213 nm $^3P_1$ 457.109 nm	375.6 nm 251.5 nm	61,671.05 cm^{-1}
Ca I	4s4p $^1P_1$ 422.544 nm $^3P_1$ 657.278 nm	389.8 nm 293.3 nm	49,305.95 cm^{-1}
Sr I	5s5p $^1P_1$ 460.733 nm $^3P_1$ 689.259 nm	412.6 nm 318.2 nm	45,932.09 cm^{-1}
Ba I	6s6p $^1P_1$ 553.548 nm $^3P_1$ 791.133 nm	340.1 nm 417.1 nm	42,034.91 cm^{-1}

In Table 5, a comparison of different experimental and theoretical results is presented for the resonance energy, width, and q-parameter of this autoionizing resonance in magnesium. The agreement is reasonably good, which shows the maturity of the experimental evidence and the theoretical models.

Table 5. Comparison of experimental and theoretical excitation energies E_r, width Γ, and line profile index q for the doubly excited $3p^2\ ^1S_0$ auto-ionization state of magnesium. The experimental uncertainty is predicted by the values in the parentheses.

Energy E_r (cm^{-1})	Width Γ (cm^{-1})	Line Profile Index, q	References
Theoretical			
68,600	137		Thompson et al. [146]
68,130	271		Chang [147]
67,936	331		Moccia and Spizzo [148]
68,700	316		Fang and Chang [149]
68,231			Kim [150]
Experimental			
68,275 (11)	276 (11)		Bradley et al. [144]
68,268 (4)	278 (8)	24.4 (1)	Bonanno et al. [151]
68,275 (11)			Okasaka and Fukuda [152]
68,273	280		Shao et al. [153]
68,270 (5)	280 (5)	24.5 (5)	Rafiq et al. [145]

Kim [150] calculated the photoionization cross-sections for the Mg$^+$ (3s, 3p, 4s, and 3d) states from the 3s3p ^{1,3}P excited states of atomic magnesium from the first ionization threshold up to the Mg$^+$ 4p threshold limit using the enhanced non-iterative variational R-matrix approach combined with multichannel quantum defect theory at the R-matrix surface. Recently, Wang et al. [154] theoretically studied the photoionization cross-sections of the excited levels (3s3p $^3P_{0,1,2}$) of atomic Mg using both the nonrelativistic and fully relativistic R-matrix method. The calculations show significant differences (a factor of 3) from the former experimental values. More experimental measurements of the photoionization cross-sections from the (3s3p $^3P_{0,1,2}$) excited states of Mg are desirable. The gf-values of the lower members of the principal series of calcium and photoionization cross-section at the first ionization threshold were determined by Parkinson et al. [155] using the hook method.

Geiger [156] calculated the bound oscillator strength distribution and density of oscillator strength in the continuum using multichannel quantum defect theory and reported the oscillator strengths of the $4s^2\,{}^1S_0 \rightarrow 4snp\,{}^1P_1$ and $4s^2\,{}^1S_0 \rightarrow 3d4p\,{}^1P_1$ transitions. Barrientos and Martin [157] computed the oscillator strengths of the principal series of alkaline earth atoms and determined the photoionization cross-section of magnesium and calcium. The relative oscillator strengths from the $4s4p\,{}^1P_1$ excited state to the $4sns\,{}^1S_0$ (n = 6, 7), the $4snd\,{}^1D_2$ (n = 6, 7), the $4p^2\,{}^1D_2$, and the $4p^2\,{}^1S_0$ states were measured by Smith [158]. The autoionization of the Rydberg series $3dnp\,{}^1P_1$ in calcium was studied by Karamatskos et al. [159] using an atomic beam apparatus and the synchrotron radiation facility at BESSY. Subsequently, Griessmann et al. [30] measured the cross-sections of the doubly excited resonances in calcium using a hot-wire diode detector and the synchrotron radiation facility at Hamburg. Lucas et al. [96] used resonance photoionization for the isotope-selective loading of Ca ions into a Paul trap. The $4s^2\,{}^1S_0 \rightarrow 4s4p\,{}^1P_1$ transition was driven by a 423 nm laser and the atoms were photoionized by a second laser at 389 nm. The lower limit for the absolute photoionization cross-section was reported as 170 Mb. Daily et al. [160] used two-photon photoionization of the Ca $4s3d\,{}^1D_2$ level in an optical dipole trap and reported the $4s4f\,{}^1F_3$ photoionization cross-section as 230 Mb. Sato et al. [161] studied the single and double photoionization of Ca atoms between 35 and 42 nm. The absolute values of the photoionization cross-sections for the $4s4p\,{}^1P_1$ and $4s4p\,{}^3P_1$ excited states of calcium at the 4s ionization threshold were measured by Haq et al. [162] as 140 ± 20 Mb and 117 ± 20 Mb, respectively. This value of the photoionization cross-section is in good agreement with that reported by Daily et al. [160].

Figure 7 shows a typical ion-signal plot, keeping the $4s4p\,{}^1P_1$ level saturated while the ionizing laser wavelength was adjusted at 389.8 nm, which corresponds to the first ionization threshold. The continuous line that passes through the experimental data points is a fit of Equation (8), which yields the value of the photoionization cross-section. Subsequently, the measured values of the photoionization cross-sections for the $4s4p\,{}^1P_1$ and $^3P_1$ levels at the threshold were used to calibrate the f-values of the transitions excited from the $4s4p\,{}^{1,3}P_1$ intermediate, terminating at the first ionization threshold. The Rydberg transitions are identified as $4s4p\,{}^1P_1 \rightarrow 4snd\,{}^1D_2$ and $4s4p\,{}^3P_1 \rightarrow 4snd\,{}^3D_2$, respectively.

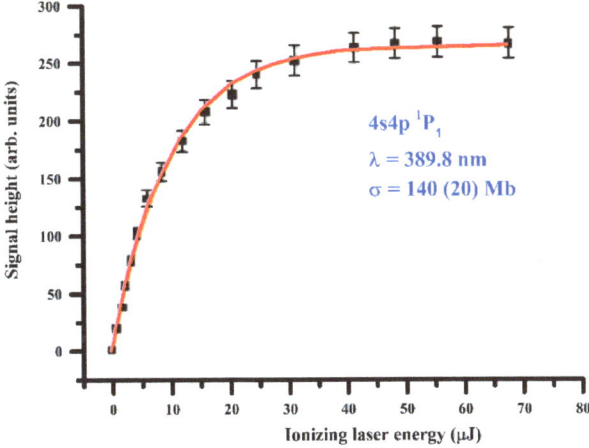

Figure 7. Photoionization signal for the $4s4p\,{}^1P_1$ level at the ionization threshold of calcium versus the ionizing laser energy.

Ewart and Purdie [163] studied the two-photon excitation of the even parity $5sns\,{}^1S_0$ and $5snd\,{}^1D_2$ Rydberg states and an autoionization level $4d^2\,{}^1D_2$ just above the first ionization threshold of strontium. The $5p_{1/2}ns_{1/2}$ and $5p_{3/2}\,ns\,J = 1$ autoionizing states were reported by Xu et al. [164], whereas the even parity $J = 0$ autoionizing spectrum below the 4d$_{5/2}$ threshold was reported by Kompitsas [165]. The absolute photoabsorption cross-sections of strontium from the 5s ionization threshold to the 5p threshold were measured by Chu et al. [35] who reported the absolute value of the cross-section for the 5s ionization threshold as 7.2 ± 0.9 Mb. Luc-Koenig et al. [166] studied the two-photon ionization processes from the ground state of strontium using the jj-coupled Eigen-channel R-matrix approach combined with the multichannel quantum defect theory. Mende and Kock [167] applied the saturation technique and determined the photoionization cross-section at the first ionization threshold as 12 ± 2.5 Mb, which was used to calibrate the f-values of the 5s^2 ^{1}S$_0$ → 5snp ^{1}P$_1$ Rydberg series. The photoionization cross-sections from the 5s5p ^{1}P$_1$ excited state of strontium just above the first ionization threshold show a giant resonance, with a peak cross-section as 56 (10) Mb, as stated by Mende et al. [139]. The spatial distribution of the photons in the laser pulse was measured with a CCD camera and applied in the evaluation procedure as a Gauss profile. Based on a 12-state non-relativistic close coupling R-matrix calculation, the giant resonance was assigned as predominantly (5p^2) ^{1}D$_2$ with a strong admixture of (4d^2) ^{1}D$_2$. Baig et al. [81] presented a comparison between the line shape of an even parity isolated autoionization resonance (4d^2 + 5p^2) ^{1}D$_2$ in strontium excited from the ground state either via two-photon non-resonant excitation or by the two-step resonant excitation processes. It was reported that the line shape q-parameter is different for the two processes, while the width of the resonance is independent of the excitation mechanism. Subsequently, Haq et al. [168] measured the photoionization cross-section from the 5s5p ^{1}P$_1$ and 5s6s ^{1}S$_0$ levels of strontium at the first ionization threshold as well as at six different wavelengths between 355 nm and 410 nm.

The dominant autoionizing resonance (4d^2 + 5p^2) ^{1}D$_2$ excited from the 5s5p ^{1}P$_1$ intermediate level is reproduced in Figure 8. The absolute value of the cross-section at the peak of the (4d^2 + 5p^2) ^{1}D$_2$ autoionizing resonance was determined as 5450 (18%) Mb. The photoionization cross-section was also measured at different ionizing laser wavelengths; 410 nm, 405 nm, 404 nm, 400 nm, and 390 nm. Interestingly, the measured values of the photoionization cross-sections completely retrace the line profile of the autoionizing resonance. The normalization was found to be very consistent, and by selecting any one of the measured points as a reference, the value of the cross-section at any other point is reproducible. Using Fano's relation for an isolated autoionization resonance, the resonance energy (E_r = 46,379 (1) cm^{-1}), width (Γ = 45 (5) cm^{-1}) and line profile index (q = 6.8 (2)) were extracted.

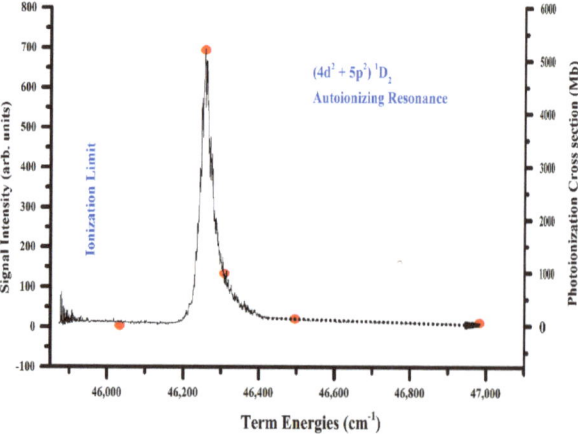

Figure 8. The photo-absorption from the 5s5p ^{1}P$_1$ level of strontium shows the (4d^2 + 2p^2) ^{1}D$_2$ autoionizing resonance just above the first ionization threshold. The absolute values of the photoionization cross-sections were determined at five different energies, marked with dots in the figure.

In Table 6, a comparison of the resonance energy and width of this domination autoionizing resonance is presented, showing a good agreement.

Table 6. Comparison of the resonance energy and width of the $(4d^2 + 5p^2)\,^1D_2$ autoionizing resonance.

Experimental Energy (cm^{-1})	Width (cm^{-1})	Calculated Energy (cm^{-1})	Width (cm^{-1})	References
46,380	60	-	-	Esherick [169]
46,376.8	56.2	-	-	Mende et al. [139]
46,380.1	59.6	-	-	Dai et al. [170]
-	-	46,433	71	Luc-Koenig et al. [166]
46,380	56 (1)	-	-	Baig et al. [81]
46,379 (1)	45 (5)	-	-	Haq et al. [168]

The absolute values of the photoionization cross-sections for the 5s5p $^1P_1$ and 5s5p $^3P_1$ excited states of strontium at the first ionization threshold were measured by Haq et al. [171] as 11.4 ± 1.8 Mb and 10.7 ± 1.7 Mb, respectively, using a thermionic diode ion detector. These threshold photoionization cross-sections values were then used to determine the oscillator strengths of the 5s5p $^1P_1 \rightarrow$ 5snd $^1D_2$ and 5s5p $^3P_1 \rightarrow$ 5snd $^3D_2$ Rydberg transitions. The oscillator strength densities in the continuum corresponding to the 5s5p $^3P_1$ excited state were determined by measuring the photoionization cross-sections at five ionizing wavelengths above the first ionization threshold. Haq et al. [172] measured the photoionization cross-section from the 5s6s $^1S_0$ excited state of strontium using a linearly polarized dye laser with the polarization vector along the direction of propagation. The total angular momentum of the excited state is $J = 0$, therefore the $M_j = 0$ sublevel is populated. The polarization vector of the ionizing laser was set parallel to that of the exciting laser, which accessed the $M_j = 0$ out of the $M_j = 0, \pm1$ sublevels of the 5s εp, $J = 1$ continuum. By adjusting the polarization vector of the ionizing laser perpendicular to that of the exciting laser, the accessible channels were 5s εp $J = 1$ and $M_j = \pm1$, and the corresponding cross-section at the threshold was determined as 1.1 ± 0.2 Mb. However, by adjusting the polarization vector of the ionizing laser at the magic angle of 54.7°, the photoionization cross-section was measured as 1.0 ± 0.2 Mb. The measured absolute value of the photoionization cross-section for the 5s6s $^1S_0$ excited state of strontium at the 5s threshold was also used to extract the f values of the 5s6s $^1S_0 \rightarrow$ 5snp $^1P_1$ Rydberg transitions.

Burkhardt et al. [76] employed multi-step laser excitation and ionization to measure the photoionization cross-sections of the resonance levels of sodium (3p $^2P_{3/2}$), potassium (4p $^2P_{3/2}$), and barium (6s6p $^1P_1$), and their atomic densities in the ground states. In this experiment, a beam of the atoms was generated, which intersected at the center of an electrically shielded cell by two collinear laser beams. One laser beam was used to excite the atoms to the resonance level and the other to photo-ionize the atoms out of that level. The diameter of the excitation laser beam was 3 mm, and the energy of the laser was 75 µJ. The ionization laser beam was the frequency tripled (353.3 nm) output of the same Nd:YAG laser (1064 nm) used to pump the dye lasers. The diameter of the ionization laser beam was much smaller than that of the exciting laser beam by a long focal length lens to make the ionizing volume as nearly cylindrical as possible over the ion collection length. The values of the photoionization cross-sections were reported as $(3.7 \pm 0.7) \times 10^{-18}$ cm^2, for the 3p $^2P_{3/2}$ level of sodium, as $(7.6 \pm 1.1) \times 10^{-18}$ cm^2 for the 4p $^2P_{3/2}$ level of potassium, and as $(17.6 \pm 2.3) \times 10^{-18}$ cm^2 for the 6s6p $^1P_1$ level of barium. Kallenbach et al. [173] measured the photoionization cross-section for the 6s6p $^1P_1$ level of barium using a thermionic diode ion detector in combination with a two-step pulsed laser excitation and ionization technique. The barium atoms were excited to the resonance level by a nitrogen laser (1MW, 4ns) pumped dye laser (energy 4 µJ and width 0.3 nm) tuned to the $6s^2\,^1S_0 \rightarrow$ 6s6p $^1P_1$ transition at 553.5 nm. The second dye laser (energy 4 µJ and width 0.08 nm) was tuned at high lying Rydberg states and across the photoionization threshold for the 6s6p $^1P_1$

level of Ba at 416 nm. The absorption oscillator strengths of the 6s6p $^1P_1 \to$ 6snd $^3D_2$ transitions were measured along with the photoionization cross-section at the threshold as $(5 \pm 2) \times 10^{-20}$ m^2. Willke and Kock [174] improved the experimental arrangements to measure the cross-sections for photoionization for the excited states of barium by using a Nd:YAG pumped dye laser system and a very narrow bandwidth dye laser (0.2 cm^{-1}) in conjunction with the thermionic diode technique. The revised value of the photoionization cross-section for the 6s6p $^1P_1$ level of barium at the threshold was reported as 80 Mb, which is in good agreement with the theoretical values, 120×10^{-18} cm^2 (Bartschat and MacLaughlin [175]) and 100×10^{-18} cm^2 (Greene and Theodosiou [176]). The barium ion signal of the high-lying Rydberg transitions up to and above the ionization threshold of the 6s continuum was also measured and the observed Rydberg series were designated as 6s6p $^1P_1 \to$ 6snd $^1D_2$ transitions. The photoionization cross-section for the 6s6p $^1P_1$ level was also measured above the ionization threshold up to 381 nm, showing numerous autoionizing resonances. The polarization of the exciting and the ionizing lasers also plays an important role in the measurement of the photoionization cross-section. Therefore, if the exciting and the ionizing lasers pulses are linearly polarized, the photoionization cross-section will depend on the angle θ between the two laser polarization directions (He et al. [77]):

$$\sigma(\theta) = \cos^2 \theta \sigma^{\parallel} + \cos^2 \theta \sigma^{\perp}, \tag{14}$$

where $\sigma^{\parallel}$ and $\sigma^{\perp}$ are the cross-sections for the parallel and perpendicular polarizations of the exciting and the ionizing laser beams, respectively. When both the lasers are linearly polarized and the polarization vectors are parallel, then only the term $\cos^2 \theta \sigma^{\parallel}$ contributes to the measured cross-section. He et al. [77] measured the absolute photoionization cross-sections for the 6s6p $^1P_1$ excited state of Ba in the threshold region and found a good agreement with the theoretically calculated value. The "magic" angle was used for the relative orientation of the linear polarization of the exciting and ionizing lasers, as, at this orientation angle, the measured cross-section is the same as if the excited state is populated isotropically. Keller et al. [177] measured the photoion spectra and angular distribution of photoelectrons using a two-color, two-photon resonant ionization of barium and probed the effect of the photoionization pathway on the autoionization processes. Using 6s6p $^1P_1$ and 5d6p $^1P_1$ as intermediate states, the 5dnd autoionizing states were studied, showing a dramatic difference in the profiles of the autoionizing resonances. Wilke and Kock [174] measured the absolute photoionization cross-sections from an excited state of Ba I using an improved thermionic diode technique with an excellent signal-to-noise ratio from the threshold of the 6s continuum at 417 nm down to 370 nm. The prominent resonances were fitted Beutler–Fano profiles, extracting the parameters that show good agreement with the experimental as well as theoretical studies. He et al. [178] reported the absolute cross-sections for photoionization of isotropically populated Ba (6s6p $^1P_1$) to each accessible final state (0, 1, and 2) from the threshold to 370 nm, using spectra obtained with four different combinations of linear and circular polarization of the exciting and ionizing radiation. The effects of de-alignment due to the hyperfine interaction of the nuclear spin I = 2 isotopes of barium were also taken into account. The results were in good agreement with R-matrix predictions of shapes and positions of autoionizing resonances as well as the magnitudes of the background (non-resonance) cross-sections, but some discrepancies with oscillator strengths of the autoionizing states were remarked. It was inferred that to obtain the correct magnitude of the background cross-section theoretically, it is important to include the electron correlation effects in the wave functions of both initial and final states. Data were also presented that verify the equivalence of obtaining the total photoionization cross-section by properly combining two sets of data, each acquired with a different angle between linear polarization vectors, e.g., parallel and perpendicular, or by making a single measurement at the magic angle, 54.74^0. Langadec et al. [179] reported the experimental and theoretical studies of the photoionization for the 6p^2 multiplet in barium, in the range 350–402 nm, by considering the 6p^2 $^3P_0$, 6p^2 $^1D_2$, and 6p^2 $^3P_2$ states, after the 6p^2 $^3P_1$ state, which was studied by Carre et al. [180]. The differential cross-sections from below to above the 6p$_{1/2}$ threshold, where several 6p$_{1/2}$nl; J and 6p$_{3/2}$nl resonances were excited using different cases of laser polarizations, and the experimental results were compared with eigen-channel R-matrix calculations. Kalyar et al. [181] measured the photoionization cross-section measurements for the 6s6p ^{1}P$_1$ and ^{3}P$_1$ excited states of barium at the first ionization threshold as 90 ± 14 Mb and 102 ± 15 Mb, respectively. Li and Budker [182] measured the photoionization cross-sections of two even parity excited states of barium, 5d6d ^{3}D$_1$ and 6s7d ^{3}D$_2$, at the 556.6 nm ionization laser wavelength, and reported that the total cross-section depends on the relative polarization of the atoms and the ionization laser light. Kalyar et al. [183] extended these studies and investigated the line shapes of the 6s7p configuration based on ^{1}P$_1$, ^{3}D$_1$, and ^{3}P$_1$ autoionizing resonances using the 6snp (n = 6, 7, 8) and 5d6p ^{1}P$_1$ as intermediate states. It was inferred, based on the extracted Fano parameters of the resonance line profiles, that the width of an autoionizing resonance is independent of the excitation path while the line profile parameter changes with the selection of different states. The even-parity autoionizing resonances in barium using two-step laser excitation via the 5d6p ^{1}P$_1$ intermediate level covering the energy region from the first ionization threshold to the 5d ^{2}D$_{5/2}$ limit using an atomic beam apparatus in conjunction with a time-of-flight mass spectrometer were studied by Kalyar et al. [184].

In Figure 9, the photoionization spectra of barium excited from the 5d6p ^{1}P$_1$ intermediate level are presented at two polarization vector orientations of the ionizing laser; parallel or perpendicular to the exciting laser polarization. The selection of the combinations of polarization vectors of the two laser beams permits access to the final states with different total angular momentum. The probability of absorption of a photon is proportional to the square of the modulus of the dipole matrix element, as stated by Elizarov and Cherepkov [185]:

$$W_{n_0 l_0 J_0 m_0 \to n_1 l_1 J_1 m_1} \propto \left| \langle n_1 l_1 J_1 M_1 | D_q | n_0 l_0 J_0 M_0 \rangle \right|^2. \tag{15}$$

Here J and M_j are the total angular momentum and its projection, respectively, and D_q are the spherical components; $q = 0$ applies to the linearly polarized light and ± 1 to the circularly polarized light. The Wigner–Eckart theorem leads to the separation of the matrix element into a geometrical and physical part, as the angles are incorporated in the coefficients and the physical part is expressed in terms of a reduced matrix element:

$$\langle n_1 l_1 J_1 M_1 | D_q | n_0 l_0 J_0 M_0 \rangle = (-1)^{J_1 - M_1} \begin{pmatrix} J_1 & 1 & J_0 \\ -M_1 & q & M_0 \end{pmatrix} \langle n_1 l_1 J_1 \| D \| n_0 l_0 J_0 \rangle. \tag{16}$$

The dipole matrix element describing the absorption of a second photon in the second step can be described as:

$$\langle n_2 l_2 J_2 M_2 | D_q | n_1 l_1 J_1 M_1 \rangle = (-1)^{J_2 - M_2} \begin{pmatrix} J_2 & 1 & J_1 \\ -M_2 & q & M_1 \end{pmatrix} \langle n_2 l_2 J_2 \| D \| n_1 l_1 J_1 \rangle. \tag{17}$$

Only those values of the total angular momentum via the single-photon absorption are allowed, for which the two 3j-symbols in the above two equations are non-vanishing. In the first-step excitation from the ground state 6s^2 ^{1}S$_0$ (J = 0 and M_j = 0) with the linearly polarized light, the 5d6p ^{1}P$_1$ (J = 1 and M_j = 0) level is populated. When the polarization vector of the second-step laser beam is set parallel to that of the first-step laser, the final states possessing M_j = 0, J = 0, and 2 are allowed while J = 1 is forbidden. However, when the polarization vector of the second step laser is set perpendicular to that of the first laser, the excited states possessing M_j = ±1, J = 1, and 2 are allowed but J = 0 is forbidden.

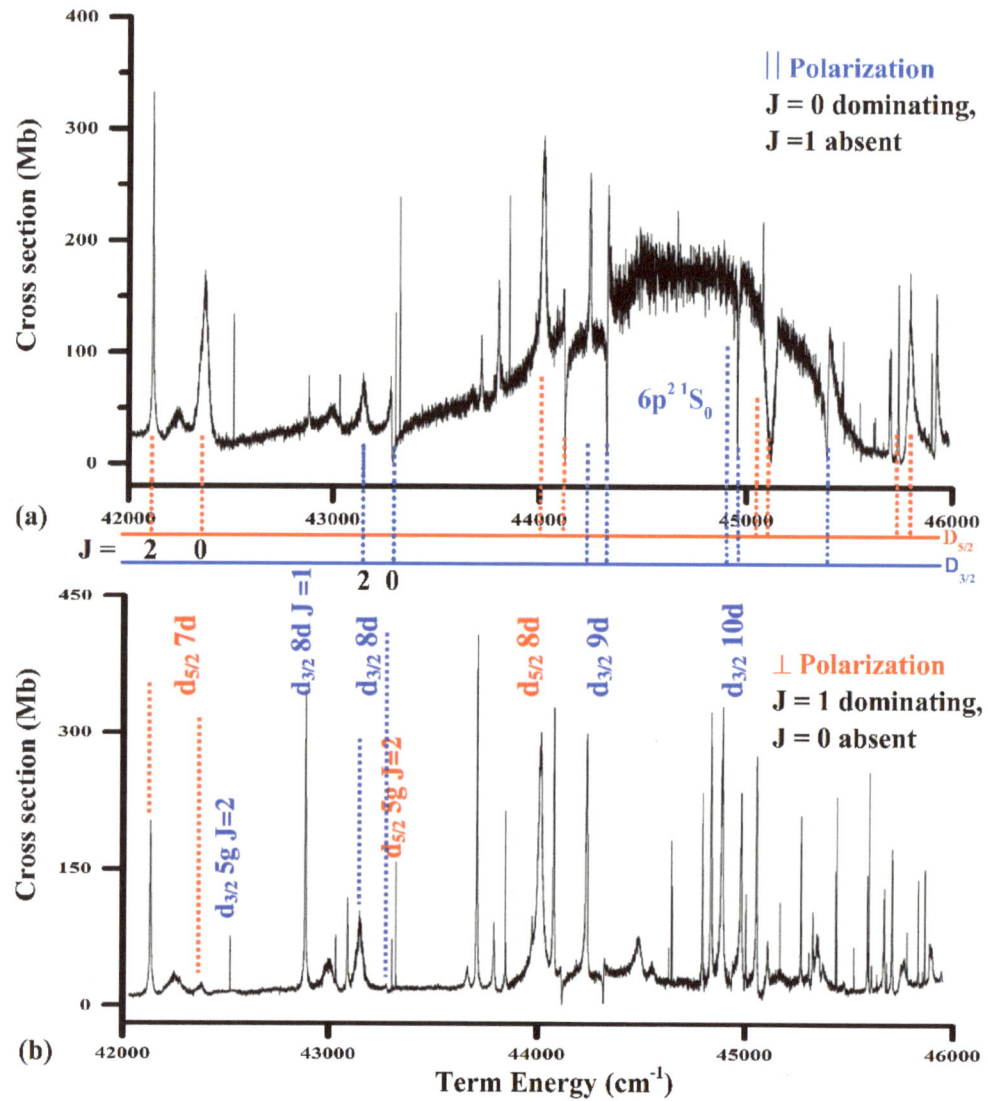

Figure 9. Photoionization spectra of barium excited from the 5d6p $^1P_1$ intermediate level at polarization vectors of the ionizing laser either parallel (**a**) or perpendicular (**b**) to that of the exciting laser that enabled the assignment of the *J*-values of the excited levels unambiguously due to excitation selection rules.

Thus, the $6p^2\,^1S_0$ excited state is only accessible from the 5d6p $^1P_1$ intermediate state when the polarization vectors of the exciting as well as that of the ionizing dye lasers are parallel to each other (Figure 9a), and this *J* = 0 transition will be forbidden in the case of the polarization vector perpendicular to that of the exciting dye laser (Figure 9b). The transitions possessing *J* = 0 are allowed but *J* = 1 are forbidden in the first arrangement. In the second arrangement, transitions with *J* = 0 are forbidden and *J* = 1 are allowed. Indeed, the broad autoionizing resonance $6p^2\,^1S_0$ possessing *J* = 0 is absent in the spectrum recorded with the polarization vectors set perpendicular to each other. Some forbidden lines also appear in the spectra, which may emerge either due to the admixture of the elliptically polarized light as the light is not 100% linearly polarized or due to the hyperfine depolarization, as reported by Wood et al. [186]. The $J = 2$ lines are present in both the spectra, and thus the assignment of the J values to the resonances is unambiguously assigned. As in the upper trace, the dominating lines possess $J = 0$; therefore, the superimposed broad feature is identified as the 6p^2 ^{1}S$_0$ autoionizing resonance, which disappears in the lower trace, in conformity with the spectra for the parallel polarization of the two lasers in the work of Lange et al. [187] and Wood et al. [186] for the photoionization from the 6s6p ^{1}P$_1$ excited state.

The line shape of the 6p^2 ^{1}S$_0$ autoionizing resonance was fitted with Fano's relation (Fano [188]), describing the interaction of a discrete state with one continuum. The photoionization cross-section of an isolated autoionizing state is represented as:

$$\sigma(\varepsilon) = \sigma_a \frac{(q+\varepsilon)^2}{1+\varepsilon^2} + \sigma_b, \qquad (18)$$

where $\varepsilon = \frac{(E-E_r)}{\Gamma/2}$ measures the departure of the incident photon energy E from the resonance energy E_r, Γ is the width of the autoionizing resonance, $\sigma(\varepsilon)$ represents the absorption cross-section for photons of energy E, whereas σ_a and σ_b are the portions of the cross-section of the continuum that interact and do not interact with the discrete level, respectively, and q is the line profile index parameter. The fitting yields the parameters: resonance energy = 44,850.50 cm^{-1}, width 980 $\pm$ 50 cm^{-1}, $q = -5.4 \pm 2$, and absolute value of photoionization cross-section = 185 $\pm$ 35 Mb. The interactions between the 6p^2 ^{1}S$_0$ broad feature and the 5d$_{5/2}$nd $J = 0$ Rydberg series were simulated using the phase-shifted multichannel quantum defect theory (MQDT), as shown by Gallagher [142], showing very interesting line shapes of the autoionizing resonances. There is clear evidence of a change in the q-values of the Rydberg series of the autoionizing resonances in the vicinity of the broad 6p^2 ^{1}S$_0$ resonance. The q-parameter describes the ratio of the transition dipole moment connecting the ground state to the discrete level and the continuum channel. The line profile is a Lorentzian shape when q is very large, and it will be asymmetric when q is small, with a minimum either below or above the resonance energy when q is either positive or negative. A window resonance appears when the value of q is very small or nearly zero.

In Figure 10a, the line profile of the 5d$_{3/2}$ 8d ($J = 0$) autoionizing resonance is reproduced along with the Fano profile fitting of Equation (16). Due to its interaction with the 6p^2 ^{1}S$_0$ broad resonance, the line profile is asymmetric, with a minimum above the resonance energy and the q-value is -1.1. The experimentally observed and the simulated line profile of the 5d$_{3/2}$ 9d ($J = 0$) autoionizing resonance is shown in Figure 10b. The interaction with the 6p^2 ^{1}S$_0$ broad resonance reveals an asymmetric line profile, with a minimum below the resonance energy and the extracted q-value is $+1.0$.

In Figure 10c, the line profile of the 5d$_{3/2}$ 10d ($J = 0$) autoionizing resonance is reproduced. The interaction with the 6p^2 ^{1}S$_0$ broad resonance results in a window-type line profile. The Fano line profile fitting yields the value of $q = 0.05$.

Tolsma et al. [189] calculated the one and two-photon ionization cross-sections of the aligned 6s6p ^{1}P$_1$ state of barium in the energy range between the $5_{d3/2}$ and $5_{d5/2}$ states of Ba+. These photoionization spectra were also measured in the same energy region, driving the one- or two-photon processes with the second or first harmonic of a tunable dye laser, respectively. The calculations were performed using the Eigen-channel R-matrix method and multichannel quantum defect theory and calculated the autoionizing resonances in this energy range. The calculations provided an absolute normalization for the experiment, reproduced the resonance structures in both the one- and two-photon cross-sections, and confirmed other aspects of the experimental observations. Afrousheh et al. [190] investigated the spectroscopic behavior of the 5d$_{3/2}$nd ($J = 0$ and 2) autoionizing Rydberg series of barium under collision with rare gases using two-photon excitation of the two valence electrons in the 6s^2 ^{1}S$_0$ ground state of barium. The barium vapor was produced in a heat-pipe-like oven, and a tunable dye laser pumped by an excimer laser was used as the excitation source. The spectral behavior of the autoionizing resonances was investigated in the presence of inert gases Ar, Kr, and Xe at different pressures. The collision-induced line shifts were measured and the shift parameters for the even-parity $5d_{3/2}\,nd$ ($J = 0$) and $5d_{3/2}\,nd$ ($J = 2$) autoionizing states were extracted from the data.

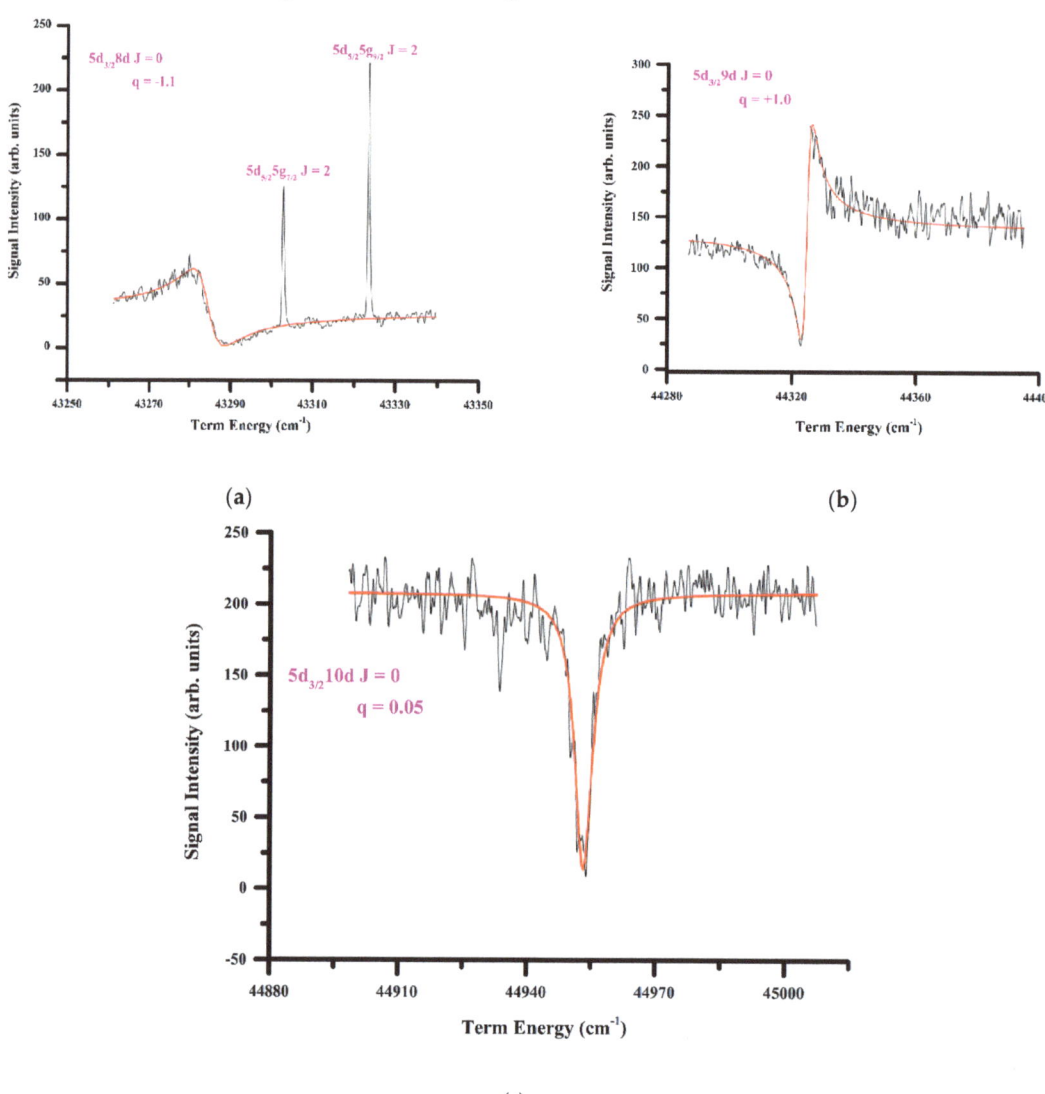

Figure 10. (**a**,**b**,**c**). Line profiles of the $5d_{3/2}$ 8d ($J = 0$) and $5d_{3/2}$ 9d ($J = 0$) autoionizing resonances in barium show different signs of q-parameter as the principal quantum number is increased by one unit. (**c**). Line profile of the $5d_{3/2}$ 10d ($J = 0$) autoionizing resonance along with the fit of Fano relation for an isolated autoionizing resonance.

In Table 7, we enlist the values of the photoionization cross-sections for the resonance lines of alkaline earth atoms.

Table 7. Measured values of photoionization cross-sections for the resonance and inter-combination transitions in alkaline earth atoms.

Atom	First Step (Excitation) $ms^2\,^1S_0 \to msnp\,^1P_1, ^3P_1$	Second Step (Ionization) $msmp\,^1P_1 \to ms\,\varepsilon s,\,\varepsilon d$ $^3P_1 \to ms\,\varepsilon s,\,\varepsilon d$	Photoionization Cross-Section
Mg I	3s3p $^1P_1$ 285.213 nm $^3P_1$ 457.109 nm	375.6 nm 251.5 nm	90 ± 16 Mb
Ca I	4s4p $^1P_1$ 422.544 nm $^3P_1$ 657.278 nm	389.8 nm 293.3 nm	140 ± 20 Mb 117 ± 20 Mb
Sr I	5s5p $^1P_1$ 460.733 nm $^3P_1$ 689.259 nm	412.6 nm 318.2 nm	11.4 ± 1.8 Mb 10.7 ± 1.7 Mb
Ba I	6s6p $^1P_1$ 553.548 nm $^3P_1$ 791.133 nm	340.1 nm 417.1 nm	90 ± 14 Mb 102 ± 15 Mb

In the next section, the autoionizing resonances in the spectra of inert gases are presented.

5. Inert Gases

Helium is the simplest atomic system, and therefore the photoionization of helium atoms is of considerable importance as it provides a prospect to evaluate different atomic models and comparison with the experimental observation. Stebbings et al. [191] determined the absolute cross-section for the photoionization of helium 1s2s ^1S and 1s2s ^3S metastable atoms from the threshold to 240 nm and compared the results with the theoretical calculations. The ultraviolet radiation was generated by frequency, doubling the output of a nitrogen pumped dye laser. Subsequently, Dunning and Stebbings [192] used a molecular beam in conjunction with a tunable laser and measured the absolute cross-sections for the photoionization of helium 3p $^{1,3}P$, 4p $^{1,3}P$, and 5p $^{1,3}P$ atoms excited optically from the 1s2s $^{1,3}S$ metastable levels at the wavelength used for its excitation from the metastable states. Marr [193] reported the cross-section data for the photoionization of the ground-state helium atoms from threshold to 200 eV and compared it with the RPAE calculations. Domke et al. [194] (and references therein) reported an extensive study on the double-excitation resonances of helium using a synchrotron radian facility (BESSY) and measured the resonance energies, linewidths, Fano q-parameters, and quantum defects of the various Rydberg series that were compared with the theoretical calculations. Schulz et al. [195] measured the double-excitation resonance of helium at a much improved spectral resolution in the soft x-ray range. The resonance parameters of the photo doubly excited helium were determined by Rost et al. [196], who systematically analyzed the doubly excited Rydberg series in helium, taking into account the theoretical results and experimental data on photoionization cross-sections, the quantum defects, the width, the oscillator strengths, and the shape parameters of the ^1P Rydberg series. Chan et al. [197] described an alternative method for the measurement of absolute optical oscillator strengths (cross-sections) for electronic excitation of free atoms and molecules throughout the discrete region of the valence-shell spectrum at high energy resolution. The absolute scale was obtained from the Thomas–Reich–Kuhn sum-rule normalization of the Bethe–Born transformed electron-energy-loss spectrum without involving the difficult determinations of photon flux or target density. The measured dipole oscillator strengths for helium excitation $1s^2\,^1S \to np\,^1P$ (n = 2–7) were in excellent quantitative agreement with the earlier calculations. The absolute measurements were also compared with other experimental and theoretical oscillator strength determinations for photoexcitation and photoionization processes in helium up to 180 eV, including the 2snp and 3snp autoionizing resonances in the 59–72-eV energy region. Gisselbrecht et al. [198] reported the absolute photoionization cross-sections of the excited, short-lived He* 1s2p ^1P and 1s3p ^1P states in the region close to the He$^+$ 1s ^2S threshold (from 0 to 2 eV). The intermediate He* $^{1,3}p$ states were populated by photoabsorption of a high-order harmonic of an intense picosecond tunable laser and subsequently ionized by absorption of photons of several fixed frequencies ranging from the near-infrared to the ultraviolet (752 nm, 532 nm, 355 nm, 266 nm). The results confirmed the earlier experimental theoretical work. Sahoo and Ho [199] reported the theoretical photoionization cross-sections for He 1s2s ^{1}S and He 1s2p ^{1}P states in a Debye plasma environment by the complex coordinate rotation method, using a finite L2 basis set constructed from one electron Laguerre orbitals. The plasma environment was found to appreciably influence the photoionization cross-sections near the ionization threshold. The photoionization cross-sections of helium were compared with other theoretical and experimental results, showing good agreement. A new minimum in the photoionization cross-section curve for the metastable 1s2s ^{1}S state was also predicted.

An alternate experimental technique to access the highly excited states of helium and autoionizing resonances in inert gases is to use a mild discharge, either a DC or RF, that populates the metastable states. One laser is used to excite the atoms from the metastable state to an intermediate upper level and a second tunable dye laser is employed to access the autoionizing states. The advantage of such an experimental arrangement is that the dye lasers working in the visible region can be used for excitation and ionization studies. The photoionization cross-sections for the excited states can also be measured using the saturation technique, as described for the case of alkali atoms and alkaline earth atoms. Figure 11a,b shows the photoionization cross-sections for the 1s3p ^{1}P and 1s3p ^{3}P excited states of helium, at and near-ionization threshold region (0–0.2 Ry) measured by Hussain et al. [200]. The 1s2s ^{1}S metastable level was populated by running a DC glow discharge, the 1s3p ^{1}P level was populated by tuning the dye laser at 310.7 nm, and the 1s3p ^{3}P level by the dye laser tuned at 388.97 nm, whereas the photoionization cross-sections were measured at different ionizing laser wavelengths: 826.4, 752, 630, 532, 501, and 355 nm for the 1s3p ^{1}P level and at 784.6, 752, 630, 532, 389, and 355 nm for the 1s3p ^{3}P level. Smooth frequency dependence of the cross-section was observed for both the excited states following the theoretical calculations. The measured values of the photoionization cross-sections are in close agreement with the reported theoretical values by Jacobs [201] and experimental values by Gisselbrecht et al. [198] and Dunning and Stebbings [192].

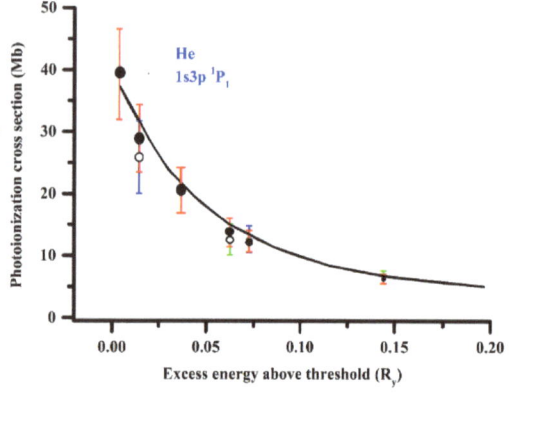
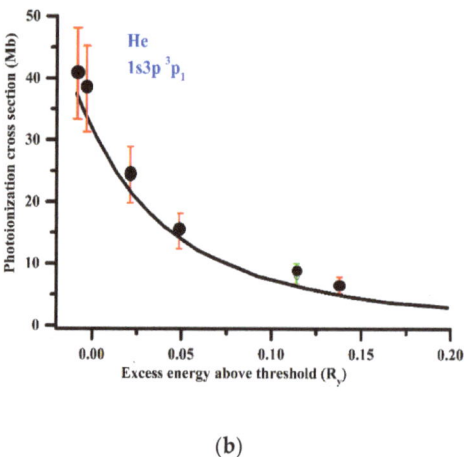

(a) (b)

Figure 11. (a,b). Show the dependence of the photoionization cross-sections for the 1s3p ^{1}P and 1s3p ^{3}P state of helium as a function of excess energy above the first ionization threshold. The experimental data points are by Dunning and Stebbings [192] and Gisselbrecht et al. [198]. The solid line is the theoretical calculation by Jacobs [201].

The oscillator strength distribution in the discrete and continuous regions of the spectrum of helium from the 2s $^1S_0$ metastable using a low-pressure RF glow discharge was measured by Hussain et al. [130]. The measured value of the photoionization cross-section for the 2s $^1S_0$ excited state at the first ionization threshold was used to determine the f values for the 2s $^1S_0 \rightarrow np$ $^1P_1$ Rydberg series from $n = 10$ to $n = 52$. The f values of the Rydberg series decrease smoothly with an increase of the principal quantum number. In the continuum region, the oscillator strength densities were determined by measuring the photoionization cross-sections from the 2s $^1S_0$ excited state at five ionizing laser wavelengths above the first ionization threshold. The discrete f values smoothly merge into the continuous oscillator strength densities across the ionization threshold.

The ionization potentials of inert gases are much higher than that of the alkali atoms or alkaline earth atoms, and therefore one needs a vacuum-ultraviolet radiation source to observe the Rydberg series and the ionization continuum. The ground state configuration of the inert gases is a filled p-subshell and the ionization of an outer p-electron from the ground state yields two ionic levels; p^5 ($^2P_{3/2}$ and $^2P_{1/2}$). The $p^5(^2P_{1/2})$ ns, nd Rydberg levels lying above the first ionization threshold ($^2P_{3/2}$) are degenerated with the p^5 $\varepsilon\ell$ continuum above the ($^2P_{3/2}$) ionization threshold, and consequently the line shapes are broad and asymmetric due to autoionization, due to the interaction of the discrete levels with the adjacent continuum, as shown by Fano [188]. The autoionizing resonance in Ar, Kr, and Xe from the ground state has been studied by Yoshino [41–43], who used conventional light sources, and by synchrotron radiation (Maden and Codling [202], Baig and Connerade [45], and Ueda [203,204]).

The photoionization cross-sections and photoelectron angular distribution parameters across the $4p^5(^2P_{1/2})$ ns', and nd' autoionization resonances in Ar, Kr, and Xe were reported by Wu et al. [205], and subsequently, Maeda et al. [206] measured their photoabsorption cross-sections in the autoionization regions. Using the ionization of inert gases from the ground state, only the $J = 1$ upper levels are accessible via single-photon excitation. The higher angular momentum states can be approached either via multi-photon excitation, as shown by Pratt et al. [207], Koeckhoven et al. [208,209], and Blazewicz et al. [210], or by multi-step excitation from the ground state (Rundel et al. [39], Grandin and Husson [211], Wada et al. [212], Wang and Knight [213], King and Latimer [214], Harth et al. [215], and Klar et al. [216]). Alternatively, the $mp^5(m+1)s$ $[1/2]_0$ and $[3/2]_2$ metastable states can be populated by running a discharge enabling one to approach the $\ell = 0, 2$ excited states via two-photon excitation (Knight and Wang [217], L'Huillier et al. [218], Piracha et al. [85,219], and Ahmed et al. [220,221]).

A schematic diagram for the two-step excitation and ionization for the argon atom is shown in Figure 12.

The excited states are represented in the j_cK-coupling scheme (Racah [222], Cowan [1]) as $n\ell$ $[K]_J$, where ℓ is the orbital angular momentum of the excited electron and K is the vector sum of the orbital angular momentum of the excited electron and the angular momentum of the core electrons. The total angular momentum J is the vector sum of K and the spin quantum number of the excited electron. The prime refers to the terms associated with the $4p^5(^2P_{1/2})$ ionic state. According to the electric dipole selection rules in the jK-coupling scheme: $\Delta j = 0$, $\Delta K = 0, \pm 1$ and $\Delta J = 0, \pm 1$, the Rydberg series can be studied. Since the first excited states of all the inert gages lie more than 10 eV above the ground state, a radiation source emitting in the VUV region is therefore required to access these states. However, the metastable state can be populated by running a mild discharge, either in a DC or RF environment. The excitation of an electron from the 3p sub-shell yields four levels: $3p^54s$ $[3/2]_2$, $3p^54s$ $[3/2]_1$, $3p^54s'$ $[1/2]_0$, $3p^54s'$ $[1/2]_1$. The 4s $[3/2]_2$, and 4s' $[1/2]_0$ are the metastable states while the 4s $[3/2]_1$ and 4s' $[l/2]_1$ states are short-lived and combine instantly with the ground state. The $3p^5(^2P_{3/2})$ and $3p^5(^2P_{1/2})$ are the parent ion levels that combine with the added electron and yield the Rydberg series converging to two ionization limits. Dunning and Stebbings [192] used a molecular beam apparatus in conjunction with a pulsed tunable UV laser to study the photoionization of argon and krypton in the wavelength range 325–269 nm and reported the $p^5(^2P_{1/2})$ nf and np levels, which autotomize into the underlying $^2P_{3/2}$ continuum. Gornik et al. [223] reported the $5p^5$ 5d $[K = 5/2]_{J=3}$ and $5p^5$ 5d $[K = 1/2]_{J=3}$ atomic states of xenon excited by two-photon absorption from the ground state. Wu et al. [205] reported the photoionization cross-sections and photoelectron angular distribution parameters across the autoionizing resonances for Ar, Kr, and Xe using the synchrotron radiation-based photoelectron spectroscopy. Samson and Stolte [224] reported the absolute photoionization cross-sections for neon, argon, krypton, and xenon from threshold to 125 eV with an accuracy of 61 to 3% in both tabular and graphical form. These data were compared with the optical oscillator strength measurements of Bonin et al. [44], who used the Dipole (e, e) collision technique. The autoionizing resonances in Ar, Kr, and Xe were also studied via three-photon excitation from the ground state by Prat et al. [207] and Koechhoven et al. [208], who reported the $p^5(^2P_{3/2})$ ns $[1/2]_1$, nd $[3/2]_1$, nd $[5/2]_3$, and ng $[7/2]_3$ levels. Subsequently, Koechhoven et al. [209] reported the $p^5(^2P_{3/2})$ np $[1/2]_0$, np $[3/2]_2$, nf $[5/2]_2$, and nf $[7/2]_4$ autoionizing resonances in inert gases using four-photon excitation from the ground state. The spectra of argon, krypton, and xenon in the autoionization region using a two-step resonant laser excitation and Optogalvanic detection technique were reported by Baig et al. [225]. By selecting $(m)p^5$ $(m + 1)p$ $[3/2]_2$ as an intermediate state (m = 4, 5, and 6 for Ar, Kr, and Xe, respectively), the $(m)p^5$ nd $[5/2]_3$ autoionizing resonances were studied. The Multichannel Quantum Defect Theory (MQDT) parameters were derived from the analysis of the series perturbations among the $(m)p^5$ nd $[5/2]_3$, $(m)p^5$ nd $[7/2]_3$, and $(m)p^5$ nd $[5/2]_3$ series in the discrete region using the phase-shifted formulation of the three-channel quantum defect theory Gallagher [142] and from the line profile analysis of the autoionizing resonances above the first ionization threshold. The predicted reduced widths for the autoionizing resonances based on the series perturbation analysis showed good agreement with those of the experimentally observed profiles.

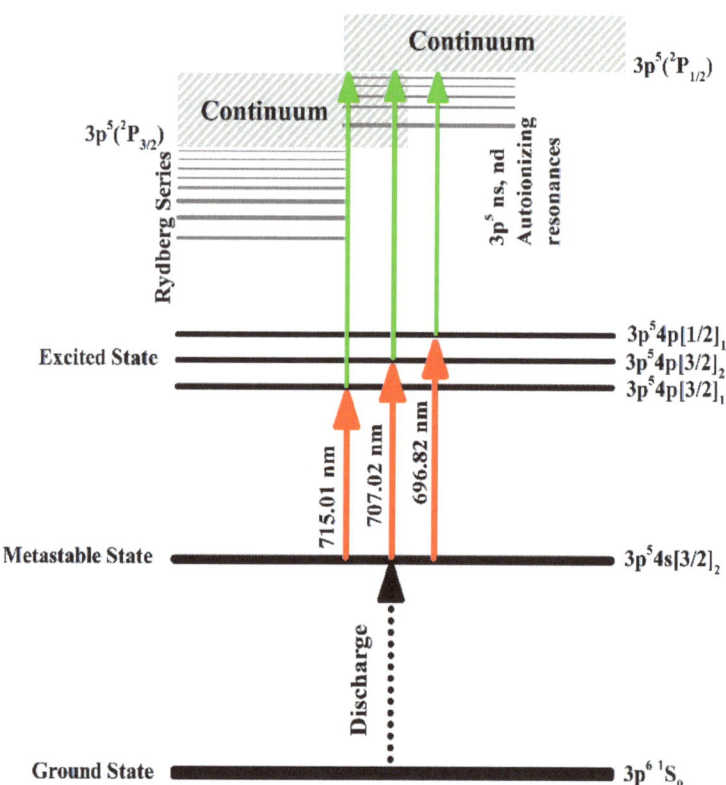

Figure 12. Excitation scheme to access the above threshold photoionization in argon.

The $2p^5(^2P_{1/2})$ np [J = 0, 1, 2] and $2p^5(^2P_{1/2})$ nf [J = 2] autoionizing Rydberg states of Ne by two-step two-color photoionization of ground-state neon atoms via the intermediate states $2p^5$ $(^2P_{3/2})3s$ J = 1 and $2p^5(^2P_{1/2})3s$ J = 1, followed by time-of-flight ion detection and strong interference between the direct and virtual intershell photoexcitation channels in autoionizing resonances production, especially for the np J = 0 series, were reported by Petrov et al. [226]. The reduced width Γr = 6431 cm^{-1} and the quantum defect μ_p = 0.7634 of the $2p^5{}_{1/2}$ 13p J = 0 resonance were found to be in good agreement with the experimental results Γ = 5334 cm^{-1} and μp = 0.7662 determined with a Fano-type line shape analysis. The even parity $mp^5(^2P_{1/2})$ np' and mp^5 $(^2P_{1/2})$ nf' autoionizing resonances of Ar, Kr, and Xe (m = 3, 4, 5) were studied experimentally and theoretically by Petrov et al. [227] by one-photon excitation from the lower-lying intermediate levels. The high-resolution measurements for the Ar(nf), Kr(12p, 8f), and Xe(8p) resonances were reported and the line shape parameters for these resonances were derived by a Fano-type analysis. The experimental spectra and the resonance parameters were compared with theoretical calculations, which were based on the configuration interaction Pauli–Fock approach, including core polarization.

The oscillator strengths of the $2p^5$ $(^2P_{1/2})$ nd' J = 2, 3 autoionizing resonances in neon were studied by Mahmood et al. [228] using a DC discharge, multistep laser excitation/ionization, and Optogalvanic detection technique. The excited states were approached using two-step laser excitation via $2p^5$ 3p' $[1/2]_1$, $2p^5$ 3p' $[3/2]_1$, and $2p^5$ 3p' $[3/2]_2$ intermediate levels, which were accessed from the $2p^5$ 3s $[1/2]_2$ metastable level. The f-values were determined for the nd' $[3/2]_2$, nd' $[5/2]_2$, and nd' $[5/2]_3$ series following the K = J = +1 selection rules. The photoionization cross-section at the $2p^5\ ^2P_{1/2}$ ionization threshold was determined as 5.5(6) Mb. Claessens et al. [229] reported the photoionization-cross-section for the $2p^53p$ $^3D_3$ state of neon at the wavelengths of 351 and 364 nm by monitoring the decay of the fluorescence of atoms trapped in a magneto-optical atom trap under the presence of a photo-ionizing laser. The absolute photoionization cross-sections were obtained as $(2.05 \pm 0.25) \times 10^{-18}$ cm^2 at 351 nm and $(2.15 \pm 0.25) \times 10^{-18}$ cm^2 at 364 nm ionizing laser wavelength. The photoionization of excited ($2p^5$ 3p, J = 3) atoms over the photoelectron energy range 0–2.5 eV was calculated by Petrov et al. [69] using the configuration interaction Pauli–Fock with a core polarization method. Subsequently, Petrov et al. [230] calculated the photoionization cross-sections for Ne atoms in the excited $2p^53p$ [K]J (J = 0 − 3) levels at near-threshold energies within the configuration interaction Pauli–Fock approach, including core polarization. The computed spectra and the line shape parameters of the odd parity $2P^5$ $(^2P_{1/2})$ ns and nd autoionizing resonances were in good agreement with high-resolution laser spectroscopic results. In addition, the absolute partial photoionization cross-sections for the $^2P_{3/2}$ and $^2P_{1/2}$ channels at photoelectron energies up to 7 eV were determined. Except for the highest lying $2p1(^1S_0)$ level, these cross-sections monotonically decrease with energy (as reported earlier in single-electron calculations for the Ne($2p^53p$) configuration) with branching ratios that essentially reflect the core compositions of the 2px levels. For the 2p1 level, the resonance structure and the partial cross-sections were reported to be strongly influenced by a Cooper–Seaton minimum in the $d_{3/2}$ channel, located just above the $^2P_{1/2}$ ionization limit. Mahmood et al. [231] reported the photoionization cross-section from three intermediate levels $2p^5$ 3p' $[1/2]_1$, $2p^5$ 3p' $[3/2]_2$ and $2p^5$ 3p' $[5/2]_3$ of neon up to the $2p^5\ ^2P_{1/2}$ ionization threshold using the optogalvanic technique. Baig et al. [232] measured the excitation spectra from the $2p^53p$ $[5/2]_{3,2}$ levels in neon using two-step laser excitation and ionization in conjunction with an optogalvanic detection in DC and RF discharge cells. The $2p^53p$ $[5/2]_{3,2}$ intermediate levels were approached via the collisionally populated $2p^53s$ $[3/2]_2$ metastable levels. The Rydberg series $2p^5(^2P_{3/2})nd$ $[7/2]_4$, $2p^5(^2P_{3/2})ns$ and the parity forbidden transitions $2p^5(^2P_{3/2})np$ $[5/2]_3$ were observed from the $2p^53p$ $[5/2]_3$ level, whereas the $2p^5(^2P_{3/2})nd$ $[7/2]_3$, $2p^5(^2P_{3/2})ns$ $[3/2]_2$, and $2p^5(^2P_{1/2})nd$ $[5/2]_3$ series were observed from the $2p^53p$ $[5/2]_2$ level in accordance with the $J = K = \pm 1$ selection rules. The photoionization cross-sections from the $2p^53p'$ $[5/2]_3$ intermediate levels were measured at eight ionizing laser wavelengths (399, 395, 390, 385, 380, 370, 364, and 355 nm), which is in good agreement with that reported by Petrov et al. (2008) and that from the $2p^53p'$ $[5/2]_2$ level at 401.8 nm.

In a high-resolution study of odd argon ($^2P_{3/2}$ ns, nd, $J = 2, 3$), Rydberg states using transverse resonant two-photon CW laser excitation of metastable (4s $^3P_2$) atoms in a collimated beam via the (4p $^3D_3$) level was performed by Weber et al. [233]. The perturbations in the spectra associated with the ($^2P_{1/2}$ $8d_0$, $9d_0$, $J = 2, 3$) levels and widths of the low-lying (nd_0, $J = 2, 3$) autoionization resonances were determined. The multichannel quantum defect theory (MQDT) analysis of the measured bound Rydberg levels yielded reliable Eigen-channel quantum defect parameters to describe the odd $J = 2$ and 3 spectra of Ar in the bound as well as in the autoionization region. Using a mild DC discharge system coupled with a multistep laser excitation and ionization scheme and optogalvanic detection, Aslam et al. [234] studied the $3p^5nd$ $[3/2]_2$ and nd $[5/2]_{2,3}$ Rydberg series, converging to the $3p^5(^2P_{3/2})$ and ($^2P_{1/2}$) limits, excited from the $3p^54p'$ $[1/2]_1$, $3p^54p'$ $[3/2]_1$, and $3p^54p'$ $[3/2]_2$, and the intermediate levels that were populated from the $3p^54s$ $[3/2]_2$ metastable levels of argon. Three perturbers, $3p^5(^2P_{1/2})8d'$ $[3/2]_2$, $3p^5(^2P_{1/2})8d'$ $[5/2]_2$, and $3p^5(^2P_{1/2})8d'$ $[5/2]_3$, were identified, as reported earlier by Weber et al. [233], and their interaction with the $3p^5(^2P_{3/2})$ nd $J = 2, 3$ Rydberg series were studied using the multichannel quantum defect theory (MQDT). The parameters obtained by fitting the lines shapes of the corresponding autoionizing resonances adjacent to the first ionization threshold $3p^5(^2P_{1/2})9d$ $[3/2]_2$, $3p(^2P_{1/2})$ $9d$ $[5/2]_2$, and $3p(^2P_{1/2})$ $9d$ $[5/2]_3$ show good correlation with those determined from the MQDT analysis in the discrete region. Qian et al. [235] studied the ionization spectrum of Ar in the energy region between the ionization thresholds for Ar$^+$ ($^2P_{3/2}$) and Ar$^+$ ($^2P_{1/2}$) using VUV synchrotron radiation and high-repetition-rate IR optical parametric oscillator (OPO) laser source.

The photoionization cross-section for the $3p^54p'$ $[3/2]_1$, $3p^54p'$ $[3/2]_2$, and $[1/2]_1$ intermediate levels were measured by Baig et al. [236] at the $3p^5$ $^2P_{1/2}$ ionization threshold as 34 ± 5, 31 ± 5, and 28 ± 4 Mb, respectively. The absolute values of the cross-section at the $3p^5$ $^2P_{1/2}$ threshold were used to determine the f-values for the $4p'$ $[3/2]_1 \rightarrow nd$ $[5/2]$, $4p'$ $[3/2]_2 \rightarrow nd$ $[5/2]_3$, and $4p'$ $[1/2]_1 \rightarrow nd$ $[3/2]_2$ Rydberg transitions. In Figure 13, we reproduce the $3p^5(^2P_{1/2})nd$ $[5/2]_3$ Rydberg series due to excitation from the $3p^54p'$ $[3/2]_2$ intermediate level showing the linear behavior of the optogalvanic detector, as the intensities of the higher members of the series decreases nearly to $(1/n^3)$ and the calibration of the f-values was straight forward. In Table 8, we enlist the results of the line profile analysis of the leading autoionizing resonances in Ar, Kr, and Xe, revealing the resonance energies, widths, line profile parameters, effective quantum number, and reduced widths.

Wright et al. [237] reported an experimental and theoretical study of the photoionization of even parity autoionizing Rydberg series of argon. The 4s' $[1/2]_0$ and 4s $[3/2]_2$ metastable states were used to access the even-parity autoionizing resonances between the first and second ionization limits and the widths and energies of the overlapping resonances data were determined by fitting to a sum of Fano-type or Shore-type profiles. A combination of synchrotron radiation and lasers was exploited by Lee et al. [238], who used synchrotron radiation to excite the $3p^5(^2P_{1/2})$ {3d' $[3/2]_1$, 5d' $[3/2]_1$, 5s' $[3/2]_1$, 5s' $[1/2]_1$, 7s' $[1/2]_1$}, and $3p^5(^2P_{3/2})$ {6d $[1/2]_1$, 6d $[3/2]_1$, and 8s $[3/2]_1$} intermediate levels and then the autoionizing Rydberg series $3p^5(^2P_{1/2})np'$ { $[1/2]_{0,1}$, $[3/2]_{1,2}$, and nf' $[5/2]_2$} were excited using lasers. The spectral line shapes of the autoionizing resonances were analyzed with Beuler–Fano profiles. Petrov et al. [70] (reported the $3p^5_{1/2}$ np $J = 0, 1, 2$ and $3p^5_{1/2}nf$ $J = 2$ autoionizing Rydberg states of Ar by two-step two-color photoionization of ground-state argon atoms via the intermediate levels $3p^5_{1/2}4s$ $[1/2]_1$, $3p^5_{3/2}5s$ $[3/2]_1$, and $3p^5_{3/2}3d$ $[1/2]_1$, followed by time-of-flight ion detection. Accurate energy positions and widths were determined for the 15p $[1/2]_0$, 15p $[3/2]_2$, and 13f $[5/2]_2$ resonances. The measured spectra and earlier results obtained for the intermediate levels 5s $[1/2]_1$, 7s $[1/2]_1$, and 8s $[3/2]_1$ were compared with absolute cross-sections and calculated with the configuration interaction Pauli–Fock core-polarization method.

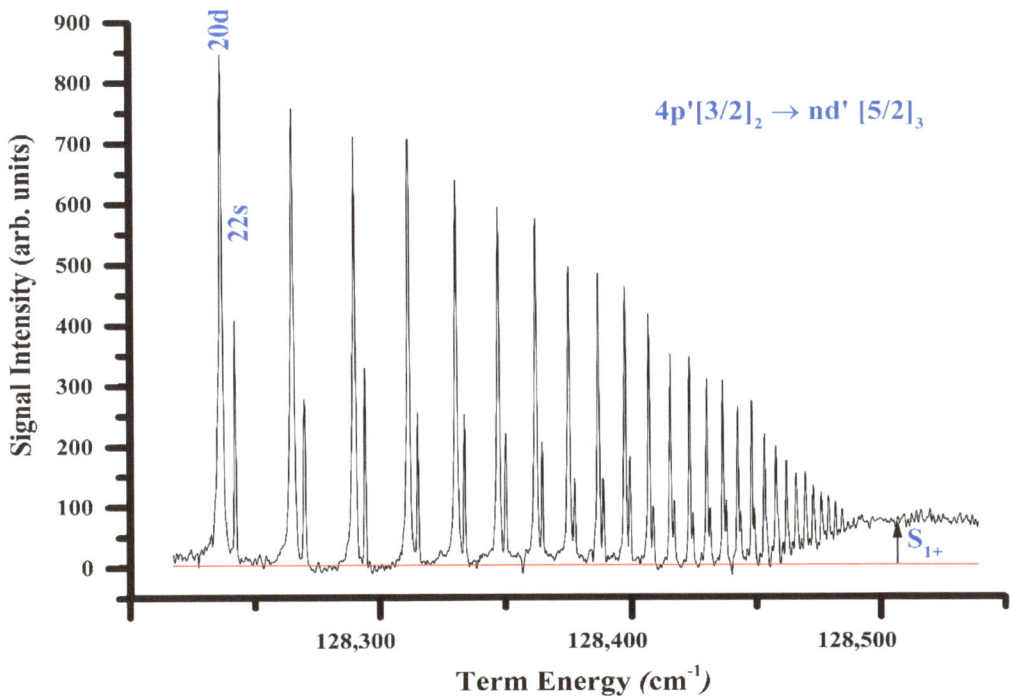

Figure 13. The $3p^5(^2P_{1/2})nd$ $[5/2]_3$ Rydberg series excited from the $3p^54p'$ $[3/2]_2$ intermediate level in argon. The nd series is dominating in intensity, as compared with the ns series.

Table 8. Line profile analysis of the autoionizing resonances in Ar, Kr, and Xe.

Atom	Resonance	E_r (cm^{-1})	Γ (cm^{-1})	q	$\nu_{1/2}$	$\Gamma_r = (\Gamma \nu_{1/2}^3)$
Ar	$9d'$ $[5/2]_2$	127 075 (1)	8 (1)	−30 (5)	8.651 (2)	5180
	$9d'$ $[5/2]_3$	127 090 (1)	9 (2)	−30 (5)	8.695 (2)	5915
Kr	$6d'$ $[5/2]_2$	113 205 (3)	45 (5)	−35 (5)	4.648 (2)	4520
	$6d'$ $[5/2]_3$	113 285 (3)	50 (5)	−30 (5)	4.685 (2)	5140
Xe	$7d'$ $[5/2]_2$	103 008 (3)	40 (8)	−20 (5)	4.524 (2)	3700
	$7d'$ $[5/2]_3$	103 110 (3)	55 (8)	−20 (5)	4.567 (2)	5240

The pioneering work on the absorption spectrum of krypton was reported by Beutler [239] in which the $4p^5$ ns and $4p^5$ nd odd-parity J = 1 states were observed. Yoshino and Tanaka [42] extended these studies by observing the $4p^5$ nd series up to a much higher principal quantum number and extracting the value of the first ionization potential of krypton. Klar et al. [240] used two-step laser excitation of metastable Kr ($4p^55s$ $^3P_{0,2}$) atoms via selected Kr ($4p^55s$ J = 1, 2) levels, and recorded the low-lying autoionizing resonances Kr($^2P_{1/2}$ nd, J = 2, 3) (n = 6, 7) at high resolution in two different experiments (atomic beam spectroscopy and opt-galvanic spectroscopy in a discharge). Accurate values for the resonance positions (quantum defects) and the (reduced) widths were determined, and the profile indices were reported. For comparison, multichannel quantum defect theory (MQDT) analyses, using energy-dependent MQDT parameters, were carried out for the odd Kr (J = 1, 2, 3) levels, yielding good agreement in the quantum defects and partial agreement in the predicted reduced widths for the nd resonances with the experimental values. Moreover, the simplified phase-shifted quantum defect theory analyses of the measured bound odd Kr (J = 2 and 3) Rydberg levels were carried out to provide improved insight into the dominant decay mechanism of the resonances. A summary and comparative discussion of the reduced widths for the nd [K]J resonances of all the rare-gas atoms Ne, Ar, Kr, and Xe was also presented. The odd-parity autoionizing resonances in krypton $4p^5ns$ $[1/2]_{0,1}$, $4p^5nd$ $[3/2]_2$, and $4p^5nd$ $[5/2]_{2,3}$ excited from three intermediate levels $4p^5\,5p'$ $[1/2]_1$, $4p^5\,5p'$ $[3/2]_1$, and $4p^5\,5p'$ $[3/2]_2$ were studied by Baig et al. [241] using resonant two-photon excitation from the $4p^5\,5s$ $[3/2]_2$ metastable level in a mild DC discharge and an optogalvanic detection technique.

To demonstrate the utility of the selected polarization vectors of the first and second laser, the autoionizing resonance for $n = 8$ in krypton, $4p^58d'$ $[3/2]_2$, $4p^58d'$ $[5/2]_2$, and $4p^58d'$ $[5/2]_3$ excited from the $4p^55p'$ $[1/2]_1$, $4p^55p'$ $[3/2]_1$, and $4p^55p'$ $[3/2]_2$ intermediate levels using both the lasers linearly polarized were studied. Within the framework of the transitions selection rules, only the $5p^58d'$ $[3/2]_2$ transitions are allowed from the $4p^55p'$ $[1/2]_1$ intermediate level and the line profile is shown in Figure 14a. The $5p^58d'$ $[5/2]_2$ is only accessible from the $4p^55p'$ $[3/2]_1$ intermediate level and, indeed, an isolated resonance was observed, as shown in Figure 14b. However, both the $4p^58d'$ $[5/2]_2$ and $4p^58d'$ $[5/2]_3$ states are allowed from the $4p^55p'$ $[3/2]_2$ intermediate level, but $4p^58d'$ $[5/2]_3$ dominates in intensity. The line profiles are shown in Figure 14c. Since both the resonances possess different J-values, they are non-interacting, and therefore the individual profiles were fitted and the sum of the two profiles yields the full curve. The fitted Fano profile for the $4p^58d'$ $[5/2]_2$ resonance is with q-parameter = -90, $\Gamma = 14$ cm^{-1} and $E_r = 115{,}792.5$ cm^{-1} and that of $4p^58d'$ $[5/2]_3$ is q-parameter = -50, $\Gamma = 17$ cm^{-1} and $E_r = 115{,}819$ cm^{-1}. The even parity autoionizing resonances in krypton were studied by Li et al. [242] from the metastable states $4p^55s$ $[3/2]_2$ and $4p^55s'$ $[1/2]_0$ using a pulsed DC discharge and subsequent single-photon excitations to the $4p^5np'$ $[3/2]_{1,2}$, $[1/2]_1$ and $4p^5nf'$ $[5/2]_3$ Rydberg states.

The work on the Rydberg states in Xe below the $5p^5\,^2P_{3/2}$ or $5p^5\,^2P_{1/2}$ thresholds accessed from the Xe $5p^6\,^1S_0$ ground state includes one-photon excitation of the odd parity ns $J = 1$ and nd $J = 1$ states and multiphoton excitation of odd or even Rydberg states by non-resonant or resonant multiphoton excitation. In addition, single-photon or resonant or non-resonant two-photon excitation from the metastable levels of Xe has been exploited to study even and odd Rydberg states. Using two different experimental set-ups involving optogalvanic spectroscopy and atomic beam spectroscopy with mass spectrometric ion detection, respectively, Hanif et al. [243] reported the $5p^5\,^2P_{1/2}\,nd$ [K = 3/2]$J = 2$ and [K = 5/2]$J = 2,3$ autoionizing resonances of xenon by means of two-step resonant laser excitation from the metastable Xe ($J = 2, 0$) levels. By selecting a particular intermediate level, 6p $[1/2]_1$ and 6p $[3/2]_{1,2}$, autoionizing resonances with specified K and J values were addressed. Level energies and resonance widths were derived by a Beutler–Fano-type line shape analysis. Extended calculations for the Xe (ns $J = 0, 1$), Xe (nd $J = 1, 2, 3$), and Xe (ng $J = 3$) autoionizing resonances were also performed based on the Pauli–Fock approach and include core polarization and electron correlation effects at a high level. Alois et al. [244] studied the photoionization of short-lived Rydberg states in Xe using resonant atomic excitation by synchrotron radiation and subsequent ionization by a tunable dye laser. By combining circular and linear polarization of the synchrotron as well as of the laser photons, the partial photoionization cross-sections were separated in the region of overlapping autoionizing resonances of different symmetry and the parameters of the resonances were extracted. Hanif et al. [243] observed the even parity autoionizing resonances built on the $5p^5\,7p$ and $5p^5\,4f$ configurations in xenon using the laser optogalvanic detection technique in conjunction with a DC discharge cell. The autoionizing resonances $5p^5\,7p$ $[1/2]_1$ and $5p^5\,7p$ $[3/2]_1$ excited from the $5p^5\,6s$ $[1/2]_0$ metastable level and the $5p^5\,7p$ $[3/2]_2$ resonance excited from the $5p^5\,6s$ $[1/2]_1$ level were studied. In addition, the $5p^5(^2P_{1/2})nf$ ($n = 4$ and 5) autoionizing resonances excited from the collisionally populated levels of the $5p^55d$ configuration were investigated. The resonance energies, quantum defects, line-shape parameters, and resonance widths were determined by fitting the observed resonance profiles to Fano's formula for the photoionization cross-section. Petrov et al. [245]

studied the near-threshold photoionization from the excited states of p^5p (J = 0–3) levels of Ar, Kr, and Xe. The absolute total and partial cross-sections for photoionization of excited $mp^5(m+1)p$ J = 0–3 levels in Ar, Kr, and Xe (m = 3–5) near the threshold was calculated using the configuration interaction Pauli–Fock, including the core polarization (CIPFCP) approach. The line shapes of the mp^5 ($^2P_{1/2}$) ns and nd autoionizing resonances and their variation with the character of the intermediate excited J = 0–3 state were also studied. Sukhorokov et al. [53] presented a review of the experimental and theoretical studies of the threshold photoionization of the heavier rare-gas atoms with particular emphasis on the autoionizing resonances in the spectral region between the lowest two ionization thresholds $^2P_{3/2}$ and $^2P_{1/2}$, accessed from the ground or excited states. The observed trends in the positions, widths, and shapes of the autoionizing resonances depending on the atomic number, the principal quantum number n, the orbital angular momentum quantum number, and further quantum numbers specifying the fine- and hyperfine-structure levels were summarized and supplemented with the ab initio and multichannel quantum defect theory calculations. Besides, the effects of various approximations in the theoretical treatment of photoionization in these systems were analyzed. The studies on the measurements of the photoionization cross sections for the excited states of atoms, molecules, and ions are still very actively being carried out at different synchrotron radiation facilities, and new data are being provided to compare with the theoretical models.

The ionization dynamic of atoms exposed to high power lasers, (intensities of 10^{20} W/cm^2, short pulse duration of 10 fs, and high repetition rates) has been extensively studied, as the presence of strong laser fields that drastically alters the nature of atomic systems. The interaction of laser field with atoms in the regime from multiphoton to tunneling ionization has been comprehensively studied by several groups that are summarized in a review by DiMauro et al. [246]. Yamakawa et al. [247] exposed Xe atoms to an IR laser (800 nm laser wavelength, 20 fs pulse duration) having an intensity range of 10^{13}–10^{18} W/cm^2 and observed the ion yields of Xe$^+$–Xe^{20+} as a function of laser intensity. The experimental data were compared with the results from a single active electron-based Ammosov–Delone–Krainov model. A double laser-plasma technique was used by Lu et al. [248] to measure the photoabsorption spectrum of Bi$^+$ covering the wavelength range from 37 and 60 nm. One laser was used to produce the vacuum ultraviolet continuum radiation and the second laser-produced plasma was used as the sample of atoms to be probed. The observed structure was identified due to 5d–6p transitions with the help of the Cowan suit to the atomic codes. The synchrotron radiation (SR) facilities have been extensively used to study the photoionization from the ground state of the atoms. A combination of synchrotron radiation with a tunable laser further enhanced the capabilities to study the photoionization from the excited states. Aloise et al. [244] used the resonant atomic excitation by synchrotron radiation and subsequent ionization by a tunable dye laser to study the photoionization of short-lived Rydberg states. The 5d (J = 1) level of xenon at 10.401 eV and 7s (J = 1) at 10.562 eV were excited from the ground state with SR. The 4f' (J = 2) autoionizing resonance from the 5d J = 1 and the 8p' J = 0, 1, 2 autoionizing resonances from the 7s J = 1 level were investigated. By combining the circular and linear polarization of the synchrotron radiation as well as the laser photons, the partial photoionization cross-sections were measured.

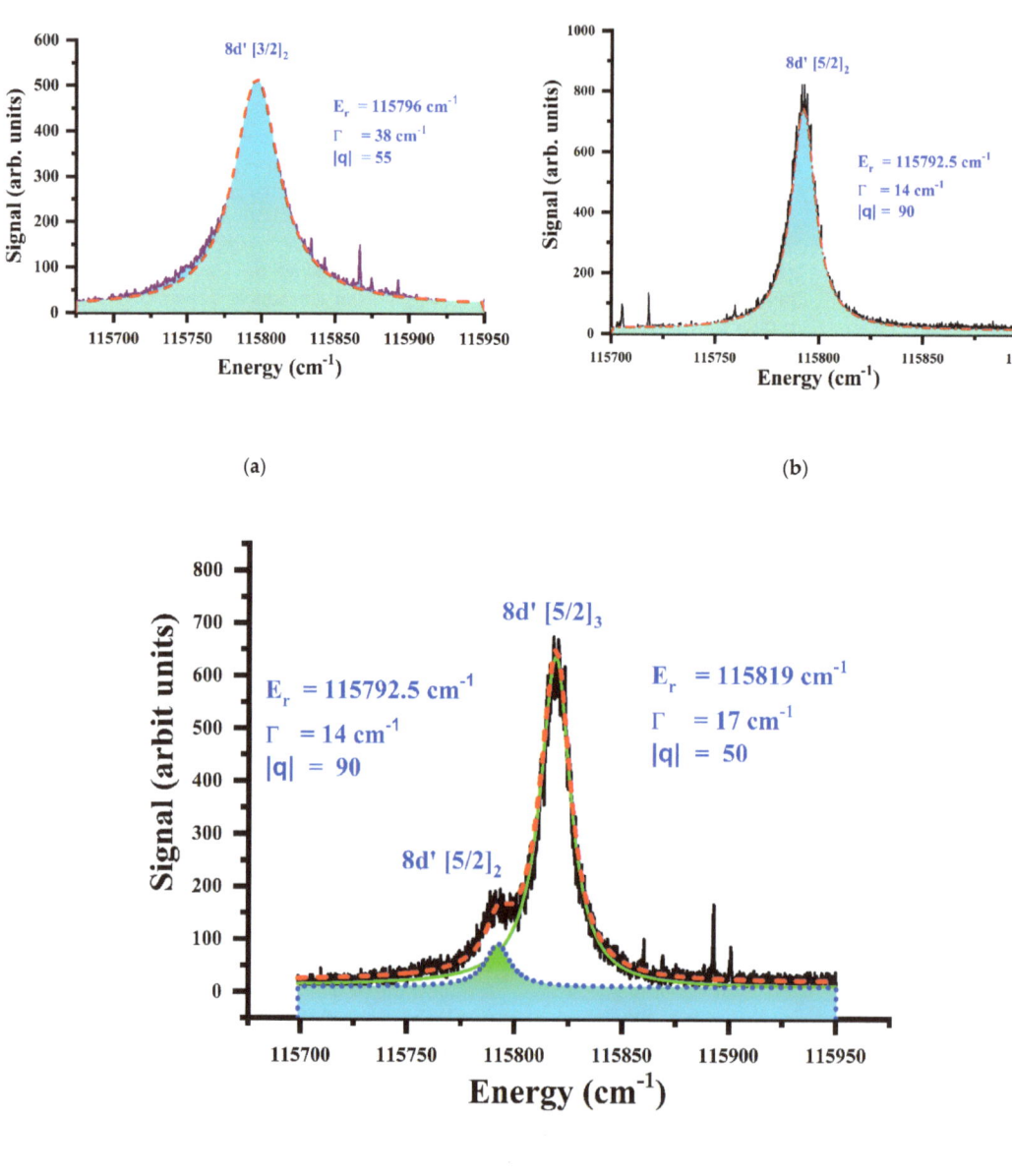

Figure 14. (**a**). The $5p^58d'$ $[3/2]_2$ autoionizing resonances in krypton excited from the $4p^55p'$ $[1/2]_1$ intermediate level in krypton. (**b**). The $5p^58d'$ $[5/2]_2$ autoionizing resonances excited from the $4p^55p'$ $[3/2]_1$ intermediate level in krypton. (**c**). The $5p^58d'$ $[5/2]_2$ and $5p^58d'$ $[5/2]_3$ autoionizing resonances excited from the $4p^55p'$ $[5/2]_2$ intermediate level in krypton.

The advent of powerful soft X-ray sources based on storage rings with insertion devices, free-electron laser (FEL), and high-harmonic generation techniques, has opened up new avenues for the photoionization studies of small quantum systems. The first FEL facility (FLASH) was developed at DESY, Germany, in 2005, which provides ultra-short pulses at high fundamental photon energies (13–48 nm) with high peak and average power. For the pump-probe experiments, an optical laser facility was also set up, which provides femtosecond infrared (110 fs, 800 nm) and picosecond visible (12 ps, 532 nm) pulses, electronically synchronized with the FEL (pulse duration 19–50 fs) pulses. Mayer et al. [249] measured the two-photon ionization of atomic helium by combining the femtosecond extreme ultraviolet pulses from the free-electron laser (FLASH) with the intense light pulses from a synchronized neodymium-doped yttrium lithium fluoride laser. The two-photon ionization measurements were in good agreement with the theoretical predictions. Sorokin et al. [250] studied the multiphoton ionization of neon and helium by the ion mass-to-charge spectroscopy using a focused beam at 42.8 nm, 38.4 eV photon energy, and irradiance of up to 10^{14} W/cm^2 and observed direct, sequential, and resonant two-, three-, and four-photon excitations as a function of absolute photon intensity. It was inferred that the atomic and ionic photoionization cross-sections are dominantly due to sequential excitation as compared to direct multiphoton processes. Martins et al. [251] studied the multiphoton ionization of neon atoms using the PG0 branch of the FLASH monochromator beamline to observe the Ne$^+$ 2p $\rightarrow$ nℓ resonances in the energy region 41 and 42 eV. Richter et al. [252] studied the photoionization of different rare gases acquired at FLASH by applying ion spectroscopy at the wavelength of 13.7 nm and irradiance levels of thousands of terawatts per square centimeter. The degree of nonlinear photoionization was found to be significantly higher than for neon, argon, and krypton. The collective giant 4d resonance of Xe may be responsible for this effect, which arises in this spectral range. Schippers et al. [55] presented the use of X-rays from a synchrotron radiation source and the photo-ion merged-beams technique to study the photoionization of mass/charge selected ionized atoms, molecules, and clusters. Examples for photoionization of atomic ions were discussed by going from the outer shell ionization of simple few-electron systems to the inner-shell ionization of complex electron ions. The unique capabilities of the photon-ion merged-beams technique for the study of photoabsorption by nanoparticles were demonstrated by the examples of endohedral fullerene ion. Young et al. [56] reported a roadmap of ultrafast X-ray atomic and molecular physics, highlighting the contributions into four categories: ultrafast molecular dynamics, multidimensional x-ray spectroscopies, high-intensity x-ray phenomena, and attosecond x-ray science. The development of X-ray free-electron lasers (XFELs) and table-top sources of x-rays based upon high harmonic generation (HHG) have revolutionized the field of ultrafast x-ray atomic and molecular physics. XFELs provide very intensity (10^{20} W cm^{-2}) of X-rays at wavelengths down to $\sim$10 nm, and HHG provides high time resolution ($\sim 10^{-18}$ s) and a large coherent bandwidth at longer wavelengths. Thus, one can focus on individual atoms and view electronic and nuclear motion on their intrinsic scales using these X-ray sources. The XFEL (European XFEL, Swiss-FEL, and PAL-FEL in Korea) has opened new fields of multiphoton and nonlinear X-ray physics, where the behavior of matter under extreme conditions can be explored. The time resolution and pulse synchronization provided by HHG sources are used to study the time delays in photoionization, and charge migration in molecules. At the 60th anniversary of the ICPEAC conference celebrations, three roadmaps on photonic, electronic, and atomic collision physics were published. Ueda et al. [57] presented Roadmap I, in which the focus was on light-matter interaction. With the advent of new light sources, such as X-ray free-electron lasers and attosecond lasers, studies of ultrafast electronic and molecular dynamics are rapidly growing. Besides, experiments with the established synchrotron radiation sources and femtosecond lasers using state-of-the-art detection systems are enlightening new scientific areas that have never before been explored. The photon-target experiments are growing from atoms and small molecules to complex systems such as biomolecules, fullerene, clusters, and solids. This roadmap contains several exciting contributions by experts in the fields, such as photoionization of ions, photo-double ionization of atoms and molecules, molecular-frame photoelectron angular distributions from static to time-resolved, and photo-induced reaction dynamics probed by COLTRIMS.

A review of the recent work on the photoionization of atomic ions of astrophysical interest was presented by Schippers and Muller [58] based on the experimental work performed at the photon-ion merged-beam setup PIPE, installed at the XUV beamline of PETRA III synchrotron radiation source operated by DESY, Hamburg, Germany. Results on single and multiple L-shell photoionizations of Fe^+, Fe^{2+}, and Fe^{3+} ions, and single and multiple K-shell photoionizations of C^-, C^+, C^{4+}, Ne^+, and Si^{2+} ions were discussed in an astrophysical context. Interestingly, it was inferred that the results of the photon-ion merged-beams method at the world's brightest synchrotron radiation light source has led to a breakthrough in the experimental study of atomic inner-shell photoionization processes with ions. Katravulapally et al. [253] studied the photoionization of lithium-ion via its double-excited states 2s2p ^1P using a free-electron laser (FEL) radiation and also calculated the ionization yield using the perturbative statistical description of the atomic dynamics. It was inferred that the FEL temporal fluctuations affect the line shape, and is strongly dependent on intensity and pulse's coherence time. A further increase in intensity enhances the Li^{2+} ionization, causing further distortion in the line shape of Li^+. The two-photon ionization cross sections and asymmetry parameters within the independent-particle approximation and relativistic second-order perturbation theory were calculated by Hofbrucker et al. [254]. The dependence of the asymmetry parameters on the polarization and energy of the incident radiation as well as on the angular momentum properties of the ionized electrons were studied. These studies have also been expanded for measuring the photoionization cross-section of the simplest enol and vinyl alcohol, as shown by Rosch [255], who employed multiplexed photoionization mass spectrometry at 308 nm and reported the cross-sections 7.5 ± 1.9 Mb and 10.005 eV and 8.1 ± 1.9 Mb at 10.205 eV. There are several papers addressing the photoionization cross sections for the excited states of atoms published every year, and this field of research is still very active.

6. Conclusions

The photoionization cross-sections for the excited states of alkali atoms, alkaline earth atoms, and inert gases have been presented as studied by several researchers. It is observed that, by in large, the measured photoionization cross-sections by different groups are in reasonable agreement and also agree with the theoretical calculations. The situation in the alkali atoms is very satisfactory, as the issue of normalization that was persistent in the early measurements has been overcome by employing many advanced experimental techniques such as multi-step laser excitation, the saturation technique, and the technique of ion trapping and cooling to monitor the intensity variations. Three experimental techniques are commonly used; thermionic diode-ion detector, the optogalvanic detection-based system, and atomic beam apparatus coupled with a time-of-flight mass spectrometer. The measurement using a thermionic diode ion detector has to be done very carefully as it is prone to suffer from the linearity of the detection system and accurate determination of the interaction area. However, with careful alignment and proper adjustment of the exciting and ionizing laser beams, reliable measurements can be performed. The situation in the optogalvanic-based detection system is relatively better, as the interaction area is very small and can easily be determined at the crossing point of the exciting and the ionizing laser beams in a discharge. The preferable experimental setup is an atomic beam apparatus coupled with a time-of-flight mass spectrometer, as the interaction area is well defined at the intersection of the atomic beam and the exciting and ionizing laser beams. The TOF mass spectrometer has the advantage over the thermionic diode and the atomic beam apparatus, as it enables simultaneous measurement of different spectroscopic properties of the selected isotopic masses. It also ensures that the emergence of the photoion signal is purely due to the isotope of interest. Further, it is possible to study the singly, doubly, triply, etc., photoionization states of atoms. The measurements of the photoionization cross-sections for different excited states of the two isotopes of lithium and three isotopes of magnesium have demonstrated the advantage of this experimental setup. There have been extensive calculations on the highly ionized atomic systems, whereas experimental data are available on the photoionization from the excited states of atoms and molecules. With the availability of intense VUV/soft X-ray radiation sources such as storage rings along with insertion devices and free-electron lasers, the field of measuring the photoionization cross-section from the excited states of atoms, molecules, and ions using improved experimental techniques and supplemented by the more advanced theoretical calculations will remain an active area of research in the coming years.

Funding: The financial assistance to acquire the Lasers and other relevant equipment for these studies was provided by the Higher Education Commission (HEC) of Pakistan, the Pakistan Science Foundation (PSF), the Pakistan Academy of Sciences (PAS) and the Quaid-i-Azam University (QAU), Islamabad.

Institutional Review Board Statement: Not applicable.

Informed Consent Statement: Not applicable.

Data Availability Statement: Not applicable.

Acknowledgments: I am greatly indebted to Sultana Nahar for providing me with the opportunity for this contribution. My thanks are to the Higher Education Commission of Pakistan and the Pakistan Science Foundation for the financial assistance to develop the facility for the measurement of photoionization cross-section for the excited states of atoms at the Quaid-i-Azam University, Islamabad, Pakistan. Much of the work presented here was performed by my students. I am also grateful to JP Connerade at Imperial College, London, TF Gallagher at Virginia University, J. Hormes at Bonn University, W. Demtoeder, and Hotop at the Kaiserslautern University Germany, K. Ueda, Japan and Ulf Griesmann at NIST for their continued support and guidance to remain active in this field of research. The financial and technical support from the National Center for Physics is highly appreciated. I am particularly thankful to my students who did the experimental work, M. Aslam, M. Hanif, Aslam Zia, Naveed Piracha, Yasir Jamil, Nasir Amin, Yasir Jameel, Shaukat Mahmood, Raheel Ali, M. Yasin, M. Sultan, M. Saleem, Shahid Hussain, Sami-ul-Haq, M. Rafiq, Ahmad Yar, Rizwan Ahmed, Zeshan Adeel Umar, Haroon Asghar, Amir Fayyaz and many more colleagues.

Conflicts of Interest: The author declares no conflict of interest.

References

1. Cowan, R.D. *The Theory of Atomic Structure and Spectra*; University of California Press: Berkeley, CA, USA, 1981; Volume 3.
2. Axner, O.; Gustafsson, J.; Omenetto, O.; Winefordner, J.D. Line strengths, A-factors and absorption cross-sections for fine-structure lines in multiplets and hyperfine structure components in lines in atomic spectroscopy-a user's guide. *Spectrochim. Acta Part B* **2004**, *59*, 1–39. [CrossRef]
3. Demtroder, W. *Laser Spectroscopy*; Springer: Berlin/Heidelberg, Germany, 2006.
4. Paradhan, A.K.; Nahar, S.N. *Atomic Astrophysics and Spectroscopy*; Cambridge University Press: Cambridge, UK, 2011.
5. Samson, J.A.R. *Techniques of Vacuum Ultraviolet Spectroscopy*; Wiley: New York, NY, USA, 1967.
6. Berkowitz, J. *Photoabsorption, Photoionization and Photoelectron Spectroscopy*; Academic Press: New York, NY, USA, 1979.
7. Berkowitz, J. *Atomic and Molecular Photoabsorption, Absolute Total Cross Sections*; Academic Press: New York, NY, USA, 2002.
8. Schmidt, V. *Electron Spectrometry of Atoms Using Synchrotron Radiation*; Cambridge University Press: Cambridge, UK, 1997.
9. Rothe, D.E. Radiative ion-electron recombination in a sodium-seeded plasma. *J. Quant. Spectrosc. Radiant Transf.* **1969**, *9*, 49–62. [CrossRef]
10. Rothe, D.E. Radiative electron-ion recombination into the first excited state of Lithium. *J. Quant. Spectrosc. Radiant Transf.* **1971**, *11*, 355–365. [CrossRef]
11. Marr, G.V.; Creek, D.M. The photoionization absorption continua for alkali metal vapours. *Proc. Roy Soc.* **1968**, *A304*, 233–244.
12. Weisheit, J.C. Photoabsorption by Ground-State Alkali-Metal Atoms. *Phys. Rev. A* **1972**, *5*, 1621–1630. [CrossRef]
13. Ambartzumian, R.V.; Furzikov, N.P.; Letokhov, V.S.; Puretsky, A.A. Measuring photoionization cross-sections of excited atomic states. *Appl. Phys.* **1976**, *9*, 335–337. [CrossRef]
14. Duong, H.T.; Pinard, J.; Vialle, J.L. Experimental separation and study of the two partial photoionisation cross sections $\sigma_{3p,s}$ and $\sigma_{3p,d}$ from the 3p state of sodium. *J. Phy. B At. Mol. Phys.* **1978**, *11*, 797–803. [CrossRef]
15. Ditchburn, R.W.; Hudson, R.D. The absorption of light by calcium vapor (2100 to 1080 Å). *Proc. R. Soc. Lond.* **1960**, *256*, 53–61.
16. Garton, W.R.S.; Codling, K. Ultra-violet Extensions of the Arc Spectra of the Alkaline Earths: The Absorption Spectrum of Barium Vapour. *Proc. Phys. Soc.* **1960**, *75*, 87–94. [CrossRef]
17. Garton, W.R.S.; Codling, K. Ultra-violet extensions of the arc spectra of the alkaline earths: The absorption spectrum of calcium vapour. *Proc. Phys. Soc.* **1965**, *86*, 1067–1075. [CrossRef]

18. Garton, W.R.S.; Codling, K. Ultra-violet extensions of the arc spectra of the alkaline earths: The absorption spectrum of strontium vapour. *J. Phys. B* **1966**, *1*, 106–113. [CrossRef]
19. Hudson, R.D.; Carter, V.L.; Young, P.A. Absorption spectrum of Ba I in the region of autoionization from 238.2 to 170 nm. *Phys. Rev. A* **1970**, *2*, 643–648. [CrossRef]
20. Carter, V.L.; Hudson, R.D.; Brieg, E.L. Autoionization in the uv photoabsorption of Atomic Calcium. *Phys. Rev. A* **1971**, *4*, 821–825. [CrossRef]
21. Brown, C.M.; Tilford, S.G.; Ginter, M.L. Absorption spectrum of Ca I in the 1580–2090 A region. *J. Opt. Soc. Am.* **1973**, *63*, 1454–1462. [CrossRef]
22. Brown, C.M.; Ginter, M.L. Absorption spectrum of Ba I between 1770 and 1560 A. *J. Opt. Soc. Am.* **1978**, *68*, 817–825. [CrossRef]
23. Brown, C.M.; Ginter, M.L. Absorption spectrum of Ca I between 143 and 134 nm. *J. Opt. Soc. Am.* **1980**, *70*, 87–93. [CrossRef]
24. Wynne, J.J.; Hermann, J.P. Spectroscopy of even-parity autoionizing levels in Ba. *Opt. Lett.* **1979**, *4*, 106–108. [CrossRef]
25. Ueda, K.; Ashizawa, Y.; Fukuda, K. Measurements of Oscillator Strengths for the Transitions from the Metastable 3P Levels of Alkaline-Earth Atoms. I. Strontium. *J. Phys. Soc. Jpn.* **1982**, *51*, 1936–1940. [CrossRef]
26. Ueda, K.; Karasawa, M.; Fukuda, K. Measurements of Oscillator Strengths for the Transitions from the Metastable 3P Levels of Alkaline-Earth Atoms. II. Magnesium. *J. Phys. Soc. Jpn.* **1982**, *51*, 2267–2270. [CrossRef]
27. Ueda, K.; Hamaguchi, Y.; Fukuda, K. Measurements of Oscillator Strengths for the Transitions from the Metastable 3P Levels of Alkaline-Earth Atoms. III. Calcium—Low-n Members. *J. Phys. Soc. Jpn.* **1982**, *51*, 2973–2976. [CrossRef]
28. Ueda, K.; Hamaguchi, Y.; Fujimoto, T.; Fukuda, K. Oscillator Strengths and Rare-Gas-Induced Broadening of the Principal Series Lines of Ba. *J. Phys. Soc. Jpn.* **1984**, *53*, 2501–2505. [CrossRef]
29. Baig, M.A.; Connerade, J.P. Extensions to the spectrum of doubly excited Mg I in the vacuum ultraviolet. *Proc. Roy. Soc. Lond.* **1978**, *364*, 353–366.
30. Griesmann, U.; Shen, N.; Connerade, J.P.; Sommer, K.; Hormes, J. New measurements of the cross sections of doubly excited resonances in calcium. *J. Phys. B At. Mol. Phys.* **1988**, *21*, L83–L88. [CrossRef]
31. Griesmann, U.; Esser, B.; Hormes, J. The total photoabsorption cross section of strontium atoms in the range of the 5pns doubly excited states. *J. Phys. B At. Mol. Opt. Phys.* **1994**, *27*, 3939–3944. [CrossRef]
32. Griesmann, U.; Esser, B.; Baig, M.A. The total photoionization cross section of barium including 7snp, 6dnp and 4fnd double excitations. *J. Phys. B At. Mol. Phys.* **1992**, *25*, 3475–3487. [CrossRef]
33. Griesmann, U.; Baig, M.A.; Ahmad, S.; Kaenders, W.G.; Esser, B.; Hormes, J. Photoionization cross sections of doubly excited resonances in ytterbium. *J. Phys. B At. Mol. Phys.* **1992**, *25*, 1393–1404. [CrossRef]
34. Yih, T.S.; Wu, H.H.; Chu, C.C.; Fung, H.S.; Lin, Y.P.; Hsu, S.J. Measurement of the photoabsorption cross section of divalent atoms. *J. Korean Phys. Soc.* **1998**, *32*, 405–412.
35. Chu, C.C.; Fung, H.S.; Wu, H.H.; Yih, T.S. Absolute photoabsorption cross section of Sr from the 5s ionization threshold to the 5p threshold. *J. Phys. B At. Mol. Opt. Phys.* **1998**, *31*, 3843–3859. [CrossRef]
36. Maeda, K.; Aymar, M.; Ueda, K.; Chiba, H.; Ohmori, K.; Sato, Y.; West, J.B.; Ross, K.J.; Ito, K. Absolute photoionization cross sections of the Ba ground state in the autoionization region: I. 238–218 nm. *J. Phys. B* **1997**, *30*, 3159–3171. [CrossRef]
37. Maeda, K.; Ueda, K.; Aymar, M.; Matsui, T.; Chiba, H.; Ito, K. Absolute photoionization cross sections of the Ba ground state in the autoionization region: II. 221–209 nm. *J. Phys. B* **2000**, *33*, 1943–1966. [CrossRef]
38. Huffman, R.E.; Tanaka, Y.; Larrabee, J.C. Absorption coefficient of Xe and Ar in the 600-1025 A wavelength region. *J. Chem. Phys.* **1963**, *39*, 902–909. [CrossRef]
39. Rundel, R.D.; Dunning, F.B.; Goldwire, H.C.; Stebbings, R.F. Near-threshold photoionization of xenon metastable atoms. *J. Opt. Soc. Am.* **1976**, *65*, 628–630. [CrossRef]
40. Ito, K.; Yoshino, Y.; Morioka, Y.; Namioka, T. The $1s^2\,{}^1S_0$-1 $snp\,{}^1P_1$ series of the helium spectrum. *Phys. Scr.* **1987**, *36*, 88–92. [CrossRef]
41. Yashino, K. Absorption Spectrum of the Argon Atom in the Vacuum-Ultraviolet Region. *J. Opt. Soc. Am.* **1970**, *60*, 1220–1229. [CrossRef]
42. Yoshino, K.; Tanaka, Y. Absorption spectrum of krypton in the vacuum UV region. *J. Opt. Soc. Am.* **1979**, *69*, 159–165. [CrossRef]
43. Yoshino, K. Absorption spectrum of xenon in the vacuum-ultraviolet region. *J. Opt. Soc. Am. B* **1985**, *2*, 1268–1274. [CrossRef]
44. Bonin, K.D.; McIlrath, T.J.; Yoshino, K. High-resolution laser and classical spectroscopy of xenon autoionization. *J. Opt. Soc. Am. B* **1985**, *2*, 1275–1283. [CrossRef]
45. Baig, M.A.; Connerade, J.P. Centrifugal barrier effects in the high Rydberg states and autoionizing resonances in neon. *J. Phys. B At. Mol. Phys.* **1984**, *17*, 1785–1796. [CrossRef]
46. Ito, K.; Ueda, K.; Namioka, T.; Yoshino, K.; Marioka, Y. High-resolution absorption spectrum of Ne I in the region of 565–595. *J. Opt. Soc. Am.* **1988**, *5*, 2006–2014. [CrossRef]
47. Marr, G.V. (Ed.) *Handbook on Synchrotron Radiation: Vacuumultraviolet and Soft X-ray*; Elsevier Sci. Publishers. B.V.: Amsterdam, The Netherlands, 1987.
48. Wuilleumier, F.J.; Ederer, D.L.; Picque, J.L. Photoionization and collisional ionization of excited atoms using synchrotron and laser radiations. *Adv. At. Mol. Phys.* **1988**, *23*, 197–286.
49. Sontag, B.; Zimmermann, P. XUV spectroscopy of metal atoms. *Rep. Prog. Phys.* **1992**, *55*, 911–987. [CrossRef]

50. Wuilleumier, F.J.; Mayer, M. Pump-probe experiments in atoms involving laser and synchrotron radiation: An overview. *J. Phys. B At. Mol. Opt. Phys.* **2006**, *39*, R425–R477. [CrossRef]
51. Mayer, M. Combination of lasers and synchrotron radiation in studies of atomic photoionization. *Nucl. Instrum. Methods Phys. Research. A* **2009**, *601*, 88–97. [CrossRef]
52. Kjeldsen, H. Photoionization cross sections of atomic ions from merged–beam experiments. *Phys. Rev. A* **2006**, *39*, R325. [CrossRef]
53. Sukhorukov, V.L.; Petrov, I.D.; Schafer, M.; Merkt, F.; Ruf, M.-W.; Hotop, H. Photoionization dynamics of excited Ne, Ar, Kr and Xe atoms near threshold. *J. Phys. B At. Mol. Opt. Phys.* **2012**, *45*, 092001. [CrossRef]
54. Bostedt, C.; Chapman, H.N.; Costello, J.T.; López-Urrutia, J.R.C.; Düsterer, S.; Epp, S.W.; Feldhaus, J.; Föhlisch, A.; Meyer, M.; Möller, T.; et al. Experiments at FLASH. *Nucl. Instrum. Methods Phys. Res. Sect. A Accel. Spectrometers Detect. Assoc. Equip.* **2009**, *601*, 108–122. [CrossRef]
55. Schippers, S.; Kilcoyne, A.D.; Phaneuf, R.A.; Müller, A. Photoionisation of ions with synchrotron radiation: From ions in space to atoms in cages. *Contemp. Phys.* **2016**, *57*, 215–229. [CrossRef]
56. Young, L.; Ueda, K.; Gühr, M.; Bucksbaum, P.H.; Simon, M.; Mukamel, S.; Rohringer, N.; Prince, K.C.; Masciovecchio, C.; Meyer, M.; et al. Roadmap of ultrafast x-ray atomic and molecular physics. *J. Phys. B At. Mol. Opt. Phys.* **2018**, *51*, 032003. [CrossRef]
57. Ueda, K.; Sokell, E.; Schippers, S.; Aumayr, F.; Sadeghpour, H.; Burgdörfer, J.; Lemell, C.; Tong, X.M.; Pfeifer, T.; Calegari, F.; et al. Roadmap on photonic, electronic and atomic collision physics: I. Light–matter interaction. *J. Phys. B At. Mol. Opt. Phys.* **2019**, *52*, 171001. [CrossRef]
58. Schippers, S.; Muller, A. Photoionization of Astrophysically relevant atomic ions at PIPE. *Atoms* **2020**, *8*, 45. [CrossRef]
59. Cooper, J.W. Photoionization from outer Atoms subshells—A Model Study. *Phys. Rev.* **1962**, *128*, 681–694. [CrossRef]
60. Manson, S.T.; Cooper, J.W. Photo-Ionization in the Soft I-Ray Range: Z Dependence in a Central-Potential Model. *Phys. Rev.* **1968**, *165*, 126–138. [CrossRef]
61. Kennedy, D.J.; Manson, S.T. Photoionization of the noble Gases: Cross Sections and Angular Distributions. *Phys. Rev. A* **1972**, *5*, 227–247. [CrossRef]
62. Burke, P.G.; Taylor, K.T. R-matrix theory of photoionization. Application to neon and argon. *J. Phys. B At. Mol. Phys.* **1975**, *8*, 2620–2639. [CrossRef]
63. Aymar, M.; Luc-Koenig, E.; Farnoux, F.C. Theoretical investigation on photoionization from Rydberg states of lithium, sodium and potassium. *J. Phys. B At. Mol. Phys.* **1976**, *9*, 1279–1292. [CrossRef]
64. Johnson, W.R.; Lin, C.D.; Chung, K.T.; Lee, C.M. Relativistic Random-Phase Approximation. *Phys. Scr.* **1980**, *21*, 409–422. [CrossRef]
65. Savukov, I.M. Quasicontinuum relativistic many-body perturbation theory photoionization cross sections of Na, K, Rb and Cs. *Phys. Rev. A* **2007**, *76*, 032710. [CrossRef]
66. Zatsarinny, O.; Tayal, S.S. Photoionization of potassium atoms from the ground and excited states. *Phys. Rev. A* **2010**, *81*, 043423-1. [CrossRef]
67. Kim, D.S.; Tayal, S.S. Autoionizing resonances in the photoionization of ground state atomic magnesium. *J. Phys. B At. Mol. Opt. Phys.* **2000**, *33*, 3235–3247. [CrossRef]
68. Johnson, W.R.; Cheng, K.T.; Huang, K.N.; Le Dourneuf, M. Analysis of Beutler-Fano autoionizing resonances in the rare-gas atoms using the relativistic multichannel quantum defect theory. *Phys. Rev. A* **1980**, *22*, 989–997. [CrossRef]
69. Petrov, I.D.; Sukhorukov, V.I.; Hotop, H. Photoionization of excited Ne ($2p^5$ 3p J=3) atoms near threshold. *J. Phys. B At. Mol. Opt. Phys.* **2008**, *41*, 065205. [CrossRef]
70. Petrov, I.D.; Sukhorukov, V.L.; Hollenstein, U.; Kaufmann, L.J.; Merkt, F.; Hotop, H. Autoionization dynamics of even ($3p^5_{1/2}$ np, nf) resonances: Comparison of experiment with theory. *J. Phys. B At. Mol. Opt. Phys.* **2011**, *44*, 025004. [CrossRef]
71. Nahar, S.N.; Manson, S.T. Photoionization of the 7d excited state of cesium. *Phys. Rev. A* **1989**, *41*, 6300. [CrossRef] [PubMed]
72. Nahar, S.N. Photoionization and electron recombination of P II. *MNRAS* **2017**, *469*, 3225–3231. [CrossRef]
73. Nahar, S.N. Photoionization features of the ground and excited levels of Cl II and benchmarking with experiment. *New Astron.* **2021**, *82*, 101447. [CrossRef]
74. Nahar, S.N. Photoionization and electron-ion recombination of n = 1 to very high n-vales of hydrogenic ions. *Atom* **2021**, *9*, 73. [CrossRef]
75. Letokhov, V.S. *Laser Photoionization Spectroscopy*; Academic: New York, NY, USA, 1987.
76. Burkhardt, C.E.; Libbert, J.L.; Xu, J.; Leventhal, J.J.; Kelley, J.D. Absolute measurement of photoionization cross sections of excited atoms: Application to determination of atomic beam densities. *Phys. Rev. A.* **1988**, *38*, 5949–5952. [CrossRef]
77. He, L.-W.; Burkhardt, C.E.; Ciocca, M.; Leventhal, J.J.; Manson, S.T. Absolute cross sections for the photoionization of the 6s6p 1P excited state of barium. *Phys. Rev. Lett.* **1991**, *67*, 2131–2134. [CrossRef]
78. Saleem, M.; Hussain, S.; Rafiq, M.; Baig, M.A. Simultaneous measurements of photoionization cross sections of lithium isotopes from the 3p $^2P_{1/2,3/2}$. *J. Phys. B At. Mol. Opt. Phys.* **2006**, *39*, 5025–5033. [CrossRef]
79. Niemax, K. Spectroscopy using thermionic diode detectors. *Appl. Phys. B* **1985**, *38*, 147–157.
80. Baig, M.A.; Akram, M.; Piracha, N.K.; Mahmood, M.S.; Bhatti, S.A.; Ahmad, N. Resonantly enhanced ns and nd Rydberg spectroscopy of sodium. *J. Phys. B* **1995**, *28*, 1421–1432. [CrossRef]
81. Baig, M.A.; Yaseen, M.; Ali, R.; Nadeem, A.; Bhatti, S.A. Near-threshold photoionization spectra of strontium. *Chem. Phys. Lett.* **1998**, *296*, 403–407. [CrossRef]

82. Yaseen, M.; Ali, R.; Nadeem, A.; Bhatti, S.A.; Baig, M.A. Two-color three-photon resonant excitation spectrum of strontium in the autoionization region. *Eur. Phys. J. D* **2002**, *20*, 177–189. [CrossRef]
83. Barbieri, B.; Beverini, N.; Sasso, A. Optogalvanic spectroscopy. *Rev. Mod. Phys.* **1990**, *62*, 603–644. [CrossRef]
84. Babin, F.; Gagné, J.-M. Hollow cathode discharge (HCD) dark space diagnostics with laser photoionization and galvanic detection. *J. Phys. B* **1992**, *54*, 35–45. [CrossRef]
85. Piracha, N.K.; Suleman, B.; Khan, S.H.; Baig, M.A. Two-photon optogalvanic Rydberg spectra of neon. *J. Phys. B At. Mol. Phys.* **1995**, *28*, 2525–2538. [CrossRef]
86. Stockhausen, G.; Mende, W.; Kock, M. Laser-induced photoionization in the dark space of a hollow cathode. *J. Phys. B* **1996**, *29*, 665–676. [CrossRef]
87. Hanif, M.; Aslam, M.; Ali, R.; Nadeem, A.; Riaz, M.; Bhatti, S.A.; Baig, M.A. Laser optogalvanic spectroscopy of 5p5nf $J = 1–5$ even-parity Rydberg levels of xenon. *J. Phys. B At. Mol. Phys.* **2000**, *33*, 4647–4655. [CrossRef]
88. Zia, M.A.; Suleman, B.; Baig, M.A. Two-photon laser optogalvanic spectroscopy of the Rydberg states of mercury by RF discharge. *J. Phys. B At. Mol. Opt. Phys.* **2003**, *36*, 4631–4639. [CrossRef]
89. Zia, M.A.; Baig, M.A. Two-photon laser optogalvanic studies of the 6snf $^3F_4$ Rydberg states of mercury by RF discharge. *Spectrochim. Acta Part B* **2005**, *60*, 1545–1551. [CrossRef]
90. Wiley, W.C.; McLaren, I.H. Time-of-Flight Mass Spectrometer with Improved Resolution. *Rev. Sci. Instrum.* **1955**, *26*, 1150–1157. [CrossRef]
91. Saleem, M.; Hussain, S.; Zia, M.A.; Baig, M.A. An efficient pathway for Li^6 isotope enrichment. *Appl. Phys. B* **2007**, *87*, 723–726. [CrossRef]
92. Gilbert, S.L.; Noecker, M.C.; Wieman, C.E. Absolute measurement of the photoionization cross section of the excited 7s state of cesium. *Phys. Rev. A* **1984**, *29*, 3150–3153. [CrossRef]
93. Wippel, V.; Binder, C.; Huber, W.; Windholz, L.; Allegrini, M.; Fuso, F.; Arimondo, E. Photoionization cross-sections of the first excited states of sodium and lithium in a magneto-optical trap. *Eur. Phys. J. D* **2001**, *17*, 285–292. [CrossRef]
94. Madsen, D.N.; Thomsen, J.W. Measurement of absolute photo-ionization cross sections using magnesium magneto-optical traps. *J. Phys. B At. Mol. Opt. Phys.* **2002**, *35*, 2173–2181. [CrossRef]
95. Dinneen, T.P.; Wallace, C.D.; Tan, K.-Y.N.; Gould, P.L. Use of trapped atoms to measure absolute photoionization cross sections. *Opt. Lett.* **1992**, *17*, 1706–1708. [CrossRef]
96. Lucas, D.M.; Ramos, A.; Home, J.P.; McDonnel, M.J.; Nakayama, S.; Stacey, J.-P.; Webster, S.C.; Stacey, D.N.; Steane, A.M. Isotope-selective photoionization for calcium ion trapping. *Phys. Rev. A* **2004**, *69*, 012711. [CrossRef]
97. Marago, O.; Ciampini, D.; Fuso, F.; Arimondo, E.; Gabbanini, C.; Manson, S.T. Photoionzation cross sections for the excited laser-cooled cesium atoms. *Phys. Rev. A* **1998**, *57*, R4110–R4113. [CrossRef]
98. Yang, J.-J.; Hu, X.; Wu, H.; Fan, J.; Cong, R.; Cheng, Y.; Ji, X.; Yao, G.; Zheng, X.; Cui, Z. Measuremnt of photoionization cross sections of the excited states of titanium. *Chin. J. Chem. Phys.* **2009**, *22*, 615–620. [CrossRef]
99. Cong, R.; Cheng, Y.; Yang, J.; Fan, J.; Yao, G.; Ji, X.; Zheng, X.; Cui, Z. Measurement of photoionization cross sections of the excited states of titanium, cobalt and nickel. *J. Appl. Phys.* **2009**, *106*, 013103. [CrossRef]
100. Zheng, X.; Zhou, X.; Cheng, Z.; Jia, D.; Qu, Z.; Yao, G.; Zhang, X.; Cui, Z. Photoionization cross section measurements of the excited states of cobalt in the near-threshold region. *AIP Adv.* **2014**, *4*, 107120. [CrossRef]
101. Saleem, M.; Hussain, S.; Baig, M.A. Saturation technique for the measurement of photoionization cross-section of atomic excited states—A review. *Optik* **2018**, *158*, 664–674. [CrossRef]
102. Heinzmann, U.; Schinkowski, D.; Zeman, H.D. Comment on measuring photoionization cross sections of excited atomic states. *Appl. Phys.* **1977**, *12*, 113. [CrossRef]
103. Nygaard, K.J.; Hebner, R.E.; Jones, J.D.; Crbin, R.J. Photoionization of the $6^2P_{3/2,1/2}$ fine-structure levels in cesium. *Phys. Rev. A* **1975**, *12*, 1440–1447. [CrossRef]
104. Smith, A.V.; Goldsmith, J.E.M.; Nitz, D.E.; Smith, S.J. Absolute photoionization cross-section measurements of the excited 4D and 5S states of sodium. *Phys. Rev. A* **1980**, *22*, 577–581. [CrossRef]
105. Gerwert, K.; Kollath, K.J. Measurement of photoionisation cross sections for the $7\,^2P_{3/2}$ and $6\,^2D_{3/2}$ excited states of caesium. *J. Phys. B* **1983**, *16*, L217. [CrossRef]
106. Bonin, K.D.; Gatzke, M.; Collins, C.L.; Kadar-Kallen, M.A. Absolute photoionization cross sections of the Cs $7D_{3/2}$ level measured by use of fluorescence reduction. *Phys. Rev. A* **1989**, *39*, 5624–5632. [CrossRef]
107. Baohua, F.; Zuren, Z. Absolute measurement of the photoionization cross section of the excited 7p state of potassium. *Chin. Phys. Lett.* **1993**, *10*, 217–221.
108. Maeda, H.; Ambe, F. Measurements of the absolute photoionization cross sections of the excited 7d and 8d states of cesium. *Phys. Rev. A* **1997**, *56*, 3719–3725. [CrossRef]
109. Pattersen, B.M.; Takekoshi, T.; Knize, R.J. Measurement of the photoionization cross section of the $6p_{3/2}$ state of cesium. *Phys. Rev. A* **1999**, *59*, 2508–2510. [CrossRef]
110. Petrov, I.D.; Sukhoroukov, V.L.; Leber, E.; Hotop, H. Near threshold phtoionization of excited alkali atoms Ak(np) (Ak= Na, K, Rb, Cs; n=3 – 6). *Eur. Phys. J. D* **2000**, *10*, 53–65. [CrossRef]
111. Duncan, B.C.; Sanchev-Villicana, V.; Gould, P.L.; Sadeghpour, H.R. Measurement of the Rb ($^5D_{5/2}$) photoionization cross section using trapped atoms. *Phys. Rev. A* **2001**, *63*, 043411. [CrossRef]

112. Gabbanini, C. Assessments of lifetimes and photoionization cross sections at 10.6 m of nd Rydberg states of Rb measured in a magneto-optical trap. *Spectrochim. Acta B* **2006**, *61*, 196–199. [CrossRef]
113. Amin, N.; Mahmood, S.; Saleem, M.; Kalyar, M.A.; Baig, M.A. Photoionization cross section measurements from the 2p, 3d and 3s excited states of lithium. *Eur. Phys. J. D* **2006**, *40*, 331–338. [CrossRef]
114. Amin, N.; Mahmood, S.; Anwar-ul-Haq, M.; Riaz, M.; Baig, M.A. Measurement of the photoionization cross-section of the $3P\ ^2P_{1/2,\,3/2}$ excited levels of sodium. *Eur. Phys. J. D* **2006**, *37*, 23–28. [CrossRef]
115. Amin, N.; Mahmood, S.; Anwar-ul-Haq, M.; Baig, M.A. Measurement of the 4d photoionization cross section via two-photon and two-step excitation in sodium. *J. Quant. Spectro. Radiat. Transf.* **2006**, *102*, 269–276. [CrossRef]
116. Amin, N.; Mahmood, S.; Haq, S.U.; Kalyar, M.A.; Rafiq, M.; Baig, M.A. Measurements of photoionization cross sections from the 4p, 5d and 7s excited states of potassium. *J. Quant. Spectro. Radiat. Transf.* **2008**, *109*, 863–872. [CrossRef]
117. Preses, J.M.; Burkhart, C.E.; Corey, R.L.; Earsom, D.L.; Daulton, T.L.; Garver, W.P.; Leventhal, J.J.; Msezane, A.Z.; Manson, S.T. Photoionization of the excited 3p state of sodium: Experiment and theory. *Phys. Rev. A* **1985**, *32*, 1264–1266. [CrossRef]
118. Miculis, K.; Meyer, W. Phototransition of Na ($3p_{3/2}$) into high Rydberg states and ionization continuum. *J. Phys. B At. Mol. Opt. Phys.* **2005**, *38*, 2097–2108. [CrossRef]
119. Baig, M.A.; Mahmood, S.; Kalyar, M.A.; Rafiq, M.; Amin, N.; Haq, S.U. Oscillator strength measurements of the 3p → nd Rydberg transitions of sodium. *Eur. Phys. J. D* **2007**, *44*, 9–16. [CrossRef]
120. Rafiq, M.; Kalyar, M.A.; Baig, M.A. Photoexcitation study of the 4s $^2S_{1/2}$ state of atomic sodium. *J. Phys. B At. Mol. Opt. Phys.* **2008**, *41*, 115701. [CrossRef]
121. Yar, A.; Ali, R.; Baig, M.A. Measurement of the photoionization cross section for the 6p $^2P_{3/2}$ state of potassium using a time of flight mass spectrometer. *Phys. Rev. A* **2013**, *87*, 045401. [CrossRef]
122. Yar, A.; Iqbal, J.; Ali, R.; Baig, M.A. Two-step laser excitation and ionization from the 7p $^2P_{3/2}$ state of potassium. *Laser Phys.* **2015**, *25*, 025702. [CrossRef]
123. Haq, S.U.; Nadeem, A. Photoionization from the 6p $^2P_{3/2}$ state of neutral cesium. *Phys. Rev. A* **2010**, *81*, 063432. [CrossRef]
124. Nadeem, A.; Haq, S.U. Photoionization from the 5p $^2P_{3/2}$ state of rubidium. *Phys. Rev. A* **2011**, *83*, 063404. [CrossRef]
125. Shahzada, S.; Ijaz, P.; Shah, M.; Haq, S.U.; Ahmed, M.; Nadeem, A. Photoionization studies from the 3p ^2P excited state of neutral lithium. *J. Opt. Soc. Am.* **2012**, *29*, 3386–3391. [CrossRef]
126. Saleem, M.; Amin, N.; Hussain, S.; Rafiq, M.; Mahmood, S.; Baig, M.A. Alternate technique for simultaneous measurement of photoionization cross-section of isotopes by TOF mass spectrometer. *Eur. Phys. J. D* **2006**, *38*, 277–284. [CrossRef]
127. Saleem, M.; Hussain, S.; Rafiq, M.; Baig, M.A. Laser isotope separation of lithium by two-step photoionization. *J. Appl. Phys.* **2006**, *100*, 053111. [CrossRef]
128. Hussain, S.; Saleem, M.; Baig, M.A. Angular momentum dependence of photoionization cross sections from the excited states of lithium. *Phys. Rev. A* **2006**, *74*, 052705. [CrossRef]
129. Saleem, M.; Hussain, S.; Baig, M.A. Angular momentum dependence of photoionization cross section from the excited states of lithium isotopes. *Phys. Rev. A* **2008**, *77*, 062506. [CrossRef]
130. Hussain, S.; Saleem, M.; Baig, M.A. Measurement of oscillator strength distribution in the discrete and continuous spectrum of lithium. *Phys. Rev. A* **2007**, *75*, 022710. [CrossRef]
131. Felfli, Z.; Manson, S.T. Influence of shape resonances on minimum in cross sections for photoionization of excited atoms. *Phys. Rev. A* **1990**, *41*, 1709–1710. [CrossRef] [PubMed]
132. Lahiri, J.; Manson, S.T. Oscillator-strength distributions for discrete and continuum transitions of excited states of cesium. *Phys. Rev. A* **1986**, *33*, 3151–3165. [CrossRef] [PubMed]
133. Lahiri, J.; Manson, S.T. Radiative recombination and excited-state photoionization of lithium. *Phys. Rev. A* **1993**, *48*, 3674–3679. [CrossRef]
134. Saha, H.P.; Pindzola, M.S.; Compton, R.N. Multi-configuration Hartree-Fock calculation of the photoionization of the excited Na 4d state. *Phys. Rev. A* **1988**, *38*, 128–134. [CrossRef]
135. Qi, Y.Y.; Wu, Y.; Wang, J.G.; Ding, P.Z. Calculations of Photo-Ionization Cross Sections for Lithium Atoms. *Chin. Phys. Lett.* **2008**, *25*, 3620–3623.
136. Green, J.C.; Decleva, P. Photoionization cross-sections: A guide to electronic structure. *Coordination Chem. Rev.* **2005**, *249*, 209–228. [CrossRef]
137. Sandner, W.; Gallagher, T.F.; Safinya, K.S.; Gounand, F. Photoionization of potassium in the vicinity of the minimum in the cross section. *Phys. Rev. A* **1981**, *23*, 2732–2734. [CrossRef]
138. Yar, A.; Ali, R.; Iqbal, J.; Ali, R.; Baig, M.A. Evidence of a Cooper minimum in the photoionization from the 7s $^2S_{1/2}$ excited state of potassium. *Phys. Rev. A* **2013**, *88*, 033403. [CrossRef]
139. Mende, W.; Bartschat, K.; Kock, M. Near-threshold photoionization from the Sr I (5s5p) $^1P^o_1$ state. *J. Phys. B At. Mol. Opt. Phys.* **1995**, *28*, 2385–2394. [CrossRef]
140. Kalyar, M.A.; Yar, A.; Iqbal, J.; Ali, R.; Baig, M.A. Measurements of photoionization cross section of the 4p levels and oscillator strength of the 4p nd $^2D_{3/2,5/2}$ transitions of potassium. *Opt. Laser Technol.* **2016**, *77*, 72–79. [CrossRef]
141. Collister, R.; Zhang, J.; Tandecki, M.; Aubin, S.; Gomez, E.; Gwinner, G.; Orozco, L.A.; Pearson, M.R.; Behr, J.A. Photoionization of the francium $7p_{3/2}$ state. *Can. J. Phys.* **2017**, *95*, 234–237. [CrossRef]
142. Gallagher, T.F. *Rydberg Atoms*; Cambridge University Press: Cambridge, UK, 1994.

143. Connerade, J.P. *Highly Exited Atoms*; Cambridge University Press: Cambridge, UK, 1998.
144. Bradley, D.J.; Dugan, C.H.; Ewart, P.; Purdie, A.F. Absolute photoionization cross-section measurement of selectively excited magnesium. *Phys. Rev. A* **1976**, *13*, 1416. [CrossRef]
145. Rafiq, M.; Hussain, S.; Saleem, M.; Kalyar, M.A.; Baig, M.A. Measurement of photoionization cross section from the 3s3p $^1P_1$ excited state of magnesium. *J. Phys. B At. Mol. Opt. Phys.* **2007**, *40*, 2291–2306. [CrossRef]
146. Thompson, D.G.; Hibbert, A.; Chandra, N. Photoionization of the (3s3p: ^1P) state of magnesium. *J. Phys. B At. Mol. Phys.* **1974**, *7*, 1298–1305. [CrossRef]
147. Chang, T.N. 3pnp ^1S autoionization states of the magnesium atom. *Phys. Rev. A* **1987**, *36*, 5468–5470. [CrossRef]
148. Moccia, R.; Spizzo, P. Atomic magnesium: I. Discrete and continuous energy spectrum below the 3p ionisation threshold. A valence-shell L2 CI calculation. *J. Phys. B At. Mol. Opt. Phys.* **1987**, *21*, 1121–1132. [CrossRef]
149. Fang, T.K.; Chang, T.N. Photoionization from excited Mg atoms. *Phys. Rev. A* **2000**, *61*, 052716. [CrossRef]
150. Kim, D.S. Photoionization of the excited 3s3p $^{1,3}P_1$ states of atomic magnesium. *J. Phys. B At. Mol. Opt. Phys.* **2001**, *34*, 2615–2629. [CrossRef]
151. Bonanno, R.E.; Clark, C.W.; Lucatorto, T.B. Multiphoton excitation of autoionizing states of Mg: Line-shape studies of the $3p^2\,^1S$ state. *Phys. Rev. A* **1986**, *34*, 2082–2085. [CrossRef]
152. Okasaka, R.; Fukuda, K. Doubly excited levels of alkaline-earth elements near the first ionisation limit. I. Magnesium. *J. Phys. B At. Mol. Phys.* **1982**, *15*, 347–355. [CrossRef]
153. Shao, Y.L.; Fotakis, C.; Charalambidis, D. Multiphoton ionization of Mg in the wavelength region of 300–214 nm. *Phys. Rev. A* **1993**, *48*, 3636–3643. [CrossRef] [PubMed]
154. Wang, G.; Wan, J.; Zhou, X. Photoionization cross sections of the excited 3s3p ^3P state for atomic Mg. *Radiat. Phys. Chem.* **2017**, *130*, 406–410. [CrossRef]
155. Parkinson, W.; Reeves, E.M.; Tomkins, F.S. Neutral calcium, strontium and barium: Determination of f-values of the principal series by hook method. *J. Phys. B At. Mol. Opt. Phys.* **1976**, *9*, 157–165. [CrossRef]
156. Geiger, I. The oscillator strength density of the $4s^2 \to$ 4snp, 3dnp, $^1P_1$, $^3P_1$ series of calcium as a two-channel case in quantum defect theory. *J. Phys. B At. Mol. Opt. Phys.* **1979**, *12*, 2277–2290. [CrossRef]
157. Barrientos, C.; Martin, I. Quantum defect orbital calculations on the alkaline earth elements: Oscillator strengths and photoionization cross sections. *Can. J. Phys.* **1985**, *63*, 1441–1445. [CrossRef]
158. Smith, G. Oscillator strengths for the neutral calcium lines of 2.9 eV excitation. *J. Phys. B At. Mol. Opt. Phys.* **1988**, *21*, 2827–2834. [CrossRef]
159. Karamatskos, N.; Mueller, M.; Schmidt, M.; Zimmermann, P. The double-electron resonances 3dnp $^1P_1$ in the photoionization spectrum of Ca I. *J. Phys. B At. Mol. Phys.* **1985**, *18*, L107–L109. [CrossRef]
160. Daily, J.E.; Gommers, R.; Cummings, E.A.; Durfee, D.S.; Bergeson, S.D. Two-photon photoionization of the Ca 4s3d $^1D_2$ level in an optical dipole trap. *Phys. Rev. A* **2005**, *71*, 043406. [CrossRef]
161. Sato, Y.; Hayaishi, T.; Itikawa, Y.; Itoh, Y.; Murakami, J.; Nagata, T.; Sasaki, T.; Sonntag, B.; Yagishita, A.; Yoshino, M. Single and double photoionisation of Ca atoms between 35 and 42 nm. *J. Phys. B At. Mol. Phys.* **1985**, *18*, 225–232. [CrossRef]
162. Haq, S.U.; Ali, R.; Kalyar, M.A.; Rafiq, M.; Nadeem, A.; Baig, M.A. Oscillator strengths of the 4s4p $^{1,3}P_1 \to$ 4snd $^{1,3}D_2$ transitions of neutral calcium. *Eur. Phys. J. D* **2008**, *50*, 1–8. [CrossRef]
163. Ewart, P.; Purdie, A.F. Laser ionization spectroscopy of Rydberg and autoionization levels in Sr I. *J. Phys. B At. Mol. Phys.* **1976**, *9*, L437–L441. [CrossRef]
164. Xu, E.Y.; Zhu, Y.; Mullins, O.C.; Gallagher, T.F. Sr $5p_{1/2}ns_{1/2}$ and $5p_{3/2}ns_{1/2}$ J=1 autoionizing states. *Phys. Rev. A* **1986**, *33*, 2401–2409. [CrossRef]
165. Kompitsas, M.; Goutis, S.; Aymar, M.; Camus, P. The even-parity J=0 autoionizing spectrum of strontium below the $4d_{5/2}$ threshold: Observation and theoretical analysis. *J. Phys. B At. Mol. Opt. Phys.* **1991**, *24*, 1557–1574. [CrossRef]
166. Lu-Koenig, E.; Aymar, M.; Lecomte, J.-M.; Lyras, A. Eigenchannel R-matrix study of two photon excitation of low autoionizing states in strontium: Dielectronic core-polarization effects. *J. Phys. B At. Mol. Opt. Phys.* **1998**, *31*, 727–740. [CrossRef]
167. Mende, W.; Kock, M. Oscillator strengths of Ba I and Sr I Rydberg Transitions. *J. Phys. B At. Mol. Opt. Phys.* **1996**, *29*, 655–664. [CrossRef]
168. Haq, S.U.; Mahmood, S.; Kalyar, M.A.; Rafiq, M.; Ali, R.; Baig, M.A. Photoionization cross section and oscillator strength distribution in the near threshold region of strontium. *Eur. Phys. J. D* **2007**, *44*, 439–448. [CrossRef]
169. Esherick, E. Bound, even-parity J = 0 and J = 2 Spectra of Sr. *Phys. Rev. A* **1977**, *15*, 1920–1936. [CrossRef]
170. Dai, C.J. Perturbed 5snd $^{1,3}D_2$ Rydberg series of Sr. *Phys. Rev. A* **1995**, *52*, 4416–4424. [CrossRef]
171. Haq, S.U.; Mahmood, S.; Amin, N.; Jamil, Y.; Ali, R.; Baig, M.A. Measurements of photoionization cross sections from the 5s5p 1P1 and 5s6s 1S0 excited states of strontium. *J. Phys. B At. Mol. Opt. Phys.* **2006**, *39*, 1587–1596.
172. Haq, S.U.; Kalyar, M.A.; Rafiq, M.; Ali, R.; Piracha, N.K.; Baig, M.A. Oscillator strength measurements of the 5s6s $^1S_0 \to$ 5snp $^1P_1$ Rydberg transitions of strontium. *Phys. Rev. A* **2009**, *79*, 042502. [CrossRef]
173. Kallenbach, A.; Kock, M.; Zierer, G. Absolute cross sections for Photoionization of laser excited Ba I states measured on a thermionic diode. *Phys. Rev. A* **1988**, *38*, 2356–2360. [CrossRef] [PubMed]
174. Willke, B.; Kock, M. Measurement of photoionization cross sections from the laser excited Ba I (6s6p) $^1P_1^0$ state. *J. Phys. B At. Mol. Opt. Phys.* **1993**, *26*, 1129–1140. [CrossRef]

175. Bartschat, K.; McLaughlin, M.; Hoversten, A. Photoionization and excitation of atomic barium from the (6s6p) $^1P_1^0$ state. *J. Phys. B At. Mol. Opt. Phys.* **1991**, *24*, 3359–3371. [CrossRef]
176. Greene, C.H.; Theodosiou, C.E. Photoionization of the Ba 6s6p $^1P_1^0$ state. *Phys. Rev. A* **1990**, *42*, 5773–5775. [CrossRef]
177. Keller, J.S.; Hunter, J.E.; Berry, R.S. Path dependence in resonant multiphoton excitation to autoionizing states of barium. *Phys. Rev. A* **1991**, *43*, 2270–2280. [CrossRef]
178. He, L.–W.; Burkhardt, C.E.; Ciocca, M.; Leventhal, J.J.; Zhou, H.–L.; Manson, S.T. Correlation effects in the photoionization of Ba(6s6p $^1P_1$): Determination of cross sections for production of specific final J states. *Phys. Rev. A* **1995**, *51*, 2085–2093.
179. Lagadec, H.; Carre, B.; Porterat, D.; Fournier, P.R.; Aymar, M. Photoionzation of the $6p^2$ states in barium: Measurements and R-matrix calculations. *J. Phys. B At. Mol. Opt. Phys.* **1996**, *29*, 471–486. [CrossRef]
180. Carre, B.; Fournier, P.R.; Porterat, D.; Lagadec, H.; Gounand, F.; Aymar, M. Photoionization of the $6p^2$ (LS)J state in barium: Measurement and calculation of the partial cross sections. *J. Phys. B At. Mol. Opt. Phys.* **1994**, *27*, 1027–1249. [CrossRef]
181. Kalyar, M.A.; Rafiq, M.; Haq, S.U.; Baig, M.A. Absolute Photoionization cross section form the 6s6p $^{1,3}P_1$ excited states of barium. *J. Phys. B At. Mol. Opt. Phys.* **2007**, *40*, 2307–2319. [CrossRef]
182. Li, C.-H.; Budker, D. Polarization-dependent photoionization cross sections and radiative lifetimes of atomic states of Ba I. *Phys. Rev. A* **2006**, *74*, 012512. [CrossRef]
183. Kalyar, M.A.; Rafiq, M.; Ali, R.; Piracha, N.K.; Baig, M.A. Line shape parameters study of the 6p7p ($^1P_1$, $^3D_1$ and $^3P_1$): Autoionizing resonances in barium. *Eur. Phys. J. D* **2007**, *41*, 229–236. [CrossRef]
184. Kalyar, M.A.; Rafiq, M.; Baig, M.A. Interaction of the $6p^2$ $^1S_0$ broad resonances with 5dnd J = 0 autoionizing resonances in barium. *Phys. Rev. A* **2009**, *80*, 052505. [CrossRef]
185. Elizarov, A.Y.; Cherepkov, N.A. Two-photon polarization spectroscopy of autoionizing states. *Sov. Phys. JETP* **1989**, *69*, 695–699.
186. Wood, R.D.; Greene, C.H.; Armstrong, D. Photoionization of the barium 6s 6p $^1P_1$ state: Comparison of theory and experiment including hyperfine-depolarization effects. *Phys. Rev. A* **1993**, *47*, 229–235. [CrossRef]
187. Lange, V.; Eichmann, U.; Sandner, W. Photoionization of excited barium 6s6p $^1P_1$. *Phys. Rev. A* **1991**, *44*, 4737–4739. [CrossRef] [PubMed]
188. Fano, U. Effects of Configuration Interaction on Intensities and Phase Shifts. *Phys. Rev.* **1961**, *124*, 1866–1878. [CrossRef]
189. Tolsma, J.R.; Haxton, D.J.; Greene, C.H.; Yamazaki, R.; Elliott, D.S. One- and two-photon ionization cross sectios of the laser excited 6s6p $^1P_1$ state of barium. *Phys. Rev. A* **2009**, *80*, 033401. [CrossRef]
190. Afrousheh, K.; Marafi, M.; Kojkaj, J.; Makdisi, Y.; Mathew, J. Spectroscopic studies of $5d_{3/2}$nd $^1D_{0,2}$ autoionizing lines of barium under collision with rare gases. *Phys. Rev. A* **2012**, *85*, 052517. [CrossRef]
191. Stebbings, R.F.; Dunning, F.B. Autoionization from High-Lying $3p^5$ ($^2P_{1/2}$)np' Levels in Argon. *Phys. Rev. A* **1973**, *8*, 665–667. [CrossRef]
192. Dunning, F.B.; Stebbings, R.F. Role of autoionization in the near- threshold photoionization of argon and krypton metastable atoms. *Phys. Rev. A* **1974**, *9*, 2378–2382. [CrossRef]
193. Marr, G.V. The absolute photoionisation cross section curve for atomic helium. *J. Phys. B At. Mol. Phys.* **1978**, *11*, L121–L123. [CrossRef]
194. Domke, M.; Schulz, K.; Remmers, G.; Kaindl, G. High-resolution study of $^1P_1$ series in helium. *Phys. Rev.A.* **1996**, *53*, 1424–1438. [CrossRef] [PubMed]
195. Schulz, K.; Kaindl, G.; Domke, M.; Bozek, J.D.; Heimann, P.A.; Schlachter, A.S.; Rost, J.M. Observation of new Rydberg series and resonances in doubly excited helium at ultrahigh resolution. *Phys. Rev. Lett.* **1996**, *77*, 3086–3089. [CrossRef] [PubMed]
196. Rost, J.M.; Schulaz, K.; Kaindl, G. Resonace parameters of photo doubly excited helium. *J. Phys. B At. Mol. Opt. Phys.* **1997**, *30*, 4663–4694. [CrossRef]
197. Chan, W.F.; Cooper, G.; Guo, X.; Brion, C.E. Absolute optical oscillator strengths for the electronic excitation of atoms at high resolution. The absorption spectrum of neon. *Phys. Rev. A.* **1992**, *45*, 1420–1433. [CrossRef]
198. Gisselbrecht, M.; Descamps, D.; Lynga, C.; L'Huillier, A.; Wahlstrom, C.G.; Meyer, M. Absolute Photoionization Cross Sections of Excited He States in the Near-Threshold Region. *Phys. Rev. Lett.* **1999**, *82*, 4607–4610. [CrossRef]
199. Sahoo, S.; Ho, Y.K. Photoionization of the excited He atom in Debye plasmas. *J. Quant. Spectr. Radiat. Transfer.* **2010**, *111*, 52–62. [CrossRef]
200. Hussain, S.; Saleem, M.; Rafiq, M.; Baig, M.A. Photoionization cross section measurements of the 3p $^{1,3}P$ excited states of helium in the near-threshold region. *Phys. Rev. A* **2006**, *74*, 022715. [CrossRef]
201. Jacobs, V.L. Photoionization of He(np $^{1,3}P$) Atoms Excited by Polarized Light. *Phys. Rev. Lett.* **1974**, *32*, 1399–1402. [CrossRef]
202. Madden, R.P.; Codling, K. New autoionizing atomic energy levels in He, Ne, and Ar. *Phys. Rev. Lett.* **1963**, *10*, 516–518. [CrossRef]
203. Ueda, K. Spectral line shapes of autoionizing Rydberg series of xenon. *Phys. Rev. A* **1987**, *35*, 2484–2492. [CrossRef] [PubMed]
204. Ueda, K.; Maeda, K.; Ito, K.; Namioka, T. High-resolution measurement and quantum defect analysis of the 8s' and 6d' autoionisation resonances of krypton. *J. Phys. B At. Mol. Opt. Phys.* **1989**, *22*, L481–L485. [CrossRef]
205. Wu, J.Z.; Whitfield, S.B.; Caldwell, C.D.; Krause, M.O.; van der Meulen, P.; Fahlman, A. High-resolution photoelectron spectrometry of selected ns' and nd' autoionization resonances in Ar, Kr, and Xe. *Phys. Rev. A* **1990**, *42*, 1350–1357. [CrossRef] [PubMed]
206. Maeda, K.; Ueda, K.; Ito, K. High-resolution measurement for photoabsorption cross sections in the autoionization regions of Ar, Kr and Xe. *J. Phys. B At. Mol. Opt. Phys.* **1993**, *26*, 1541–1555. [CrossRef]

207. Pratt, S.T.; Dehmer, P.M.; Dehmer, J.L. Three-photon excitation of autoionizing series of atomic xenon between the $^2P_{3/2}$ and $^2P_{1/2}$ fine-structure thresholds. *Phys. Rev. A* **1987**, *35*, 3793–3798. [CrossRef] [PubMed]
208. Koeckhoven, S.M.; Buma, W.J.; de Lange, C.A. Three-photon excitation of antoionizing states of Ar, Kr, and Xe between the $^2P_{3/2}$ and $^2P_{1/2}$ ionic limits. *Phys. Rev. A* **1994**, *49*, 3322. [CrossRef] [PubMed]
209. Koeckhoven, S.M.; Buma, W.J.; de Lange, C.A. Four-photon excitation of autoionizing states of Ar, Kr, and Xe between the $^2P_{3/2}$ and $^2P_{1/2}$ ionic limits. *Phys. Rev. A* **1995**, *51*, 1097–1109. [CrossRef]
210. Blazewicz, P.R.; Stockdale, J.A.D.; Miller, J.C.; Efthimiopoulos, T.; Fotakis, C. Four photon excitation of even-parity Rydberg states in krypton and xenon. *Phys. Rev. A* **2001**, *35*, 1092–1098. [CrossRef]
211. Grandin, J.P.; Husson, X. Even-parity Rydberg and autoionising states in xenon. *J. Phys. B At. Mol. Phys.* **1981**, *14*, 433–440. [CrossRef]
212. Wada, A.; Adachi, Y.; Hirose, C. Observation of Autoionization Spectrum of Kr 7d′-5p′ and 9s′-5p′ Visible Transitions by Opt-galvanic Spectroscopy. *J. Phys. Chem.* **1986**, *90*, 6645–6648. [CrossRef]
213. Wang, L.; Knight, R.D. Two-photon laser spectroscopy of the ns and nd autoionizing Rydberg series in xenon. *Phys. Rev. A* **1986**, *34*, 3902–3907. [CrossRef] [PubMed]
214. King, R.F.; Latimer, C.J. Autoionizing resonances in the photoionization of long lived excited xenon atoms. *J. Opt. Soc. Am.* **1982**, *72*, 306–308. [CrossRef]
215. Harth, K.; Raab, M.; Hotop, H. Odd Rydberg spectrum of ^{20}NeI: High resolution laser spectroscopy and multichannel quantum defect theory analysis. *Z. Phys. D—At. Mol. Clust.* **1987**, *7*, 213–225. [CrossRef]
216. Klar, D.; Ueda, K.; Ganz, J.; Harth, K.; Bussert, W.; Baier, S.; Weber, J.M.; Ruf, M.W.; Hotop, H. High-resolution measurement and quantum-defect analysis for the Ne nd′ J = 1, 2 and 3 autoionizing resonances. *J. Phys. B At. Mol. Opt. Phys.* **1994**, *27*, 4897–4907. [CrossRef]
217. Knight, R.D.; Wang, L.G. One photon laser spectroscopy of the np and nf Rydberg series in xenon. *J. Opt. Soc. Am. B* **1985**, *2*, 1084–1087. [CrossRef]
218. L'Huillier, A.; Lompre, L.A.; Normand, D.; Morellec, J.; Ferray, M.; Lavancier, J.; Mainfray, G.; Manus, C. Spectroscopy of the np and nf even-parity Rydberg series in xenon by two-photon excitation. *J. Opt. Soc. Am. B* **1989**, *6*, 1644–1647. [CrossRef]
219. Piracha, N.K.; Baig, M.A.; Khan, S.H.; Suleman, B. Two-photon optogalvanic spectra of argon: Odd parity Rydberg states. *J. Phys. B At. Mol. Opt. Phys.* **1997**, *30*, 1151–1162. [CrossRef]
220. Ahmed, M.; Zia, M.A.; Baig, M.A.; Suleman, B. Two-photon laser-optogalvanic spectroscopy of the odd-parity Rydberg series of krypton. *J. Phys. B At. Mol. Opt. Phys.* **1997**, *30*, 2155–2165. [CrossRef]
221. Ahmed, M.; Baig, M.A.; Suleman, B. Laser optogalvanic spectroscopic studies of xenon. *J. Phys. B At. Mol. Opt. Phys.* **1998**, *31*, 4017–4028. [CrossRef]
222. Racah, G. On a New Type of Vector Coupling in Complex Spectra. *Phys. Rev.* **1942**, *61*, 537. [CrossRef]
223. Gornik, W.; Kindt, S.; Matthias, E.; Rinneberg, H.; Schmidt, D. Off-Resonant E2 Transition Observed in Two-Photon Absorption in Xe I. *Phys. Rev. Lett.* **1980**, *45*, 1941–1944. [CrossRef]
224. Samson, J.A.R.; Stolte, W.C. Precision measurements of the total photoionization cross-sections of He, Ne, Ar, Kr, and Xe. *J. Elect. Spectro. Rel. Phenom.* **2002**, *123*, 265–276. [CrossRef]
225. Baig, M.A.; Hanif, M.; Aslam, M.; Bhatti, S.A. Laser optogalvanic observations and MQDT analysis of mp 5 nd J = 3 autoionizing resonances in Ar, Kr and Xe. *J. Phys. B At. Mol. Opt. Phys.* **2006**, *39*, 4221–4229. [CrossRef]
226. Petrov, I.D.; Peters, T.; Halfmann, T.; Alois, S.; O'Keeffe, P.; Meyer, M.; Sukhorukov, V.L.; Hotop, H. Lineshapes of the even $mp^5_{1/2}n(p'/f')$ autoionizing resonances of Ar, Kr and Xe. *Eur. Phys. J. D* **2006**, *40*, 181–193. [CrossRef]
227. Petrov, I.D.; Sukhorukov, V.L.; Peters, T.; Zehnder, O.; Worner, H.J.; Merkt, F.; Hotop, H. Autoionizing even $2p^5_{1/2}$ $nl'[K']_{0,1,2}(l'= 1, 3)$ Rydberg series of Ne: A comparison of many-electron theory and experiment. *J. Phys. B At. Mol. Opt. Phys.* **2006**, *39*, 3159–3176. [CrossRef]
228. Mahmood, S.; Amin, N.; ul Haq, S.; Shaikh, N.M.; Hussain, S.; Baig, M.A. Measurements of oscillator strengths of the $2p^5(^2P_{1/2})$ nd J = 2, 3 autoionizing resonances in neon. *J. Phys. B At. Mol. Opt. Phys.* **2006**, *39*, 2299–2313. [CrossRef]
229. Claessens, B.J.; Ashmore, J.P.; Sang, R.T.; MacGillivray, W.R.; Beijerinck, H.C.W.; Vredenbregt, E.J.D. Measurement of the photoionization cross section of the $(2p)^5(3p)$ $^3D_3$ state of neon. *Phys. Rev. A* **2006**, *73*, 012706. [CrossRef]
230. Petrov, I.D.; Sukhorukov, V.L.; Ruf, M.W.; Hotop, H. Odd autoionizing $2p_{1/2}$ $^5n(s'/d')$ resonances of Ne excited from the $2p^53p[K]_J$ states. *Eur. Phys. J. D* **2009**, *53*, 289–302. [CrossRef]
231. Mahmood, S.; Shaikh, N.M.; Kalyar, M.A.; Rafiq, M.; Piracha, N.K.; Baig, M.A. Measurements of electron density, temperature and photoionization cross sections of the excited states of neon in a discharge plasma. *J. Quant. Spectro. Radiat. Trans.* **2009**, *110*, 1840–1850. [CrossRef]
232. Baig, M.A.; Bokhari, I.A.; Rafiq, M.; Kalyar, M.A.; Hussian, T.; Ali, R. Photoexcitation and photoionization from the $2p^5$ $3p[5/2]_{2,3}$ levels in neon. *Phys. Rev. A* **2011**, *84*, 013421. [CrossRef]
233. Weber, J.M.; Ueda, K.; Klar, D.; Kreil, J.; Ruf, M.W.; Hotop, H. Odd Rydberg spectrum of ^{40}Ar(I): High-resolution laser spectroscopy and multichannel quantum defect analysis of the J = 2 and 3 levels. *J. Phys. B At. Mol. Opt. Phys.* **1999**, *32*, 2381. [CrossRef]
234. Aslam, M.; Ali, R.; Nadeem, A.; Bhatti, S.A.; Baig, M.A. Observation of $3p^5nd$ J = 2, 3 odd parity spectra of argon and MQDT analysis in the discrete and autoionizing regions. *Opt. Commun.* **1999**, *172*, 37–46. [CrossRef]

235. Qian, X.; Zhang, T.; Ng, C.Y. Two-color photoionization spectroscopy using vacuum ultraviolet synchrotron radiation and infrared optical parametric oscillator laser. *Rev. Sci. Instrum.* **2003**, *74*, 2784–2788. [CrossRef]
236. Baig, M.A.; Mahmood, S.; Mumtaz, R.; Rafiq, M.; Kalyar, M.A.; Hussain, S.; Ali. R. Oscillator strength measurements of the $3p^5(^2P_{1/2})$ nd$[3/2]_2$ and $[5/2]_{2,3}$ autoionizing resonances in argon. *Phys. Rev. A* **2008**, *78*, 032524. [CrossRef]
237. Wright, J.D.; Morgan, T.J.; Li, L.; Gu, Q.; Knee, J.L.; Petrov, I.D.; Sukhorukov, V.L.; Hotop, H. Photoionization spectroscopy of even-parity autoionizing Rydberg states of argon: Experimental and theoretical investigation of Fano profiles and resonance widths. *Phys. Rev. A* **2008**, *77*, 062512. [CrossRef]
238. Lee, Y.Y.; Dung, T.Y.; Hsieh, R.M.; Yuh, J.Y.; Song, Y.F.; Ho, G.H.; Huang, T.P.; Pan, W.C.; Chen, I.C.; Tu, S.Y.; et al. Autoionizing Rydberg series (np′, nf′ of Ar investigated by stepwise excitations with lasers and synchrotron radiation. *Phys. Rev. A* **2008**, *78*, 022509. [CrossRef]
239. Beutler, H. About absorption series of argon, krypton and xenon to terms between the two ionization limits $^2P_{3/2}$ and $^2P_{1/2}$. *Z. Phys.* **1935**, *93*, 177–196. [CrossRef]
240. Klar, D.; Aslam, M.; Baig, M.A.; Ueda, K.; Ruf, M.W.; Hotop, H. High-resolution measurements and multichannel quantum defect analysis of the Kr($4p^5(^2P_{1/2})$nd′, J = 2, 3) autoionizing resonances. *J. Phys. B At. Mol. Opt. Phys.* **2001**, *34*, 1549–1568. [CrossRef]
241. Baig, M.A.; Hanif, M.; Aslam, M. Laser-optogalvanic studies of the $4p^5$ns and nd autoionizing resonances in krypton. *J. Phys. B At. Mol. Opt. Phys.* **2008**, *41*, 035004. [CrossRef]
242. Li, C.Y.; Wang, T.T.; Zhen, J.F.; Zhang, Q.; Chen, Y. Resonance-enhanced photon excitation spectroscopy of the even-paroty autoionizing Rydberg series of Kr. *Sci. China Ser. B Chem.* **2009**, *52*, 161–168. [CrossRef]
243. Hanif, M.; Aslam, M.; Riaz, M.; Bhatti, S.A.; Baig, M.A. Laser opto galvanic measurements and line-shape analysis of $5p^57p$ and $5p^54$–$5f$ autoionizing resonances in xenon. *J. Phys. B At. Mol. Opt. Phys.* **2005**, *38*, S65–S75. [CrossRef]
244. Aloise, S.; O'Keeffe, P.; Cubaynes, D.; Meyer, M.; Grum-Grzhimailo, A.N. Photoionization of Synchrotron-Radiation-Excited Atoms: Separating Partial Cross Sections by Full Polarization Control. *Phys. Rev. Lett.* **2005**, *94*, 223002. [CrossRef] [PubMed]
245. Petrov, I.D.; Sukhorukov, V.L.; Ruf, M.W.; Klar, D.; Hotop, H. Near-threshold photoionization from the excited mp^5(m +1)p J = 0 − 3 levels of Ar, Kr and Xe (m = 3−5). Near-threshold photoionization from the excited mp^5(m + 1)p J = 0 − 3 levels of Ar, Kr and Xe (m = 3−5). *Eur. Phys. J. D* **2011**, *62*, 347–359. [CrossRef]
246. DiMauro, L.F.; Freeman, R.R.; Kulander, K.C. *Multiphoton Processes: ICOMP VIII: 8th International Conference, AIP Conference Proceedings, No. 525 [APCPCS] (No. ISBN 1-56396-946-7; ISSN 0094-243X.; CODEN APCPCS)*; American Institute of Physics: Melville, NY, USA, 2000.
247. Yamakawa, K.; Akahane, Y.; Fukuda, Y.; Aoyama, M.; Inoue, N.; Ueda, H.; Utsumi, T. Many-electron dynamics of a Xe atom in strong and superstrong laser fields. *Phys. Rev. Lett.* **2004**, *92*, 123001. [CrossRef] [PubMed]
248. Lu, H.; Varvarezos, L.; Hayden, P.; Kennedy, E.T.; Mosnier, J.-P.; Costello, J.T. The 5d-6p VUV Photoabsorption Spectrum of Bi$^+$. *Atoms* **2020**, *8*, 55. [CrossRef]
249. Meyer, M.; Cubaynes, D.; O'keeffe, P.; Luna, H.; Yeates, P.; Kennedy, E.T.; Costello, J.T.; Orr, P.; Taïeb, R.; Maquet, A.; et al. Two-color photoionization in xuv free-electron and visible laser fields. *Phys. Rev. A* **2006**, *74*, 011401. [CrossRef]
250. Sorokin, A.A.; Wellhöfer, M.; Bobashev, S.V.; Tiedtke, K.; Richter, M.; Richter, M. X-ray-laser interaction with matter and the role of multiphoton ionization: Free-electron-laser studies on neon and helium. *Phys. Rev. A* **2007**, *75*, 051402. [CrossRef]
251. Martins, M.; Wellhöfer, M.; Sorokin, A.A.; Richter, M.; Tiedtke, K.; Wurth, W. Resonant multiphoton processes in the soft-x-ray regime. *Phys. Rev. A* **2009**, *80*, 023411. [CrossRef]
252. Richter, M.; Amusia, M.Y.; Bobashev, S.V.; Feigl, T.; Juranić, P.N.; Martins, M.; Sorokin, A.A.; Tiedtke, K. Extreme ultraviolet laser excites atomic giant resonance. *Phys. Rev. Lett.* **2009**, *102*, 163002. [CrossRef]
253. Katravulapally, T.; Nikolopoulos, L.A. Effects of the FEL Fluctuations on the 2s2p Li+ Auto-Ionization Lineshape. *Atoms* **2020**, *8*, 35. [CrossRef]
254. Hofbrucker, J.; Eiri, L.; Volotka, A.V.; Fritzsche, S. Photoelectron Angular Distributions of Nonresonant Two-Photon Atomic Ionization Near Nonlinear Cooper Minima. *Atoms* **2020**, *8*, 54. [CrossRef]
255. Rosch, D.; Caravan, R.L.; Taatjes, C.A.; Au, K.; Almeida, R.; Osborn, D.L. Absolute photoionization cross section of the simplest enol, vinyl alcohol. *J. Phys. Chem. A* **2021**, *125*, 7920–7928. [CrossRef] [PubMed]

Article

Photoionization Study of Neutral Chlorine Atom

Momar Talla Gning * and Ibrahima Sakho

Department of Physics Chemistry, UFR Sciences and Technologies, University Iba Der Thiam,
Thies BP 967, Senegal; ibrahima.sakho@univ-thies.sn
* Correspondence: mtalla.gning@univ-thies.sn

Abstract: Photoionization of neutral chlorine atom is investigated in this paper in the framework of the screening constant per unit nuclear charge (SCUNC) method. Resonance energies, quantum defects and effective charges of the $3s^23p^4$ ($^3P_{2,1,0}$)ns and $3s^23p^4$ ($^3P_{1,0}$)nd Rydberg series originating from both the $^2P^0_{3/2}$ ground state and the $^2P^0_{1/2}$ excited state of chlorine atom are reported. The present study believed to be the first theoretical investigation is compared with the recent experimental measurements (Yang et al., Astrophys. J. 810:132, 2015). Good agreements are obtained between theory and experiments. New SCUNC data are tabulated as useful references for interpreting astrophysical spectra from neutral atomic chlorine.

Keywords: photoionization; resonance energy; quantum defect; Rydberg series; effective charges; ground state; excited state; SCUNC

Citation: Gning, M.T.; Sakho, I. Photoionization Study of Neutral Chlorine Atom. *Atoms* **2023**, *11*, 152. https://doi.org/10.3390/atoms11120152

Academic Editor: Yew Kam Ho, Sultana N. Nahar and Guillermo Hinojosa

Received: 11 September 2023
Revised: 3 November 2023
Accepted: 8 November 2023
Published: 6 December 2023

Copyright: © 2023 by the authors. Licensee MDPI, Basel, Switzerland. This article is an open access article distributed under the terms and conditions of the Creative Commons Attribution (CC BY) license (https://creativecommons.org/licenses/by/4.0/).

1. Introduction

Photoionization is a fundamental tool for probing our understanding of atomic structure and spectra. Knowledge of the latter is important for many derived processes and studies. Examples include understanding photon–plasma interactions, determining the abundance of chemical elements in astronomical objects [1], and modeling and diagnosing astrophysical and laboratory plasmas [2,3], to name but a few. One of the most important elements to study is chlorine which has been detected in numerous astrophysical objects, such as the planetary nebula NGC2818 [4], Jupiter's moon Io [5] and others. In addition, chlorine is used in many different applications in our daily lives, and more than that, it can be used to determine the physical conditions and chemical evolution of astronomical objects. However, as a chemical element existing both on Earth and in space, it is important to study its properties to facilitate its identification in astrophysical and laboratory plasmas, as well as its modeling for different applications. However, to date, determining its abundance remains a challenge and has been the subject of several studies over the last decade [6]. Experimental and theoretical studies were the subject of active researches as far as chlorine element is concerned. In the past, the R-matrix approach was used in the calculation of photoionization cross sections of Cl and Br [7,8]. In addition, the first absolute photoionization cross-section measurements for atomic chlorine were made from the 1S ionization threshold at 16.4 eV to 75 [9]. On the other hand, investigations were carried out in the calculations of oscillator strengths for ultraviolet resonances in Cl^+ and Cl^{2+} [7,10], in the photoionization cross section measurements of chlorine Cl^+ cation [8,11], in the experimental and theoretical photoionization of Cl^{2+} [9,12] and in the L-shell photoionization of magnesium-like ions with new results for Cl^{5+} [10,13]. In addition, theoretical investigations were carried out for Cl^+ in the framework of the dirac-coulomb R-matrix method [11,14] and for Cl^{5+} using the clean-channel R-matrix approach [15]. In the recent past, photoionization on Cl^+ was performed with the framework of the screening constant per unit nuclear charge (SCUNC) method [16] and of the relativistic Breit-Pauli R-matrix method (BPRM) [3]. Despite major efforts to understand the properties of neutral chlorine and its ions, atomic data on neutral chlorine are still scarce in the literature. The scarcity of theoretical calculations for Cl is due to its open-shell structure, which makes it a difficult atom for theorists, but also to some extent to its high reactivity [17]. In the recent past, vacuum ultraviolet photoion (VUV-PI) and VUV photoion pulsed field ionization (VUV-PFI-PI) measurements of the resonance energies in Cl were taken and the dominant series due to the $3p \to ns$ and $3p \to nd$ resonances were identified [18] to perfect and extend the earlier measurements [19] to more excited states $n = 61$. However, in the experiment carried out by Yang et al. [18], we note a number of shortcomings. The resonance energies measured for certain series of resonances are uncertain, and for all the series considered by these authors in their experiments, the quantum defects derived from experimental measurements of resonance energies are imprecise, and their variation in all directions is considered unsatisfactory. The goal of the present study is to report accurate resonance energies and quantum defects belonging to the identified $3p \to ns$ and $3p \to nd$ resonances in Cl [18]. For this purpose, we apply the SCUNC formalism [16,20,21]. Section 2 gives a brief overview of the calculation methodology. In Section 3, we present and discuss the results obtained, and we draw conclusions in Section 4.

2. Theory

For a given $(^{2S+1}L_J)nl$—Rydberg series, the general expression for the resonance energies E_n is given by (in Rydberg) [16,20,21]

$$E_n = E_\infty - \frac{Z^2}{n^2}\left[1 - \beta(nl; s; \mu, \nu;\, ^{2S+1}L_J; Z)\right]^2 \quad (1)$$

In Equation (1), ν and μ ($\mu > \nu$) denote the principal quantum numbers of the $(^{2S+1}L_J)nl$ Rydberg series used in the empirical determination of the f_k—screening constants, s represents the spin of the nl-electron ($s = 1/2$), E_∞ is the energy value of the series limit and Z stands for the atomic number. The β-parameters are screening constants by unit nuclear charge expanded in inverse powers of Z and given by the following equation:

$$\beta\left(Z,\,^{2S+1}L_J, n, s, \mu, \nu\right) = \sum_{k=1}^{q} f_k \left(\frac{1}{Z}\right)^k \quad (2)$$

where $f_k = f_k(^{2S+1}L_J, n, s, \mu, \nu)$ are screening constants to be evaluated empirically. In Equation (2), q stands for the number of terms in the expansion of the β–parameter. The resonance energy (E_n) is in the following form:

$$E_n = E_\infty - \frac{Z^2}{n^2}\left\{1 - \frac{f_1(^{2S+1}L_J^\pi)}{Z(n-1)} - \frac{f_2(^{2S+1}L_J^\pi)}{Z} \pm \sum_{k=1}^{q}\sum_{k'=1}^{q'} f_1^{k'} F(n,\mu,\nu,s) \times \left(\frac{1}{Z}\right)^k\right\}^2. \quad (3)$$

In Equation (3), $f_1(^{2S+1}L_J^\pi)$ and $f_2(^{2S+1}L_J^\pi)$ are screening constants to be evaluated. $\pm \sum_{k=1}^{q}\sum_{k'=1}^{q'} f_1^{k'} F(n,\mu,\nu,s) \times \left(\frac{1}{Z}\right)^k$ is a corrective term introduced to stabilize the resonance energies with increasing the principal quantum number n.

In general, resonance energies are analyzed from the standard quantum-defect expansion formula

$$E_n = E_\infty - \frac{RZ_{core}^2}{(n-\delta)^2}. \quad (4)$$

In Equation (4), E_∞ denotes the converging limit, R is the Rydberg constant, here, $R = R_{Cl} = 109{,}735.6176$ cm^{-1} represents the Rydberg constant for the Cl atom, which is obtained from the relation $R_{Cl} = R_\infty/(1 + m_e/M)$, where $R_\infty = 109{,}737.3157$ cm^{-1}, M is the mass of Cl$^+$, and m_e is the rest mass of the electron; Z_{core} represents the electric charge of the core ion and δ means the quantum defect. In addition, theoretical and measured energy positions can be analyzed by calculating the Z^*-effective charge in the framework of the SCUNC-procedure.

$$E_n = E_\infty - \frac{Z^{*2}}{n^2}R. \qquad (5)$$

Furthermore, comparing Equations (3) and (5), the effective charge is in the following form:

$$Z^* = Z\left\{1 - \frac{f_1\left(^{2S+1}L_J^\pi\right)}{Z(n-1)} - \frac{f_2\left(^{2S+1}L_J^\pi\right)}{Z} \pm \sum_{k=1}^{q}\sum_{k'=1}^{q'} f_1^{k'} F(n,\mu,\nu,s) \times \left(\frac{1}{Z}\right)^k\right\}. \qquad (6)$$

In addition, the f_2-parameter in Equation (3) is theoretically determined from Equation (6) with the following conditions:

$$\lim_{n\to\infty} Z^* = Z\left(1 - \frac{f_2\left(^{2S+1}L_J^\pi\right)}{Z}\right) = Z_{core}. \qquad (7)$$

So, we then get the following form:

$$f_2\left(^{2S+1}L_J^\pi\right) = Z - Z_{core} \qquad (8)$$

For a photoionization process from an atomic X^{p+}, we obtain the following form:

$$\gamma + X^{p+} \to X^{p+1} + e^- \qquad (9)$$

where γ is the absorbed photon. Using (9), we find $Z_{core} = p + 1$. For the neutral chlorine atom (Cl) considered in this work, Equation (9) becomes $\gamma + Cl \to Cl^+ + e^-$, therefore $Z_{core} = 1$ and $f_2\left(^{2S+1}L_J^\pi\right) = (17-1) = 16.0$. The remaining $f_1\left(^{2S+1}L_J^\pi\right)$-parameter is evaluated empirically using experimental data [18] for a given $(^{2S+1}L_J)nl$ level with $\nu = 0$ in Equation (3). The results obtained are indicated in the caption of the corresponding table. The details of the calculation are clearly explained in our previous original papers [16,20,21].

In addition, using Equations (4) and (5), we get

$$\frac{Z^{2*}}{n^2} = \frac{Z_{core}^2}{(n-\delta)^2}$$

which means

$$Z^* = \frac{Z_{core}}{\left(1 - \frac{\delta}{n}\right)}. \qquad (10)$$

Equation (10) indicates clearly that each Rydberg series must satisfy the following SCUNC conditions:

$$\begin{cases} Z^* \geq Z_{core} & if \quad \delta \geq 0 \\ Z^* \leq Z_{core} & if \quad \delta \leq 0 \\ \lim_{n\to\infty} Z^* = Z_{core} \end{cases} \qquad (11)$$

The resonance energies, quantum defects and effective charges of the $3s^2 3p^4$ $(^3P_{2,1,0})ns$ and $3s^2 3p^4$ $(^3P_{1,0})nd$ Rydberg series of Cl studied in the present work are listed in Tables 1–9, and comparisons are done with previous experimental measurements [18,19]. From (4) we obtain the following form for the quantum defect:

$$\delta = n - Z_{core}\sqrt{\frac{R}{(E_\infty - E_n)}}. \qquad (12)$$

Table 1. Resonance energies (E_n), quantum defect (δ) and effective charge (Z^*) for the $3s^23p^5\,^2P^0_{3/2} \rightarrow 3s^23p^4\,(^3P_2)ns\,(^4P_{5/2})$ Rydberg series in Cl I. The present SCUNC- calculations are compared with experiments [18]. Here, $f_1\,(^3P_2;\,^4P_{5/2};\,^2P^0_{3/2}) = -2.2115 \pm 0.0078$ with $\mu = 23$.

	Chlorine Initial State: $3s^23p^5\,^2P^0_{3/2}$				
ns	Rydberg Series $3s^23p^4\,(^3P_2)ns\,(^4P_{5/2})$				
	SCUNC	Experimental Data	SCUNC	Experimental Data	SCUNC
	E_n (cm^{-1})	E_n (cm^{-1}) [a]	δ	δ [a]	Z^*
23	104,339.78	104,339.78	2.101	2.108	1.101
24	104,362.21	104,361.03	2.100	2.157	1.096
25	104,381.78	104,380.42	2.099	2.173	1.092
26	104,398.94	104,397.54	2.098	2.185	1.088
27	104,414.07	104,413.48	2.097	2.139	1.085
28	104,427.49	104,426.07	2.096	2.207	1.082
29	104,439.43	104,438.29	2.094	2.195	1.079
30	104,450.12	104,448.99	2.093	2.204	1.076
31	104,459.71	104,458.49	2.091	2.225	1.074
32	104,468.36	104,467.04	2.090	2.250	1.071
33	104,476.18	104,476.51	2.088	2.043	1.069
34	104,483.28	104,483.55	2.086	2.045	1.067
35	104,489.74	104,490.51	2.084	1.958	1.065
36	104,495.63	104,495.51	2.083	2.104	1.063
37	104,501.02	104,500.80	2.081	2.124	1.061
38	104,505.97	104,505.65	2.079	2.147	1.060
39	104,510.53	104,510.02	2.077	2.193	1.058
40	104,514.72	104,514.27	2.076	2.187	1.057
41	104,518.60	104,518.21	2.074	2.178	1.055
42	104,522.19	104,522.23	2.072	2.060	1.054
43	104,525.51	104,525.23	2.071	2.159	1.053
44	104,528.61	104,528.27	2.069	2.182	1.051
45	104,531.49	104,531.11	2.067	2.202	1.050
46	104,534.17	104,533.69	2.066	2.250	1.049
47	104,536.68	104,536.29	2.064	2.222	1.048
48	104,539.02	104,538.66	2.062	2.220	1.047
49	104,541.21	104,540.83	2.061	2.241	1.046
50	104,543.27	104,542.87	2.059	2.261	1.045
51	104,545.21	104,544.89	2.058	2.227	1.044
52	104,547.03	104,546.65	2.056	2.269	1.043
53	104,548.74	104,548.45	2.055	2.228	1.043
54	104,550.35	104,550.06	2.053	2.240	1.042
55	104,551.88	104,551.57	2.052	2.259	1.041
56	104,553.32	104,552.97	2.050	2.297	1.040
57	104,554.68	104,554.38	2.049	2.274	1.039
58	104,555.97	104,555.67	2.048	2.284	1.039
59	104,557.19	104,556.95	2.046	2.247	1.038
60	104,558.35	104,558.08	2.045	2.282	1.037
61	104,559.45	104,559.14	2.044	2.330	1.037
62	104,560.49		2.042		1.036
63	104,561.49		2.041		1.036
64	104,562.44		2.040		1.035
65	104,563.34		2.039		1.035
66	104,564.20		2.038		1.034
67	104,565.02		2.036		1.034
68	104,565.80		2.035		1.033
69	104,566.55		2.034		1.033
70	104,567.27		2.033		1.032
71	104,567.95		2.032		1.032
72	104,568.61		2.031		1.031
73	104,569.23		2.030		1.031

Table 1. Cont.

	Chlorine Initial State: $3s^23p^5\,^2P^0_{3/2}$				
ns	Rydberg Series $3s^23p^4\,(^3P_2)ns\,(^4P_{5/2})$				
	SCUNC	Experimental Data	SCUNC	Experimental Data	SCUNC
	E_n (cm^{-1})	E_n (cm^{-1}) [a]	δ	δ [a]	Z^*
74	104,569.83		2.029		1.030
75	104,570.41		2.028		1.030
76	104,570.97		2.027		1.029
77	104,571.50		2.026		1.029
78	104,572.01		2.025		1.029
79	104,572.50		2.024		1.028
80	104,572.97		2.023		1.028
⋮	...	...	...		...
∞	104,591.02	104,591.02			1.000

[a] Ref. [18].

Table 2. Resonance energies (E_n), quantum defect (δ) and effective charge (Z^*) for the $3s^23p^5\,^2P^0_{3/2} \to 3s^23p^4\,(^3P_1)nd\,(^2D_{5/2})$ Rydberg series in Cl I. The present SCUNC- calculations are compared with experiments [18,19]. Here, $f_1\,(^3P_1;\,^2D_{5/2};\,^2P^0_{3/2}) = -0.3042 \pm 0.0015$ with $\mu = 13$.

	Chlorine Initial State: $3s^23p^5\,^2P^0_{3/2}$						
nd	Rydberg Series $3s^23p^4\,(^3P_1)nd\,(^2D_{5/2})$						
	SCUNC	Experimental Data		SCUNC	Experimental Data		SCUNC
	E_n (cm^{-1})	E_n (cm^{-1}) [a]	E_n (cm^{-1}) [b]	δ	δ [a]	δ [b]	Z^*
13	104,604.35	104,604.35	104,606.8	0.321	0.321	0.299	1.025
14	104,700.63	104,703.87	104,705.2	0.320	0.282	0.266	1.023
15	104,777.89	104,777.92	104,780.9	0.319	0.318	0.275	1.022
16	104,840.81	104,842.25	104,845.1	0.318	0.292	0.242	1.020
17	104,892.75	104,892.28	104,892.2	0.317	0.327	0.328	1.019
18	104,936.11	104,937.67	104,939.4	0.316	0.276	0.232	1.018
19	104,972.69	104,973.23	104,973.9	0.315	0.299	0.279	1.017
20	105,003.84	105,003.28		0.314	0.334		1.016
21	105,030.57	105,031.67		0.314	0.269		1.015
22	105,053.69	105,053.38		0.313	0.327		1.014
23	105,073.81	105,073.38		0.313	0.336		1.014
24	105,091.44	105,092.21		0.312	0.266		1.013
25	105,106.97	105,107.15		0.312	0.299		1.013
26	105,120.72	105,120.57		0.311	0.323		1.012
27	105,132.95	105,132.67		0.311	0.335		1.012
28	105,143.88	105,144.51		0.311	0.250		1.011
29	105,153.69	105,153.87		0.310	0.291		1.011
30	105,162.52	105,162.61		0.310	0.300		1.010
31	105,170.51	105,170.49		0.310	0.312		1.010
32	105,177.74	105,177.63		0.309	0.326		1.010
33	105,184.33	105,184.86		0.309	0.224		1.010
34	105,190.34	105,191.94		0.309	0.026		1.009
35	105,195.83	105,196.08		0.308	0.261		1.009
36	105,200.87	105,201.08		0.308	0.264		1.009
37	105,205.50	105,205.61		0.308	0.285		1.008
38	105,209.77	105,209.96		0.308	0.261		1.008
39	105,213.71	105,213.83		0.308	0.276		1.008
40	105,217.36	105,217.38		0.307	0.301		1.008
41	105,220.74	105,220.77		0.307	0.298		1.008
42	105,223.88	105,224.01		0.307	0.265		1.007

Table 2. Cont.

	Chlorine Initial State: $3s^23p^5\,^2P^0_{3/2}$					
nd	Rydberg Series $3s^23p^4\,(^3P_1)nd\,(^2D_{5/2})$					
	SCUNC	Experimental Data		SCUNC	Experimental Data		SCUNC
	E_n (cm^{-1})	E_n (cm^{-1}) [a]	E_n (cm^{-1}) [b]	δ	δ [a]	δ [b]	Z^*
43	105,226.81	105,226.91		0.307	0.270		1.007
44	105,229.53	105,229.49		0.307	0.322		1.007
45	105,232.07	105,232.39		0.306	0.177		1.007
46	105,234.45	105,234.81		0.306	0.150		1.007
47	105,236.68	105,236.91		0.306	0.199		1.007
48	105,238.77	105,238.85		0.306	0.266		1.006
49	105,240.73	105,240.69		0.306	0.327		1.006
50	105,242.57	105,242.59		0.306	0.297		1.006
51	105,244.31	105,244.41		0.306	0.247		1.006
52	105,245.95	105,245.89		0.305	0.341		1.006
53	105,247.49	105,247.41		0.305	0.359		1.006
54	105,248.95			0.305			1.006
55	105,250.33			0.305			1.006
56	105,251.63			0.305			1.006
57	105,252.87			0.305			1.005
58	105,254.04			0.305			1.005
59	105,255.16			0.305			1.005
60	105,256.22			0.305			1.005
61	105,257.22			0.304			1.005
62	105,258.18			0.304			1.005
63	105,259.09			0.304			1.005
64	105,259.96			0.304			1.005
65	105,260.79			0.304			1.005
66	105,261.58			0.304			1.005
67	105,262.34			0.304			1.005
68	105,263.06			0.304			1.005
69	105,263.76			0.304			1.004
70	105,264.42			0.304			1.004
⋮	...			...			...
∞	105,287.01	105,287.01					

[a] Ref. [18]. [b] Ref. [19].

Table 3. Resonance energies (E_n), quantum defect (δ) and effective charge (Z^*) for the $3s^23p^5\,^2P^0_{3/2} \rightarrow 3s^23p^4\,(^3P_1)ns\,(^2P_{3/2})$ Rydberg series in Cl I. The present SCUNC- calculations are compared with experiments [18,19]. Here, $f_1\,(^3P_1;\,^2P_{3/2};\,^2P^0_{3/2}) = -2.2052 \pm 0.0016$ with $\mu = 14$.

	Chlorine Initial State: $3s^23p^5\,^2P^0_{3/2}$					
ns	Rydberg Series $3s^23p^4\,(^3P_1)ns\,(^2P_{3/2})$					
	SCUNC	Experimental Data		SCUNC	Experimental Data		SCUNC
	E_n (cm^{-1})	E_n (cm^{-1}) [a]	E_n (cm^{-1}) [b]	δ	δ [a]	δ [b]	Z^*
14	104,521.08	104,521.08	104,520.0	2.030	2.030	2.039	1.170
15	104,634.67	104,636.11	104,630.8	2.030	2.016	2.068	1.158
16	104,724.79	104,720.65	104,716.7	2.029	2.080	2.129	1.147
17	104,797.47	104,794.65	104,790.1	2.028	2.071	2.139	1.138
18	104,856.94	104,854.28	104,854.3	2.026	2.075	2.075	1.130
19	104,906.20	104,907.24		2.025	2.001		1.123
20	104,947.46	104,945.08		2.023	2.085		1.116
21	104,982.37	104,980.03		2.021	2.093		1.110
22	105,012.15	105,009.19		2.019	2.125		1.105

Table 3. Cont.

	Chlorine Initial State: $3s^2 3p^5\,{}^2P^0_{3/2}$					
ns			Rydberg Series $3s^2 3p^4\,({}^3P_1)ns\,({}^2P_{3/2})$			
	SCUNC	Experimental Data		SCUNC	Experimental Data	SCUNC
	E_n (cm^{-1})	E_n (cm^{-1}) [a]	E_n (cm^{-1}) [b]	δ	δ [a]	δ [b]	Z^*
23	105,037.78	105,036.02		2.017	2.090		1.100
24	105,059.98	105,058.38		2.015	2.091		1.096
25	105,079.34	105,079.18		2.013	2.021		1.092
26	105,096.32	105,095.29		2.011	2.075		1.088
27	105,111.31	105,109.68		2.009	2.123		1.085
28	105,124.59	105,123.33		2.007	2.107		1.082
29	105,136.42	105,135.24		2.005	2.110		1.079
30	105,147.00	-		2.004	-		1.076
31	105,156.51	105,155.48		2.002	2.115		1.074
32	105,165.08	105,163.91		2.000	2.142		1.071
33	105,172.83	105,171.71		1.999	2.148		1.069
34	105,179.86	105,179.08		1.997	2.112		1.067
35	105,186.27			1.996			1.065
36	105,192.11			1.994			1.063
37	105,197.47			1.993			1.061
38	105,202.38			1.992			1.060
39	105,206.89			1.990			1.058
40	105,211.06			1.989			1.057
41	105,214.91			1.988			1.055
42	105,218.47			1.987			1.054
43	105,221.78			1.985			1.053
44	105,224.85			1.984			1.051
45	105,227.71			1.983			1.050
46	105,230.37			1.982			1.049
47	105,232.86			1.981			1.048
48	105,235.19			1.980			1.047
49	105,237.38			1.979			1.046
50	105,239.42			1.978			1.045
51	105,241.35			1.977			1.044
52	105,243.16			1.977			1.043
53	105,244.86			1.976			1.042
54	105,246.47			1.975			1.042
55	105,247.98			1.974			1.041
56	105,249.42			1.973			1.040
57	105,250.77			1.972			1.039
58	105,252.05			1.972			1.039
59	105,253.27			1.971			1.038
60	105,254.42			1.970			1.037
61	105,255.52			1.970			1.037
62	105,256.56			1.969			1.036
63	105,257.55			1.968			1.036
64	105,258.49			1.968			1.035
65	105,259.39			1.966			1.034
66	105,260.25			1.966			1.034
67	105,261.06			1.966			1.033
68	105,261.84			1.965			1.033
69	105,262.59			1.965			1.032
70	105,263.30			1.965			1.032
⋮							
∞	105,287.01	105,287.01					

[a] Ref. [18]. [b] Ref. [19].

Table 4. Resonance energies (E_n), quantum defect (δ) and effective charge (Z^*) for the $3s^23p^5\,^2P^0_{3/2} \to 3s^23p^4\,(^3P_1)nd\,(^2D_{5/2})$ Rydberg series in Cl I. The present SCUNC- calculations are compared with experiments [18]. Here, $f_1\,(^3P_1;\,^2D_{5/2};\,^2P^0_{3/2}) = -0.0881 \pm 0.0016$ with $\mu = 13$.

nd	Chlorine Initial State: $3s^23p^5\,^2P^0_{3/2}$				
	Rydberg Series $3s^23p^4\,(^3P_1)nd\,(^2D_{5/2})$				
	SCUNC	Experimental Data	SCUNC	Experimental Data	SCUNC
	E_n (cm^{-1})	E_n (cm^{-1}) [a]	δ	δ^a	Z^*
13	104,628.11	104,628.11	0.095	0.095	1.007
14	104,719.51	104,709.09	0.094	0.220	1.007
15	104,793.13	104,788.03	0.094	0.170	1.006
16	104,853.30	104,845.55	0.094	0.234	1.006
17	104,903.10	104,903.72	0.093	0.079	1.006
18	104,944.79	104,939.93	0.093	0.219	1.005
19	104,980.05	104,973.23	0.093	0.299	1.005
20	105,010.12	105,006.44	0.092	0.223	1.005
21	105,035.97	105,031.67	0.092	0.269	1.004
22	105,058.37	105,054.03	0.092	0.297	1.004
23	105,077.90	105,077.62	0.092	0.107	1.004
24	105,095.03		0.092		1.004
25	105,110.14		0.092		1.004
26	105,123.53		0.092		1.004
27	105,135.46		0.091		1.003
28	105,146.12		0.091		1.003
29	105,155.70		0.091		1.003
30	105,164.34		0.091		1.003
31	105,172.15		0.091		1.003
32	105,179.23		0.091		1.003
33	105,185.69		0.091		1.003
34	105,191.57		0.091		1.003
35	105,196.96		0.091		1.003
36	105,201.91		0.091		1.003
37	105,206.46		0.091		1.002
38	105,210.65		0.091		1.002
39	105,214.53		0.091		1.002
40	105,218.11		0.090		1.002
41	105,221.44		0.090		1.002
42	105,224.53		0.090		1.002
43	105,227.41		0.090		1.002
44	105,230.09		0.090		1.002
45	105,232.60		0.090		1.002
46	105,234.95		0.090		1.002
47	105,237.14		0.090		1.002
48	105,239.20		0.090		1.002
49	105,241.14		0.090		1.002
50	105,242.96		0.090		1.002
51	105,244.67		0.090		1.002
52	105,246.29		0.090		1.002
53	105,247.81		0.090		1.002
54	105,249.25		0.090		1.002
55	105,250.61		0.090		1.002
56	105,251.91		0.090		1.002
57	105,253.13		0.090		1.002
58	105,254.29		0.090		1.002
59	105,255.39		0.090		1.002
60	105,256.44		0.090		1.001
61	105,257.43		0.090		1.001

Table 4. Cont.

nd	Chlorine Initial State: $3s^23p^5\,^2P^0_{3/2}$				
	Rydberg Series $3s^23p^4\,(^3P_1)nd\,(^2D_{5/2})$				
	SCUNC	Experimental Data	SCUNC	Experimental Data	SCUNC
	E_n (cm^{-1})	E_n (cm^{-1}) [a]	δ	δ [a]	Z^*
62	105,258.38		0.090		1.001
63	105,259.28		0.090		1.001
64	105,260.14		0.090		1.001
65	105,260.97		0.090		1.001
66	105,261.75		0.090		1.001
67	105,262.50		0.090		1.001
68	105,263.22		0.090		1.001
69	105,263.90		0.090		1.001
70	105,264.56		0.090		1.001
⋮					
∞	105,287.01	105,287.01			

[a] Ref. [18].

Table 5. Resonance energies (E_n), quantum defect (δ) and effective charge (Z^*) for the $3s^23p^5\,^2P^0_{3/2} \rightarrow 3s^23p^4\,(^3P_1)nd\,(^2D_{5/2})$ Rydberg series in Cl I. The present SCUNC- calculations are compared with experiments [18,19]. Here, $f_1\,(^3P_0;^2P_{1/2};\,^2P^0_{3/2}) = -2.3766 \pm 0.0273$ with $\mu = 10$.

ns	Chlorine Initial State: $3s^23p^5\,^2P^0_{3/2}$						
	Rydberg Series $3s^23p^4\,(^3P_0)ns\,(^2P_{1/2})$						
	SCUNC	Experimental Data		SCUNC	Experimental Data		SCUNC
	E_n (cm^{-1})	E_n (cm^{-1}) [a]	E_n (cm^{-1}) [b]	δ	δ [a]	δ [b]	Z^*
10	103,834.02	103,834.01	103,831.8	2.089	2.089	2.094	1.264
11	104,205.62	104,204.13	104,203.7	2.089	2.093	2.094	1.238
12	104,470.57	-		2.088	-		1.216
13	104,666.06	104,665.63		2.087	2.089		1.198
14	104,814.39	104,813.74		2.086	2.090		1.183
15	104,929.59	104,928.56		2.085	2.093		1.170
16	105,020.83	105,019.83		2.084	2.094		1.158
17	105,094.33	-		2.083	-		1.149
18	105,154.40	-		2.082	-		1.140
19	105,204.12	105,203.12		2.081	2.099		1.132
20	105,245.75	-		2.080	-		1.125
21	105,280.94	105,281.56		2.080	2.055		1.119
22	105,310.97			2.079			1.113
23	105,336.79			2.078			1.108
24	105,359.15			2.077			1.103
25	105,378.65			2.077			1.099
26	105,395.75			2.076			1.095
27	105,410.83			2.076			1.091
28	105,424.20			2.075			1.088
29	105,436.11			2.075			1.085
30	105,446.76			2.074			1.082
31	105,456.33			2.074			1.079
32	105,464.95			2.074			1.077
33	105,472.75			2.073			1.074
34	105,479.83			2.073			1.072
35	105,486.27			2.073			1.070
36	105,492.15			2.072			1.068

Table 5. Cont.

ns	Chlorine Initial State: $3s^23p^5\,^2P^0_{3/2}$						
	Rydberg Series $3s^23p^4\,(^3P_0)ns\,(^2P_{1/2})$						
	SCUNC	Experimental Data		SCUNC	Experimental Data		SCUNC
	E_n (cm^{-1})	E_n (cm^{-1}) [a]	E_n (cm^{-1}) [b]	δ	δ [a]	δ [b]	Z^*
37	105,497.53			2.072			1.066
38	105,502.47			2.072			1.064
39	105,507.01			2.072			1.063
40	105,511.20			2.071			1.061
41	105,515.07			2.071			1.059
42	105,518.65			2.071			1.058
43	105,521.97			2.071			1.057
44	105,525.06			2.071			1.055
45	105,527.94			2.070			1.054
46	105,530.62			2.070			1.053
47	105,533.12			2.070			1.052
48	105,535.46			2.070			1.051
49	105,537.66			2.070			1.050
50	105,539.71			2.070			1.049
51	105,541.65			2.069			1.048
52	105,543.46			2.069			1.047
53	105,545.18			2.069			1.046
54	105,546.79			2.069			1.045
55	105,548.31			2.069			1.044
56	105,549.75			2.069			1.043
57	105,551.11			2.069			1.042
58	105,552.40			2.069			1.042
59	105,553.62			2.068			1.041
60	105,554.78			2.068			1.040
61	105,555.88			2.068			1.040
62	105,556.93			2.068			1.039
63	105,557.92			2.068			1.038
64	105,558.87			2.068			1.038
65	105,559.77			2.068			1.037
66	105,560.63			2.068			1.037
67	105,561.45			2.068			1.036
68	105,562.24			2.068			1.035
69	105,562.99			2.068			1.035
70	105,563.70			2.068			1.034
⋮	...			...			...
∞	105,587.48	105,587.48					

[a] Ref. [18].; [b] Ref. [19].

Table 6. Resonance energies (E_n), quantum defect (δ) and effective charge (Z^*) for the $3s^23p^5\,^2P^0_{3/2} \rightarrow 3s^23p^4\,(^3P_0)nd\,(^2P_{3/2})$ Rydberg series in Cl I. The present SCUNC- calculations are compared with experiments [18,19]. Here, $f_1\,(^3P_0,^2P_{3/2};\,^2P^0_{3/2}) = -0.2916 \pm 0.0028$ with $\mu = 16$.

nd	Chlorine Initial State: $3s^23p^5\,^2P^0_{3/2}$						
	Rydberg Series $3s^23p^4\,(^3P_0)nd\,(^2P_{3/2})$						
	SCUNC	Experimental Data		SCUNC	Experimental Data		SCUNC
	E_n (cm^{-1})	E_n (cm^{-1}) [a]	E_n (cm^{-1}) [b]	δ	δ [a]	δ [b]	Z^*
16	105,142.00	105,142.00	105,148.0	0.305	0.302	0.195	1.019
17	105,193.81	105,194.50		0.304	0.286		1.018

Table 6. Cont.

nd	Chlorine Initial State: $3s^23p^5\,^2P^0_{3/2}$						
	Rydberg Series $3s^23p^4\,(^3P_0)nd\,(^2P_{3/2})$						
	SCUNC	Experimental Data		SCUNC	Experimental Data		SCUNC
	E_n (cm^{-1})	E_n (cm^{-1}) [a]	E_n (cm^{-1}) [b]	δ	δ [a]	δ [b]	Z^*
18	105,237.08	105,237.77		0.303	0.282		1.017
19	105,273.59	105,274.04		0.302	0.284		1.016
20	105,304.67	105,305.21		0.302	0.277		1.015
21	105,331.35	105,331.58		0.301	0.285		1.015
22	105,354.43	105,354.70		0.301	0.280		1.014
23	105,374.52	105,374.94		0.300	0.269		1.013
24	105,392.12	105,392.28		0.299	0.280		1.013
25	105,407.63	105,407.79		0.299	0.276		1.012
26	105,421.36	105,421.57		0.299	0.269		1.012
27	105,433.57	105,433.69		0.298	0.273		1.011
28	105,444.49	105,444.61		0.298	0.269		1.011
29	105,454.28	105,454.48		0.297	0.257		1.010
30	105,463.10	105,463.45		0.297	0.235		1.010
31	105,471.07	105,471.25		0.297	0.251		1.010
32	105,478.30			0.296			1.009
33	105,484.88			0.296			1.009
34	105,490.88			0.296			1.009
35	105,496.37			0.296			1.009
36	105,501.40			0.295			1.008
37	105,506.03			0.295			1.008
38	105,510.29			0.295			1.008
39	105,514.23			0.295			1.008
40	105,517.87			0.295			1.007
41	105,521.25			0.294			1.007
42	105,524.39			0.294			1.007
43	105,527.31			0.294			1.007
44	105,530.03			0.294			1.007
45	105,532.58			0.294			1.007
46	105,534.95			0.294			1.006
47	105,537.18			0.293			1.006
48	105,539.26			0.293			1.006
49	105,541.22			0.293			1.006
50	105,543.07			0.293			1.006
51	105,544.80			0.293			1.006
52	105,546.44			0.293			1.006
53	105,547.98			0.293			1.006
54	105,549.44			0.293			1.006
55	105,550.81			0.292			1.005
56	105,552.12			0.292			1.005
57	105,553.36			0.292			1.005
58	105,554.53			0.292			1.005
59	105,555.64			0.292			1.005
60	105,556.70			0.292			1.005
61	105,557.70			0.292			1.005
62	105,558.66			0.292			1.005
63	105,559.57			0.292			1.005
64	105,560.44			0.292			1.005
65	105,561.27			0.291			1.005
66	105,562.06			0.291			1.004
67	105,562.82			0.291			1.004
68	105,563.54			0.291			1.004
69	105,564.24			0.291			1.004

Table 6. Cont.

	Chlorine Initial State: $3s^23p^5\,^2P^0_{3/2}$					
nd	Rydberg Series $3s^23p^4\,(^3P_0)nd\,(^2P_{3/2})$					
	SCUNC	Experimental Data		SCUNC	Experimental Data		SCUNC
	E_n (cm^{-1})	E_n (cm^{-1}) [a]	E_n (cm^{-1}) [b]	δ	δ [a]	δ [b]	Z^*
70	105,564.90			0.291			1.004
⋮							
∞	105,587.48	105,587.48					

[a] Ref. [18]. [b] Ref. [19].

Table 7. Resonance energies (E_n), quantum defect (δ) and effective charge (Z^*) for the $3s^23p^5\,^2P^0_{1/2} \to 3s^23p^4\,(^3P_0)nd\,(^2P_{3/2})$ Rydberg series in Cl I. The present SCUNC- calculations are compared with experiments [18]. Here, $f_1\,(^3P_0;\,^2P_{3/2};\,^2P^0_{1/2}) = -0.2711 \pm 0.0040$ with $\mu = 18$.

	Chlorine Initial State: $3s^23p^5\,^2P^0_{1/2}$				
nd	Rydberg Series $3s^23p^4\,(^3P_0)nd\,(^2P_{3/2})$				
	SCUNC	Experimental Data	SCUNC	Experimental Data	SCUNC
	E_n (cm^{-1})	E_n (cm^{-1}) [a]	δ	δ [a]	Z^*
18	104,355.55	104,355.55	0.283	0.282	1.016
19	104,391.93	104,391.13	0.282	0.305	1.015
20	104,422.91	104,422.71	0.281	0.288	1.014
21	104,449.51	104,449.01	0.281	0.300	1.014
22	104,472.52	104,472.08	0.280	0.300	1.013
23	104,492.55	104,492.02	0.280	0.308	1.012
24	104,510.11	104,509.52	0.279	0.314	1.012
25	104,525.57	-	0.279	-	1.011
26	104,539.27	-	0.278	-	1.011
27	104,551.45	104,551.11	0.278	0.307	1.010
28	104,562.34	104,562.22	0.278	0.289	1.010
29	104,572.12	104,572.18	0.277	0.269	1.010
30	104,580.92	104,581.29	0.277	0.231	1.009
31	104,588.87	104,589.20	0.277	0.232	1.009
32	104,596.09		0.277		1.009
33	104,602.65		0.276		1.008
34	104,608.64		0.276		1.008
35	104,614.12		0.276		1.008
36	104,619.15		0.276		1.008
37	104,623.77		0.275		1.008
38	104,628.02		0.275		1.007
39	104,631.95		0.275		1.007
40	104,635.59		0.275		1.007
41	104,638.97		0.275		1.007
42	104,642.10		0.275		1.007
43	104,645.02		0.274		1.006
44	104,647.74		0.274		1.006
45	104,650.27		0.274		1.006
46	104,652.65		0.274		1.006
47	104,654.87		0.274		1.006
48	104,656.95		0.274		1.006
49	104,658.91		0.274		1.006
50	104,660.75		0.273		1.006
51	104,662.48		0.273		1.005

Table 7. Cont.

	Chlorine Initial State: $3s^2 3p^5\,^2P^0_{1/2}$				
nd	Rydberg Series $3s^2 3p^4\,(^3P_0)nd\,(^2P_{3/2})$				
	SCUNC	Experimental Data	SCUNC	Experimental Data	SCUNC
	E_n (cm^{-1})	E_n (cm^{-1}) [a]	δ	δ [a]	Z^*
52	104,664.12		0.273		1.005
53	104,665.66		0.273		1.005
54	104,667.11		0.273		1.005
55	104,668.49		0.273		1.005
56	104,669.79		0.273		1.005
57	104,671.03		0.273		1.005
58	104,672.20		0.273		1.005
59	104,673.31		0.272		1.005
60	104,674.37		0.272		1.005
61	104,675.37		0.272		1.005
62	104,676.33		0.272		1.004
63	104,677.24		0.272		1.004
64	104,678.11		0.272		1.004
65	104,678.94		0.272		1.004
66	104,679.73		0.272		1.004
67	104,680.49		0.272		1.004
68	104,681.21		0.272		1.004
69	104,681.90		0.272		1.004
70	104,682.56		0.272		1.004
⋮	...		...		...
∞	104,705.13	104,705.13			

[a] Ref. [18].

Table 8. Resonance energies (E_n), quantum defect (δ) and effective charge (Z^*) for the $3s^2 3p^5\,^2P^0_{1/2} \rightarrow 3s^2 3p^4\,(^3P_1)ns\,(^2P_{3/2})$ Rydberg series in Cl I. The present SCUNC- calculations are compared with experiments [18]. Here, $f_1\,(^3P_1;^2P_{3/2};\,^2P^0_{1/2}) = -2.3230 \pm 0.0100$ with $\mu = 25$.

	Chlorine Initial State: $3s^2 3p^5\,^2P^0_{1/2}$				
ns	Rydberg Series $3s^2 3p^4\,(^3P_1)ns\,(^2P_{3/2})$				
	SCUNC	Experimental Data	SCUNC	Experimental Data	SCUNC
	E_n (cm^{-1})	E_n (cm^{-1}) [a]	δ	δ [a]	Z^*
25	104,193.41	104,193.41	2.206	2.209	1.097
26	104,210.79	104,210.83	2.206	2.206	1.093
27	104,226.11	104,225.45	2.206	2.254	1.089
28	104,239.68	104,238.49	2.206	2.301	1.086
29	104,251.77	104,245.29	2.206	2.759	1.083
30	104,262.58	104,263.32	2.205	2.135	1.080
31	104,272.28	104,272.34	2.204	2.201	1.077
32	104,281.02	104,280.49	2.203	2.271	1.075
33	104,288.93	104,288.40	2.202	2.276	1.073
34	104,296.10	104,295.65	2.201	2.271	1.070
35	104,302.62	104,302.52	2.200	2.221	1.068
36	104,308.57	104,309.47	2.199	2.045	1.066
37	104,314.02	104,314.25	2.197	2.159	1.065
38	104,319.02	104,318.65	2.196	2.279	1.063
39	104,323.61	104,323.57	2.194	2.211	1.061
40	104,327.85	104,327.56	2.193	2.271	1.060

Table 8. *Cont.*

	Chlorine Initial State: $3s^23p^5\,^2P^0_{1/2}$				
	Rydberg Series $3s^23p^4\,(^3P_1)ns\,(^2P_{3/2})$				
ns	SCUNC	Experimental Data	SCUNC	Experimental Data	SCUNC
	E_n (cm^{-1})	E_n (cm^{-1}) [a]	δ	δ [a]	Z^*
41	104,331.76	104,331.42	2.191	2.289	1.058
42	104,335.38	104,335.11	2.190	2.276	1.057
43	104,338.74	104,338.65	2.188	2.224	1.055
44	104,341.85	104,341.48	2.187	2.321	1.054
45	104,344.76	104,344.34	2.185	2.344	1.053
46	104,347.46	-	2.184	-	1.052
47	104,349.99	-	2.182	-	1.051
48	104,352.35	-	2.181	-	1.049
49	104,354.56	104,354.87	2.179	2.049	1.048
50	104,356.64	104,356.60	2.178	2.211	1.047
51	104,358.59	104,358.53	2.176	2.221	1.046
52	104,360.42		2.175		1.046
53	104,362.14		2.173		1.045
54	104,363.77		2.172		1.044
55	104,365.30		2.170		1.043
56	104,366.75		2.169		1.042
57	104,368.12		2.167		1.041
58	104,369.42		2.166		1.041
59	104,370.65		2.165		1.040
60	104,371.82		2.163		1.039
61	104,372.92		2.162		1.039
62	104,373.97		2.160		1.038
63	104,374.97		2.159		1.037
64	104,375.93		2.158		1.037
65	104,376.83		2.156		1.036
66	104,377.70		2.155		1.036
67	104,378.52		2.154		1.035
68	104,379.31		2.152		1.035
69	104,380.06		2.151		1.034
70	104,380.78		2.150		1.034
⋮	...		...		...
∞	104,404.62	104,404.62			1.000

[a] Ref. [18].

Table 9. Resonance energies (E_n), quantum defect (δ) and effective charge (Z^*) for the $3s^23p^5\,^2P^0_{1/2} \rightarrow 3s^23p^4\,(^3P_2)ns\,(^4P_{3/2})$ Rydberg series in Cl I. The present SCUNC- calculations are compared with experiments [18]. Here, $f_1\,(^3P_2;\,^4P_{3/2};\,^2P^0_{1/2}) = -2.3603 \pm 0.0128$ with $\mu = 27$.

	Chlorine Initial State: $3s^23p^5\,^2P^0_{1/2}$				
	Rydberg Series $3s^23p^4\,(^3P_2)ns\,(^4P_{3/2})$				
ns	SCUNC	Experimental Data	SCUNC	Experimental Data	SCUNC
	E_n (cm^{-1})	E_n (cm^{-1}) [a]	δ	δ [a]	Z^*
27	103,529.65	103,529.65	2.247	2.247	1.091
28	103,543.29	103,542.03	2.247	2.345	1.087
29	103,555.42	103,553.29	2.247	2.432	1.084
30	103,566.28	103,564.01	2.247	2.465	1.081
31	103,576.02	103,575.44	2.247	2.309	1.079
32	103,584.79	103,583.78	2.246	2.367	1.076

Table 9. Cont.

	Chlorine Initial State: $3s^2 3p^5\ {}^2P^0_{1/2}$				
ns	Rydberg Series $3s^2 3p^4\ (^3P_2)ns\ (^4P_{3/2})$				
	SCUNC	Experimental Data	SCUNC	Experimental Data	SCUNC
	E_n (cm^{-1})	E_n (cm^{-1}) [a]	δ	δ [a]	Z^*
33	103,592.73	103,594.21	2.246	2.048	1.074
34	103,599.93	103,600.99	2.245	2.089	1.072
35	103,606.48	103,607.12	2.244	2.140	1.069
36	103,612.45	103,612.94	2.243	2.157	1.067
37	103,617.92	103,618.14	2.242	2.199	1.066
38	103,622.93	103,623.11	2.241	2.204	1.064
39	103,627.54	103,627.60	2.240	2.227	1.062
40	103,631.79	103,631.81	2.238	2.234	1.061
41	103,635.72	103,634.41	2.237	2.580	1.059
42	103,639.35	103,639.85	2.236	2.092	1.058
43	103,642.72	103,642.77	2.234	2.218	1.056
44	103,645.85	103,645.66	2.233	2.294	1.055
45	103,648.76	103,648.54	2.231	2.309	1.054
46	103,651.47	103,651.21	2.230	2.329	1.052
47	103,654.00	103,653.63	2.229	2.381	1.051
48	103,656.37	103,656.01	2.227	2.385	1.050
49	103,658.59	103,658.36	2.226	2.334	1.049
50	103,660.67	103,660.33	2.224	2.394	1.048
51	103,662.63	103,662.35	2.223	2.369	1.047
52	103,664.46	103,664.03	2.221	2.464	1.046
53	103,666.19		2.220		1.045
54	103,667.82		2.218		1.045
55	103,669.36		2.217		1.044
56	103,670.82		2.215		1.043
57	103,672.19		2.214		1.042
58	103,673.49		2.213		1.041
59	103,674.72		2.211		1.041
60	103,675.89		2.210		1.040
61	103,677.00		2.208		1.039
62	103,678.06		2.207		1.039
63	103,679.06		2.206		1.038
64	103,680.01		2.204		1.037
65	103,680.92		2.203		1.037
66	103,681.79		2.201		1.036
67	103,682.62		2.200		1.036
68	103,683.41		2.199		1.035
69	103,684.16		2.198		1.035
70	103,684.88		2.196		1.034
⋮					
∞	103,708.75	103,708.75			

[a] Ref. [18].

3. Results and Discussion

Let us first determine the sign of the quantum defect (δ) using the SCUNC analysis conditions (11), considering the lowest resonance corresponding to the first entry of the Rydberg series under study. For example, for the $3s^2 3p^4\ (^3P_2)ns\ (^4P_{5/2})$ Rydberg series originating from the $3s^2 3p^5\ (^2P^0_{3/2})$ ground state of Cl (Table 1), the lowest Resonance corresponds to $n_{low} = 23$. From Table 1, we deduce $f_1\ (^3P_2;\ {}^4P_{5/2};\ {}^2P^0_{3/2}) = -2.2115$. From Equation (6), we derive the expression for the effective nuclear charge Z^*_{max} as follows:

$$Z^*_{max} = Z\left\{1 - \frac{f_1}{Z(n_{low}-1)} - \frac{16.0}{Z}\right\} = 17\left\{1 + \frac{2.2115}{17(23-1)} - \frac{16.0}{17}\right\} = 1.101. \quad (13)$$

As $Z_{core} = 1.000$, $Z^*_{max} = 1.101 > Z_{core}$. Then, the quantum defect is positive. So, for all the series analyzed in this work, positive quantum defects are allowed according to the according to the SCUNC analysis conditions (11). This is verified for all the data quoted in Tables 1–9. Table 1 presents Resonance energies, quantum defects and effective charges of the $3s^23p^4$ ($^3P_2$)ns ($^4P_{5/2}$) Rydberg series originating from the $3s^23p^5\,^2P^0_{3/2}$ ground stateCl and converging to the $^3P_2$ series limit in Cl$^+$. For this Rydberg Resonance, only the VUV-PI and VUV-PFI-PI measurements [18] are available in the literature to our knowledge. Comparison of resonance energies shows excellent agreement between theoretical data (SCUNC) and (VUV-PI and VUV-PFI-PI) measurements [18] up to $n = 61$, as well highlightedin Figure 1. In contrast, for quantum defects, the present SCUNC calculations provide good quantum defect behavior that is virtually constant or decreases slightly with increasing principal quantum number n up to $n = 80$, while experimental quantum defects vary anomalously in all directions, as shown in Figure 2. Table 2 lists resonance energies, quantum defects and effective chargesof the $3s^23p^4$ ($^3P_1$)nd ($^2D_{5/2}$) Rydberg resonance series from the $3s^23p^5\,^2P^0_{3/2}$ ground state of atomic chlorine. As shown in Figure 3, there is an excellent agreement between the SCUNC resonance energies and the quoted measurements [18,19]. But if the SCUNC quantum defect is constant or decreases slightly with increasing principal quantum number n, the measured quantum defect varies considerably [18,19], as shown in Figure 4. It is well known that quantum defects must be constant or decrease with increasing n. Especially when $n \to \infty$, we obtain a hydrogen-like system for which the quantum defect is zero. Tables 3 and 4 present the resonance energies, quantum defects and effective charges of the $3s^23p^4$ ($^3P_1$)ns ($^2P_{3/2}$) and of the $3s^23p^4$ ($^3P_1$)nd ($^2D_{5/2}$) Rydberg originating from the $3s^23p^5\,^2P^0_{3/2}$ ground state of Cl and converging to the $^3P_1$ series limit in Cl$^+$. Once again, the SCUNC results agree excellently with all the experimental resonance energies up to $n = 34$ (Table 3) and up to $n = 23$ (Table 4). These good agreements allow us to consider as accurate the extrapolated SCUNC data up to $n = 70$. Comparisons of quantum defects indicate irregular behavior of the experimental values in contrast with the SCUNC quantum defect varying correctly up to $n = 70$. Tables 5 and 6 list resonance energies, quantum defects and effective charges calculated for the $3s^23p^5\,^2P^0_{3/2} \to 3s^23p^4$ ($^3P_0$)ns ($^2P_{1/2}$) and the $3s^23p^5\,^2P^0_{3/2} \to 3s^23p^4$ ($^3P_0$)nd ($^2P_{3/2}$) Rydberg resonance series. For the $3s^23p^4$ ($^3P_0$)ns ($^2P_{1/2}$) series, experimental data are presented in Table 5 up to $n = 21$ with missing energy positions for $n = 12, 17$ 18, 20. SCUNC data associated with a nearly constant quantum defect at around 2.08 are provided for the missing experimental resonance energies [18,19]. In addition, the good behavior of the SCUNC quantum defect is observed up to $n = 70$ allowing us to consider the extrapolated news resonance energies as correct. Table 6 indicates resonance parameters of the $3s^23p^4$ ($^3P_0$)nd ($^2P_{3/2}$) series. Here again, very good consistency is obtained between the theoretical and the experimental resonance energies up to $n = 31$ as shown in Figure 5. New resonance energies associated with an almost constant quantum defect are tabulated for high-lying states $n = 32$–70 (see Figure 6). Tables 7–9 compare resonance energies and quantum defects, respectively of the $3s^23p^4$ ($^3P_0$)nd ($^2P_{3/2}$), $3s^23p^4$ ($^3P_1$)ns ($^2P_{3/2}$) and $3s^23p^4$ ($^3P_2$)ns ($^4P_{3/2}$) Rydberg series. Comparison shows reasonably good agreement between resonance energies for all the considered series as highlighted in Figures 7 and 8. For the $3s^23p^4$ ($^3P_0$)nd ($^2P_{3/2}$) series presented in Table 7, the absent experimental resonance energies for $n = 25$ and 26 [18] and that for the $3s^23p^4$ ($^3P_1$)ns ($^2P_{3/2}$) series quoted in Table 8 for $n = 46, 47$ and 48 were calculated via the present formalism. As far as quantum defects are concerned, for the above series, the SCUNC data remain again virtually constant up to $n = 70$ in contrast with the measured [18] as shown in Figures 9 and 10. For all the series investigated in this work, the effective nuclear charge decreases the monotony toward the value of the electric charge of the core ion $Z_{core} = 1.0$.

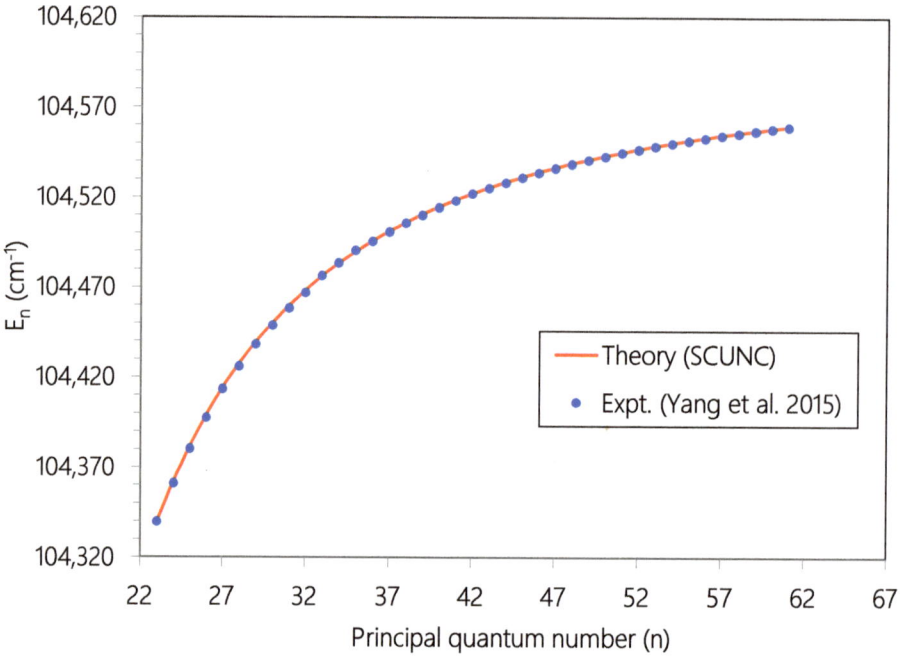

Figure 1. Plot of Resonance energies (E_n, cm^{-1}) versus principal quantum number (n) for the $3s^23p^4$ ($^3P_{2,1}$)ns ($^4P_{5/2}$) Rydberg series of resonances originating from the Cl ($^2P^0_{3/2}$) ground state. Experimental data (solid blue circles, Ref. [18] and theoretical estimates (solid red line, SCUNC).

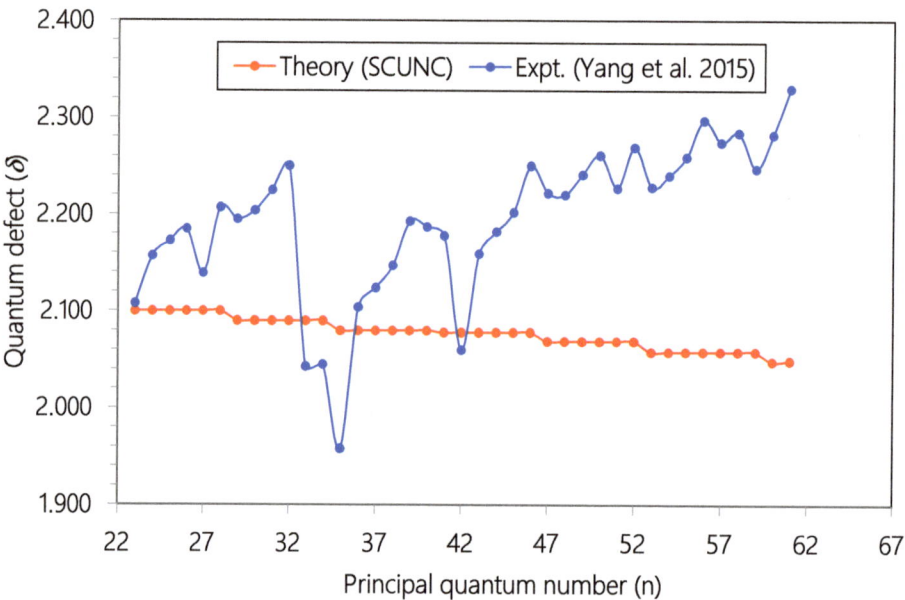

Figure 2. Plot of quantum defects (δ) versus principal quantum number (n) for the $3s^23p^4$ ($^3P_{2,1}$)ns ($^4P_{5/2}$) Rydberg series of resonances originating from the Cl ($^2P^0_{3/2}$) ground state. Experimental data (solid blue circles, Ref. [18]) and theoretical estimates (solid red circles, SCUNC).

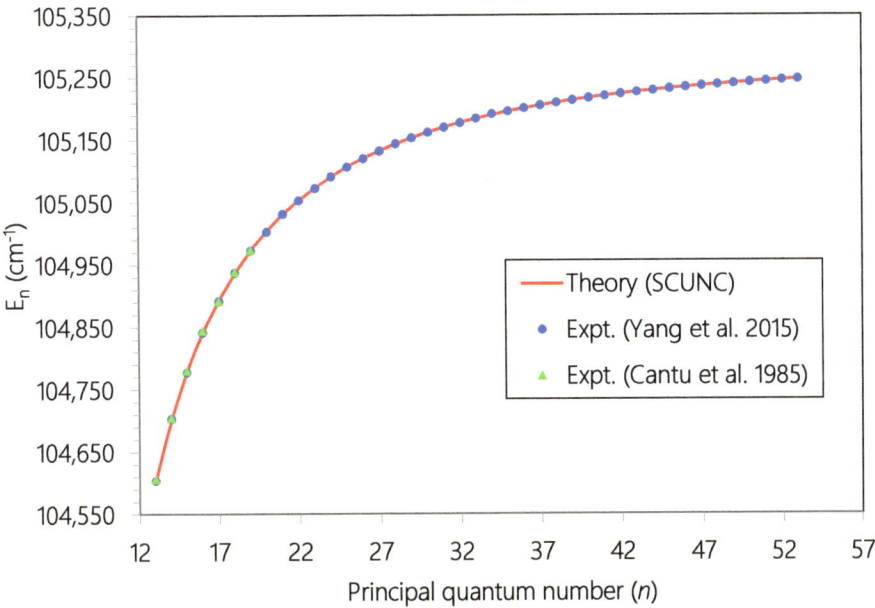

Figure 3. Plot of resonance energies (E_n, cm^{-1}) versus principal quantum number (n) for the $3s^23p^4$ ($^3P_1$)nd ($^2D_{5/2}$) Rydberg series of resonances originating from the Cl ($^2P^0{}_{3/2}$) ground state. Experimental data (solid blue circles Ref. [18] and solid green triangles Ref. [19]) and theoretical estimates (solid red line, SCUNC).

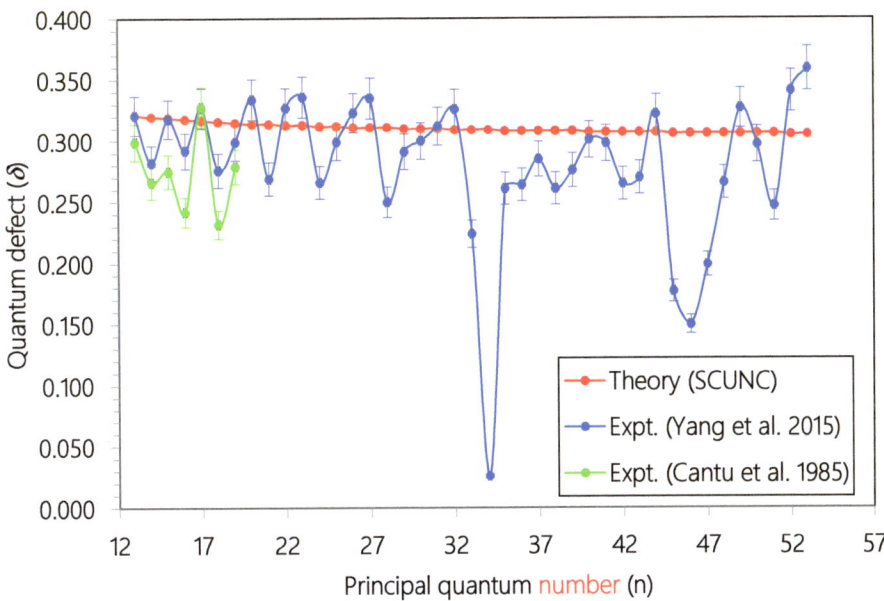

Figure 4. Plot of quantum defects (δ) versus principal quantum number (n) for the $3s^23p^4$ ($^3P_1$)nd ($^2D_{5/2}$) Rydberg series of resonances originating from the Cl ($^2P^0{}_{3/2}$) ground state. Experimental data (solid blue circles Ref. [18] and solid green circles Ref. [19]) and theoretical estimates (solid red circles, SCUNC).

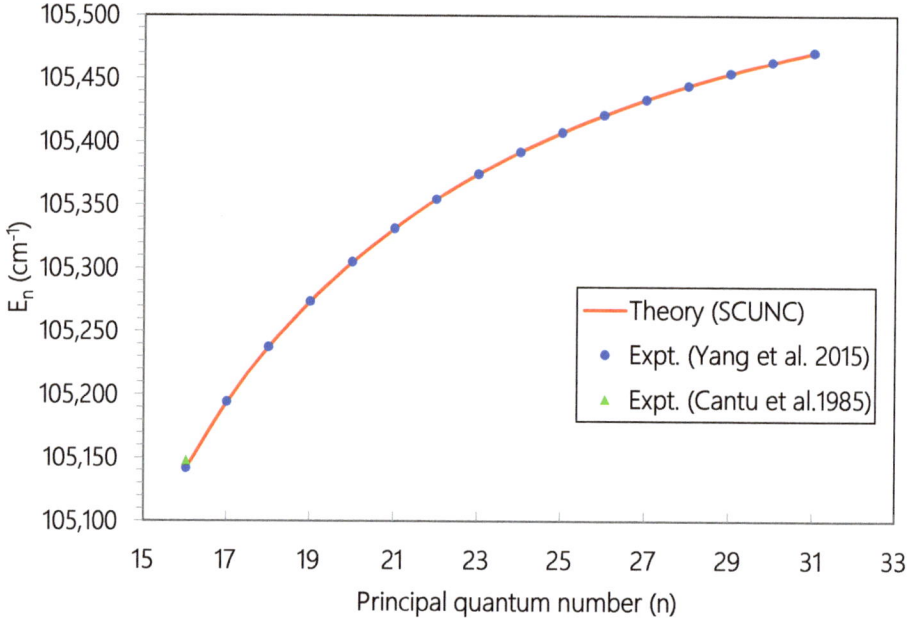

Figure 5. Plot of resonance energies (E_n, cm^{-1}) versus principal quantum number (n) for the $3s^23p^4$ ($^3P_0$)nd ($^2P_{3/2}$) Rydberg series of resonances originating from the Cl ($^2P^0_{3/2}$) ground state. Experimental data (solid blue circles Ref. [18] and solid green triangles Ref. [19]) and theoretical estimates (solid red line, SCUNC).

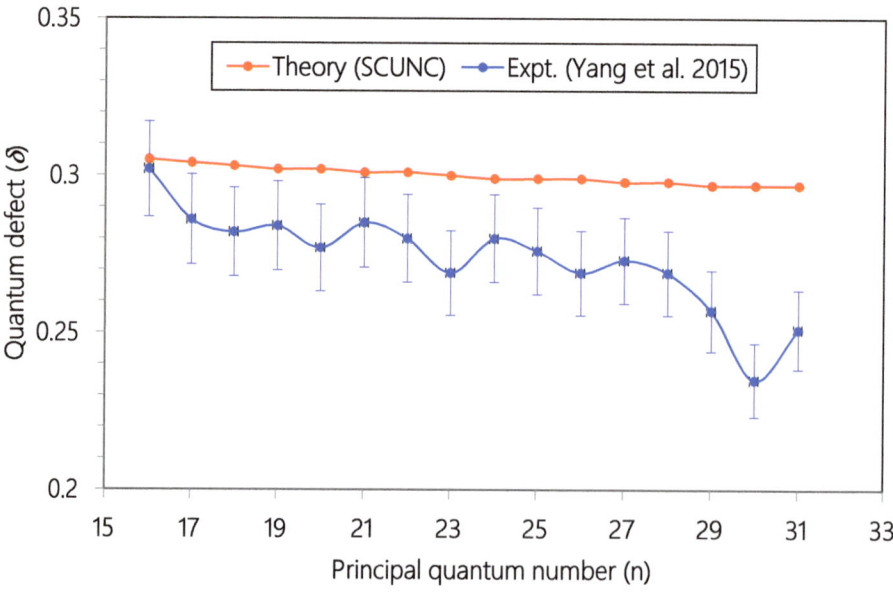

Figure 6. Plot of quantum defects (δ) versus principal quantum number (n) for the $3s^23p^4$ ($^3P_0$)nd ($^2P_{3/2}$) Rydberg series of resonances originating from the Cl ($^2P^0_{3/2}$) ground state. Experimental data (solid blue circles Ref. [18]) and theoretical estimates (solid red circles, SCUNC).

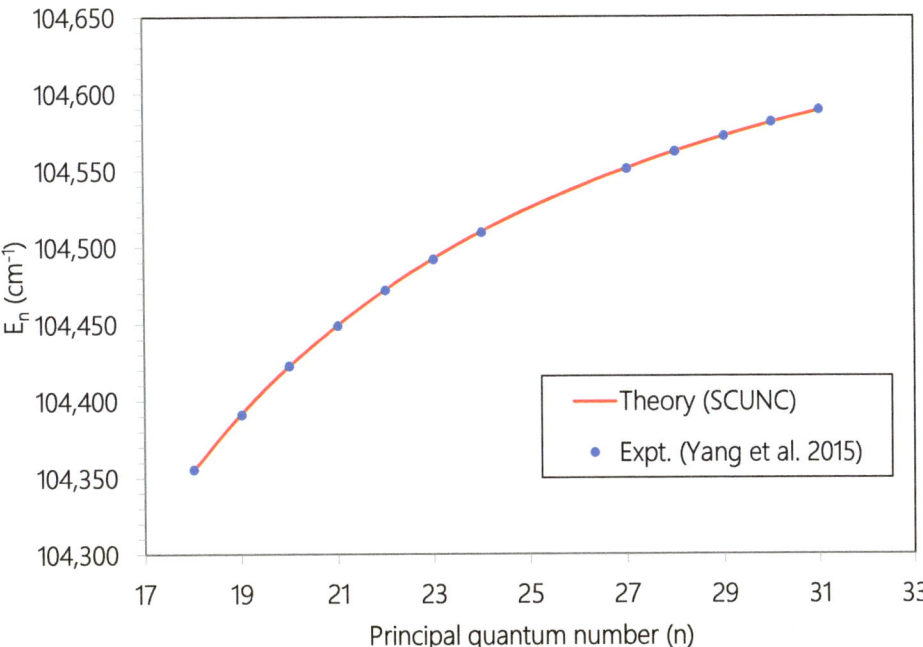

Figure 7. Plot of resonance energies (E_n, cm^{-1}) versus principal quantum number (n) for the $3s^23p^4$ ($^3P_0$)nd ($^2P_{3/2}$) Rydberg series of resonances originating from the Cl ($^2P^0_{1/2}$) excited state. Experimental data (solid blue circles Ref. [18]) and theoretical estimates (solid red line, SCUNC).

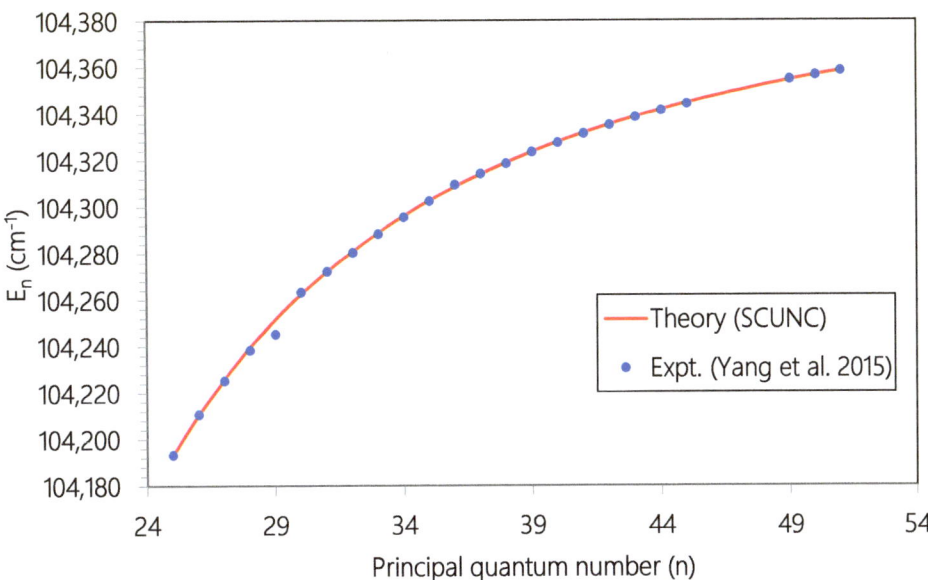

Figure 8. Plot of resonance energies (E_n, cm^{-1}) versus principal quantum number (n) for the $3s^23p^4$ ($^3P_1$)ns ($^2P_{3/2}$) Rydberg series of resonances originating from the Cl ($^2P^0_{1/2}$) excited state. Experimental data (solid blue circles Ref. [18]) and theoretical estimates (solid red line, SCUNC).

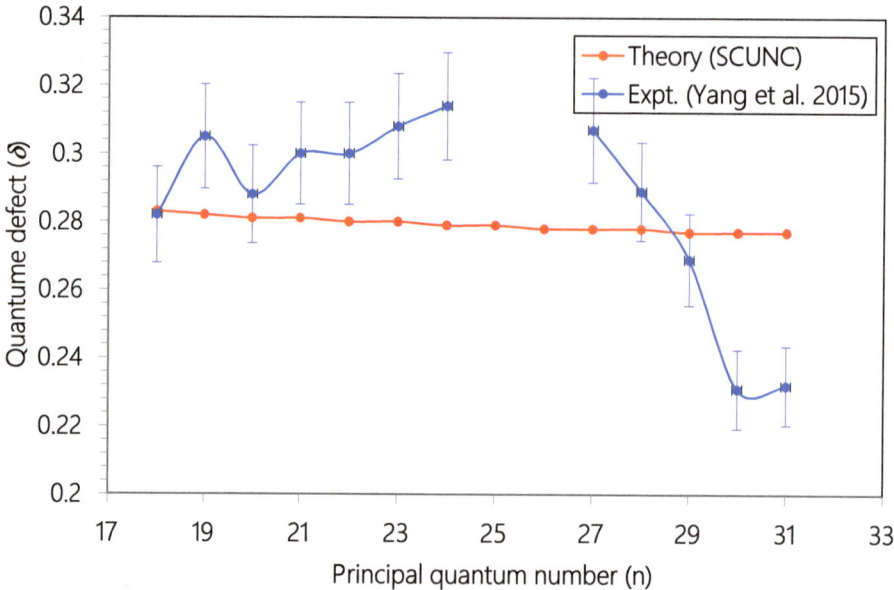

Figure 9. Plot of quantum defects (δ) versus principal quantum number (n) for the $3s^23p^4$ ($^3P_0$)nd ($^2P_{3/2}$) Rydberg series of resonances originating from the Cl ($^2P^0_{1/2}$) excited state. Experimental data (solid blue circles Ref. [18]) and theoretical estimates (solid red circles, SCUNC).

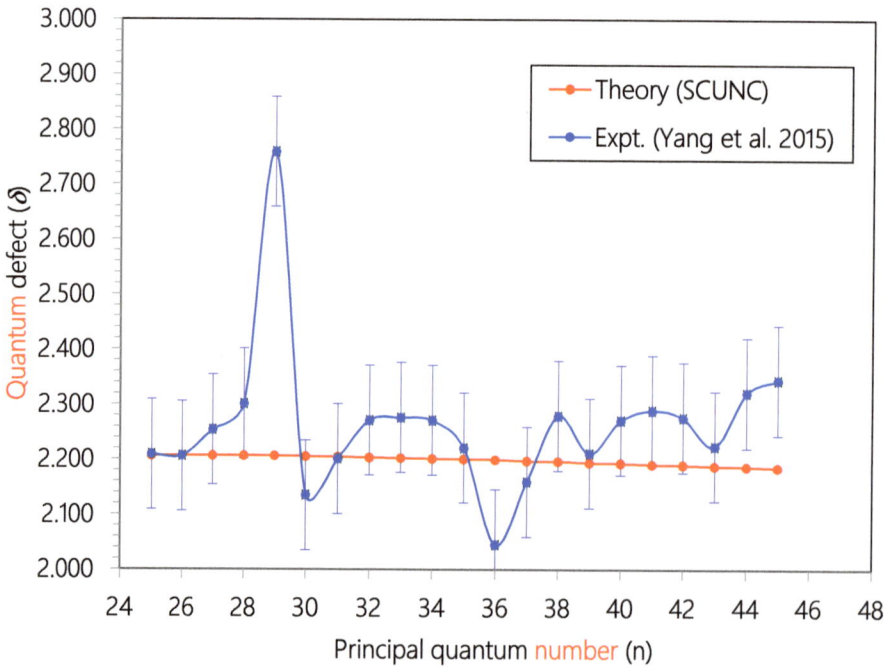

Figure 10. Plot of quantum defects (δ) versus principal quantum number (n) for the $3s^23p^4$ ($^3P_1$)ns ($^2P_{3/2}$) Rydberg series of resonances originating from the Cl ($^2P^0_{1/2}$) excited state. Experimental data (solid blue circles Ref. [18]) and theoretical estimates (solid red circles, SCUNC).

4. Summary and Conclusions

In this paper, the first calculations of resonance energies, quantum defects and effective charges of several Rydberg series resulting from the ejection of $3p$ electrons from the $^2P^0_{3/2}$ ground state and $^2P^0_{1/2}$ excited state of the neutral chlorine atom was carried out. Overall, very good agreements were obtained between the present SCUNC calculations and the available experimental data for resonance energies. In addition, for all the resonance energies associated with an experimental quantum defect that varied in all directions, an almost constant SCUNC quantum defect was tabulated up to $n = 70$. The new SCUNC data quoted in the listed tables may be of great interest for the physical community focusing their studies on the photoionization of atomic chlorine.

Author Contributions: M.T.G.: Conceptualization, Methodology, Investigation, Formal analysis, Writing—original draft, Writing—review and editing. I.S.: Methodology, Investigation, Writing—review and editing, Visualization, Formal analysis, Supervision. All authors have read and agreed to the published version of the manuscript.

Funding: This research received no funding.

Data Availability Statement: Data is contained within this article.

Conflicts of Interest: The authors declare that they have no known competing financial interest or personal relationship that could have appeared to influence the work reported in this paper.

References

1. McLaughlinand, B.M.; Balance, C.P. Photoionization Cross-Sections for the trans-iron element Se+ from 18 eV to 31 eV. *J. Phys. B At. Mol. Opt. Phys.* **2012**, *45*, 095202. Available online: https://iopscience.iop.org/article/10.1088/0953-4075/45/9/095202 (accessed on 11 September 2023). [CrossRef]
2. Bizau, J.M.; Champeaux, J.P.; Cubaynes, D.; Wuilleumier, F.J.; Folkmann, F.; Jacobsen, T.S.; Penent, F.; Blancard, C.; Kjeldsen, H. Absolute cross sections for L-shell photoionization of the ions N^{2+}, N^{3+}, O^{3+}, O^{4+}, F^{3+}, F^{4+} and Ne^{4+}. *Astron. Astrophys.* **2005**, *439*, 387–399. [CrossRef]
3. Nahar, S.N. Photoionization features of the ground and excited levels of Cl II and benchmarking with experiment. *New Astron.* **2021**, *82*, 101447. [CrossRef]
4. Dufour, R.J. The unique planetary nebula NGC 2818. *Astrophys. J.* **1984**, *287*, 341–352. Available online: https://adsabs.harvard.edu/full/1984ApJ...287..341D (accessed on 11 September 2023). [CrossRef]
5. Fegley, B.; Zolotov, M.Y. Chemistry of Sodium, Potassium, and Chlorine in Volcanic Gases on Io. *Icarus* **2000**, *148*, 193–210. [CrossRef]
6. Moomey, D.; Federman, S.R.; Sheffer, Y. Revisiting the chlorine abundance in diffuse interstellar clouds from measurements with the copernicus satellite. *Astrophys. J.* **2012**, *744*, 174. Available online: https://iopscience.iop.org/article/10.1088/0004-637X/744/2/174/meta (accessed on 11 September 2023). [CrossRef]
7. Robicheaux, F.; Green, C.H. Partial and differential photoionization cross sections of Cl and Br. *Phys. Rev. A* **1993**, *47*, 1066. Available online: https://www.physics.purdue.edu/~robichf/papers/pra47_1066.pdf (accessed on 11 September 2023). [CrossRef]
8. Lamoureux, M.; Farnoux, F.C. Photoionization cross section of the open 3p subshell in atomic chlorine. *J. Phys.* **1979**, *40*, 545. [CrossRef]
9. James, A.R.; Samson; Shefer, Y.; Angel, G.C. A Critical Test of Many-Body Theory: The Photoionization Cross Section of Cl as an Example of an Open-Shell Atom. *Phys. Rev. Lett.* **2020**, *56*, 2020. [CrossRef]
10. Schectman, R.M.; Federman, S.R.; Brun, M.; Cheng, S.; Fritts, M.C.; Irving, R.E.; Gibson, N.D. Oscillator Strengths for Ultraviolet Resonances in Cl II and Cl III. *Astrophys. J.* **2005**, *621*, 1159–1162. Available online: https://iopscience.iop.org/article/10.1086/427918/meta (accessed on 11 September 2023). [CrossRef]
11. Hernández, E.; Juárez, A.; Kilcoyne, A.; Aguilar, A.; Hernández, L.; Antillón, A.; Macaluso, D.; Morales-Mori, A.; González-Magaña, O.; Hanstorp, D.; et al. Absolute measurements of chlorine Cl+ cation single photoionization cross section. *J. Quant. Spectosc. Radiat. Transf.* **2015**, *151*, 217–223. [CrossRef]
12. Nahar, S.N.; Hernández, E.M.; Kilkoyne, D.; Antillón, A.; Covington, A.M.; González-Magaña, O.; Hernández, L.; Davis, V.; Calabrese, D.; Morales-Mori, A.; et al. Experimental and Theoretical Study of Photoionization of Cl III. *Atoms* **2023**, *11*, 28. [CrossRef]
13. Mosnier, J.P.; Kennedy, E.T.; Bizau, J.M.; Cubaynes, D.; Guilbaud, S.; Blancard, C.; Hasoğlu, M.F.; Gorczyca, T.W. L-Shell Photoionization of Magnesium-like Ions with New Results for Cl5+. *Atoms* **2023**, *11*, 66. [CrossRef]
14. McLaughlin, B.M. Photoionization of Cl+ from the $3s^23p^4\,^3P_{2,1,0}$ and the $3s^23p^4\,^1D_2,^1S_0$ states in the energy range 19–28 eV. *Mon. Not. R. Astron. Soc.* **2017**, *464*, 1990–1999. [CrossRef]

15. Kim, D.S.; Kwon, D.H. Theoretical photoionization spectra for Mg-isoelectronic Cl^{5+} and Ar^{6+} ions. *J. Phys. B At. Mol. Opt. Phys.* **2015**, *48*, 105004. Available online: https://iopscience.iop.org/article/10.1088/0953-4075/48/10/105004/meta (accessed on 11 September 2023). [CrossRef]
16. Ba, M.D.; Diallo, A.; Badiane, J.K.; Gning, M.T.; Sow, M.; Sakho, I. Photoionization of the $3s^23p$nd Rydberg series of Cl^+ ion using the Screening constant by unit nuclear charge method. *Radiat. Phys. Chem.* **2018**, *153*, 111–119. [CrossRef]
17. Nahar, S.N. Characteristic features in photoionization of Cl III. In Proceedings of the 54th Annual Meeting of The Aps Division of Atomic, Molecular And Optical Physics, Spokane, WA, USA, 5–9 June 2023. Available online: https://meetings.aps.org/Meeting/DAMOP23/Session/C11.1 (accessed on 11 September 2023).
18. Yang, L.; Gao, H.; Zhou, J.; Ng, C.Y. Vacuum ultraviolet laser photoion and pulsed field ionization–photoion study of Rydberg series of chlorine atoms prepared in the (j = 3/2 and 1/2) fine-structure states. *Astrophys. J.* **2015**, *810*, 132. Available online: https://iopscience.iop.org/article/10.1088/0004-637X/810/2/132 (accessed on 11 September 2023). [CrossRef]
19. Cantu, A.M.; Parkinson, W.H.; Grisendi, T.; Tagliaferri, G. Absorption Spectrum of Atomic Chlorine: $3p^5\,{}^2P^0_{3/2,1/2}\,3p^4$ ns,nd. *Phys. Scr.* **1985**, *31*, 579. Available online: https://iopscience.iop.org/article/10.1088/0031-8949/31/6/019 (accessed on 11 September 2023). [CrossRef]
20. Sakho, I. The Screening Constant by Unit Nuclear Charge Method. In *Description & Application to the Photoionization of Atomic Systems*; ISTE Science Publishing Ltd.: London, UK; John Wiley & Sons, Inc.: New York, NY, USA, 2018.
21. Gning, M.T.; Sakho, I. Single Photoionization Study of Br^{3+} via the Screening Constant by Unit Nuclear Charge Method. *J. At. Mol. Condens. Nano Phys.* **2019**, *6*, 131–141. [CrossRef]

Disclaimer/Publisher's Note: The statements, opinions and data contained in all publications are solely those of the individual author(s) and contributor(s) and not of MDPI and/or the editor(s). MDPI and/or the editor(s) disclaim responsibility for any injury to people or property resulting from any ideas, methods, instructions or products referred to in the content.

Article

Photoejection from Various Systems and Radiative-Rate Coefficients

Anand K. Bhatia

Heliophysics Science Division, NASA/Goddard Space Flight Center, Greenbelt, MD 20771, USA; anand.k.bhatia@nasa.gov

Abstract: Photoionization or photodetachment is an important process. It has applications in solar- and astrophysics. In addition to accurate wave function of the target, accurate continuum functions are required. There are various approaches, like exchange approximation, method of polarized orbitals, close-coupling approximation, R-matrix formulation, exterior complex scaling, the recent hybrid theory, etc., to calculate scattering functions. We describe some of them used in calculations of photodetachment or photoabsorption cross sections of ions and atoms. Comparisons of cross sections obtained using different approaches for the ejected electron are given. Furthermore, recombination rate coefficients are also important in solar- and astrophysics and they have been calculated at various electron temperatures using the Maxwell velocity distribution function. Approaches based on the method of polarized orbitals do not provide any resonance structure of photoabsorption cross sections, in spite of the fact that accurate results have been obtained away from the resonance region and in the resonance region by calculating continuum functions to calculate resonance widths using phase shifts in the Breit–Wigner formula for calculating resonance parameters. Accurate resonance parameters in the elastic cross sections have been obtained using the hybrid theory and they compare well with those obtained using the Feshbach formulation. We conclude that accurate results for photoabsorption cross sections can be obtained using the hybrid theory.

Keywords: scattering functions; photoabsorption; photoionization; radiative attachment; opacity

1. Introduction, Calculations and Results

In 1865, J. C. Maxwell proposed his theory of propagation of electromagnet waves. In 1887, experiments of H. Hertz confirmed his theory. His experiments also showed the existence of discrete energy levels and led to Einstein's photoelectric law [1]. In a photoelectric effect, photons behave like particles rather than waves, as also in the experiment on the scattering of X-rays by electrons by A. H. Compton [2]. This was also confirmed by the experiments of Bothe and Geiger [3]. Their experiments showed that the electron moved from its position in about 10^{-7} s. A wave would have taken much longer to move the electron. Photodetachment of negative hydrogen ions is given by

$$h\nu + H^- \to e + H \tag{1}$$

It was suggested by Wildt [4] that this process is an important source of opacity in the atmosphere of the Sun, in addition to processes like bound-bound transitions, free-free transitions, and Thomson scattering. The cross section, in units of Bohr radius, for this process in the length form and in the dipole approximation is given by (cf. Appendix A)

$$\sigma(a_0^2) = 4\alpha\pi k(I + k^2)\left|\left\langle \Psi_f \middle| z_1 + z_2 \middle| \Phi \right\rangle\right|^2 \tag{2}$$

In the above expression, $\alpha = 1/137.036$ is the fine structure constant, I is the ionization potential, $z_i = r_i \cos(\theta_i)$ are the dipole transition operators, and k is the momentum of the outgoing electron. Rydberg units are used in this article. The function Φ represents the bound state wave function of the hydrogen ion and Ψ_f is the wave function of the outgoing electron and the remaining hydrogen atom. Various approximations have been made for the scattering function.

The simplest approximation is the exchange approximation given by

$$\Psi(\vec{r}_1, \vec{r}_2) = u(\vec{r}_1)\phi(\vec{r}_2) \pm (1 \leftrightarrow 2) \tag{3}$$

In the above equation, $\vec{r}_1$ and $\vec{r}_2$ are the distances of the incident and bound electrons from the nucleus, assumed fixed, so that the recoil of the nucleus can be neglected, $u(\vec{r}_1)$ is the scattering function and $\phi(\vec{r}_2)$ is the target function. Exchange between similar particles is important. The plus sign refers to the singlet states and the minus sign refers to triplet states. In these equations

$$u(\vec{r}_1) = \frac{u(r_1)}{r_1} Y_{Lo}(\theta_1, \phi_1) \tag{4}$$

The angles θ_1 and ϕ_1 are the spherical polar angles, measured in radians. The ground state function is given by

$$\phi(\vec{r}_2) = 2e^{-r_2} Y_{00}(\theta_2, \phi_2) \tag{5}$$

The scattering function u of the incident particle is obtained from

$$\int Y_{lo}(\Omega_1)\phi_0(\vec{r}_2) \left| H - E \right| \Psi(\vec{r}_1, \vec{r}_2) d\Omega_1 d\vec{r}_2 = 0 \tag{6}$$

Morse and Allis [5] carried out the exchange approximation calculations in 1933. Assuming that the nucleus is of infinite mass, is fixed, and the recoil of the nucleus can be neglected, the Hamiltonian H and energy E (in Rydberg units) are given by

$$H = -\nabla_1^2 - \nabla_2^2 - \frac{2Z}{r_1} - \frac{2Z}{r_2} + \frac{2}{r_{12}} \tag{7}$$

$$E = -Z^2 + k^2 \tag{8}$$

Z is the nuclear change and k is the momentum of the incident particle. Using Equation (6), we get the equation for the scattering function

$$[\frac{d^2}{dr^2} - \frac{l(l+1)}{r^2} + v_d r_1 + k^2] u_l(r_1) \pm 4Z^2 [(Z^2 + k^2)\delta_{l0} r_1 V_l(r_1) - \frac{2}{2l+1} y_l(r_1)] = 0 \tag{9}$$

$$v_d(r) = \frac{2(Z-1)}{r} + 2e^{-Zr}(1 + \frac{1}{r}) \tag{10}$$

$$y_l(r) = \frac{1}{r^l} \int_0^r x^{l+1} \phi_0(x) u_l(x) dx + r^{l+1} \int_r^\infty \phi_0(x) \frac{u_l(x)}{x^l} dx \tag{11}$$

$$V_l = \int_0^\infty e^{-Zx} x u_l(x) dx \tag{12}$$

In Equation (9), δ_{lo} is the Dirac delta function. The scattering function behaves asymptotically like $\sin(kr - l\frac{\pi}{2} + \eta_l)$, where η_l is the phase shift for the incident electron of angular momentum l.

The target electron is distorted because of the electric field produced by the incident electron, resulting in a lowering of the energy by $\Delta E = -\frac{1}{2}\alpha E_e^2$, where $\alpha = 4.5 a_0^3$ is the polarizability of the hydrogen atom and E_e is the electric field produced by the incident electron. For a slowly moving incident electron, this distortion has been taken into account by the method of polarized orbitals of Temkin [6], assuming that the atom follows the instantaneous motion of the scattered electron. He proved that the target function, now polarized, for an incident electron at a distance r_1 is given by

$$\Phi^{pol}(\vec{r}_1, \vec{r}_2) = \phi_0(\vec{r}_2) - \frac{\varepsilon(r_1, r_2)}{r_1^2} u_{1s \to p}(\vec{r}_2) \quad (13)$$

$$u_{1s \to p}(\vec{r}_2) = \frac{\cos\theta_{12}}{(Z\pi)^{0.5}} e^{-Zr_2}\left(\frac{Z}{2}r_2^2 + r_2\right) \quad (14)$$

In Equation (13), $\varepsilon(r_1, r_2)$ is a step function which allows polarization of the target electron only when the incident electron is outside the orbit of the target because the step function is equal to 1 when r_1 is greater than r_2, zero otherwise. The integro-differential equation for the function $u(r_1)$ for all angular momenta has been given by Sloan [7]. The method of polarized orbitals has been used extensively for atoms as well as for molecules. However, this method is not variationally correct, and only the long-range correlations can be included.

The method has been modified in the hybrid theory [8] by replacing the step function $\varepsilon(r_1, r_2)$ by a cutoff function $\chi(r_1) = (1 - e^{-\beta r_1})^n$, where β and n are optimized to get the maximum phase shifts and now the polarization of the target takes place whether the incident electron is inside or outside the orbit of the target electron. Phase shifts have lower bounds, i.e., they are always below the exact phase shifts, but approaching the correct value as the number of short-range correlations is increased. Short-range correlations are also included by writing the wave function as

$$\Psi(\vec{r}_1, \vec{r}_2) = u(\vec{r}_1)\Phi^{pol}(\vec{r}_1, \vec{r}_2) + (1 \leftrightarrow 2) + \sum C_i \Phi_l^i(\vec{r}_1, \vec{r}_2) \quad (15)$$

The last term in the above equation representing correlation functions for any angular momentum l are of the Hylleraas type. The equation for the scattering function is now obtained from

$$\int d\Omega_1 d\vec{r}_2 Y_{lo}(\Omega_1) \Phi^{pol} \left| H - E \right| \Psi(\vec{r}_1, \vec{r}_2) = 0 \quad (16)$$

In the above equation, H is the Hamiltonian of the system and E is the energy.

The resulting equation is given in Ref. [8]. This formulation gives accurate phase shifts and resonance parameters of He atoms and Li$^+$ ions. The results compare well with those obtained using other approaches.

The initial state Φ in Equation (2) can be a (1s1s) ^1S state or (1s2s) $^{1,3}S$ states. This function Φ can be chosen of the Hylleraas form and is accurately known when calculating energy of the state by the Rayleigh–Ritz variational principle and the final state function can be calculated accurately using the hybrid theory or any other approach. Cross sections have been calculated using the Hylleraas functions with 364 terms for the initial state function, and when 35 short-range correlations are also included in the final state wave function, as indicated in Equation (15). These results are given in Table 1 of ref. [9] and are now given here in Table 1. We see that the inclusion of the short-range correlations does change the cross sections slightly. Bell and Kingston [10], using the method of polarized orbitals, also calculated these cross sections. Their results are also given in Table 1 along with the close-coupling results of Wishart [11], who used the close-coupling approximation for the continuum functions. We find that the results of ref. [10] differ from those calculated using the hybrid theory which provides accurate scattering functions.

Table 1. Photodetachment cross section (Mb) of H^-.

k	Cross Section without Short-Range Correlations	Cross Sections with Short-Range Correlations	Bell and Kingston, Ref. [10]	Wishart, Ref. [11]
0.01	0.0245	-	-	-
0.02	0.1959	-	-	-
0.03	0.6444	-	-	-
0.04	1.4736	1.4750	-	-
0.05	2.7480	2.7517	-	-
0.06	4.4914	4.4988	-	-
0.07	6.6844	-	-	-
0.1	15.2465	15.3024	12.34	15.937
0.2	38.3688	38.5443	40.48	37.870
0.23	39.4354	39.6366	-	38.707
0.24	39.2882	-	-	-
0.25	38.9121	39.1350	-	38.116
0.26	38.3850	-	-	-
0.3	34.9684	35.2318	36.40	34.829
0.4	24.2537	25.4709	25.296	23.858
0.5	15.8692	16.0858	16.43	15.720
0.6	10.4924	10.7410	11.29	10.431
0.7	7.1258	7.4862	-	7.101
0.74	6.1530	6.6072	-	6.139
0.8	4.9768	5.6512	5.31	4.978
0.8544	4.1421	4.1421	-	-
0.8631	4.0224	6.8976	-	-
0.8660	3.9846	7.623	-	-

We find that the maximum of the cross sections is at $k = 0.23$, which corresponds to a photon of wavelength 8406.3 Å, (using $\lambda = 911.267/\omega$ Å). As the momentum of the emitted electron, k, tends to zero, the photodetachment cross sections of the negative hydrogen ion tend to zero because the final state function is proportional to $j_1(kr)$ which goes to zero for as k tends to zero. Further, the cross section is directly proportional to k. Therefore, cross section is equal to zero at $k = 0.0$.

Ohmura and Ohmura [12], using the effective range theory and the loosely bound structure of hydrogen ion, obtained

$$\sigma = \frac{6.8475 \times 10^{-18} \gamma k^3}{(1 - \gamma \rho)(\gamma^2 + k^2)^3} \text{cm}^2 \tag{17}$$

In the above expression, $\gamma = 0.2355883$ is the square root of the binding energy and $\rho = 2.646 \pm 0.004$ is the effective range. The maximum of these cross sections occurs at $k = 0.236$ or at 8195 Å which is close to the maximum of the cross sections obtained using the hybrid theory. Miyska at el. [13], using the R-matrix approach, have obtained accurate results for the photodetachment of the negative H ion. However, their results are given in form of curves and it is difficult to get accurate results for comparison. The experimental results [14,15] are also given in the form of curves and it is difficult to get accurate results for comparison. However, they appear to be close to the present results. The maximum is around 8000 Å, which is close to 8406.3 Å obtained using the hybrid theory. The results obtained using the hybrid theory and those obtained using Equation (17) are given in Figure 1. We find that the two sets of results are very close to each other.

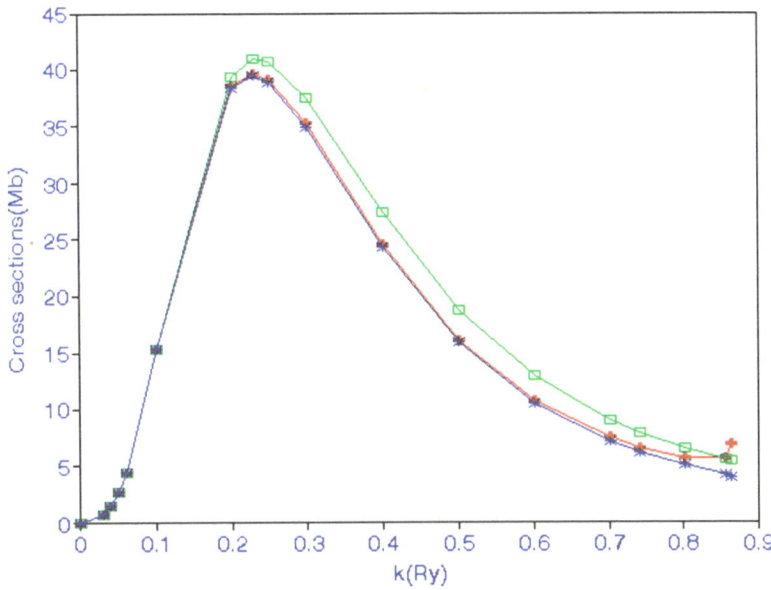

Figure 1. Photodetachment of a hydrogen negative ion. The lowest curve is obtained when only the long-range correlations in Equation (15) are included; the middle curve is obtained when the short-range and long-range correlations are included in Equation (15). The top curve is obtained for cross sections obtained by Ohmura and Ohmura using the effective range theory, Equation (17).

Similar calculations have been carried out for the photoionization of He and Li$^+$. The results have been compared with the results obtained in other calculations and also with the experimental results, the agreement is good. Cross sections for photoionization of He agree with those obtained using the R-matrix theory [16] and the experimental results [17,18]. These results are given in Table 2 and are also shown in Figure 2. All the three curves overlap. It is not possible to distinguish one curve from the other. This indicates that the hybrid theory gives results which are as accurate as those obtained using the R-matrix formulation, which is a very a versatile but is a very complicated theory. Photoabsorption in He played an important role in indicating the presence of resonances and in determining their positions and widths [19]. The line shape parameter q [20] is inversely proportional to the matrix element in the calculation of photoionization given in Equation (2). This parameter can be calculated accurately because the matrix element is known accurately. The matrix element appearing in the expression for q depends on the bound state wave function and the continuum function of the ejected electron in the photoionization cross section, and these functions can be obtained with very high accuracy.

It should be pointed out that these cross sections are finite as k goes to zero. In this case, the final continuum functions are Coulomb functions which behave like reciprocal of the square root of k. The k outside cancels with k inside the square of the matrix element giving finite cross sections as k goes to zero.

Yan et al. [21] using the accurate measurements at low energies and theoretical calculated results at high energies have calculated photoionization cross sections of He and H_2. Their interpolated results for He agree well with cross sections obtained using the hybrid theory given in Table 2. They have also calculated photoionization cross sections of H_2 as well as sum rules for He and H_2.

Table 2. Photoionization cross sections (Mb) for the ground state of He obtained with correlations.

k	Hybrid Theory [8]	R-Matrix [16]	Experiment Ref. [17]	Experiment Ref. [18]
0.1	7.3300	7.295	7.51	7.44
0.2	7.1544	7.115	7.28	7.13
0.3	6.8716	6.838	6.93	6.83
0.4	6.4951	6.474	6.49	6.46
0.5	6.0461	6.006	5.99	6.02
0.6	5.5925	5.535	5.46	5.55
0.7	5.0120	4.995	4.92	5.04
0.8	4.4740	4.482	4.38	4.51
0.9	3.9649	-	-	-
1.0	3.4654	3.476	3.38	3.48
1.1	3.0206	3.023	2.91	3.00
1.3	2.2561	2.271	2.17	2.19
1.4	1.9821	1.943	1.87	1.89
1.5	1.6817	-	-	-
1.6	1.6329	-	-	-

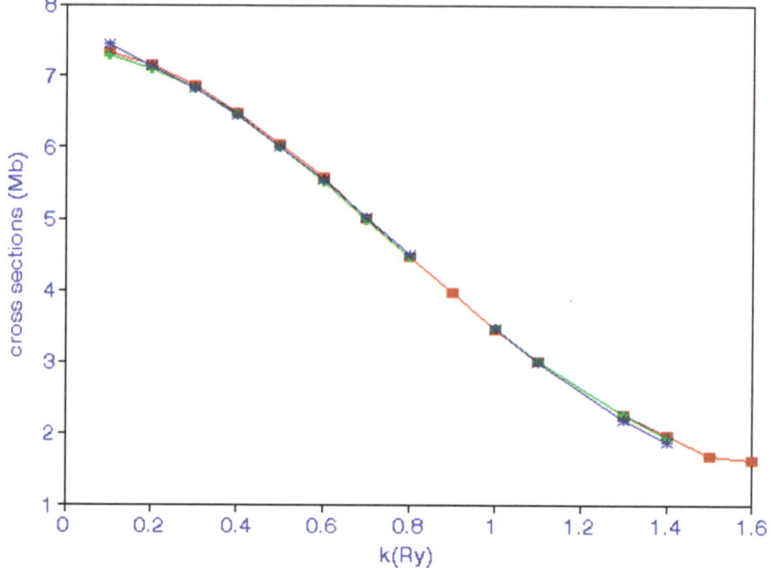

Figure 2. Photoionization of He. A comparison of cross sections calculated using the hybrid theory [8] and R-matrix approximation [16] with the experimental results [17,18] is shown.

Similar calculations [9] for photoionization of (1s2s) ^1S and ^3S states of He have been carried using the hybrid theory [8] with 455 terms in the bound state wave function. These results are shown in Table 3. The cross sections are compared with those obtained by Norcross [22], using the method of coupled equations for calculating the continuum functions. The results obtained using the hybrid theory are also compared with those of Jacobs [23], who also used pseudostates in the coupled equations.

Similar calculations [9] also have been carried for Li^+ using the hybrid theory with 165 terms for the ground state wave function. Calculations have also been carried out by the method of polarized orbits. The results obtained using the hybrid theory are given in Table 4 and they are compared with results of Bell and Kingston [10] and Daskhan and Ghosh [24]; the method of polarized orbitals [6] has been used in Refs. [10,24].

Table 3. Photoionization cross sections (Mb) for the metastable states of He.

	(1s2s) ^{1}S State of He		
k	Hybrid Theory [9]	Norcross [22]	Jacobs [23]
0.1	8.7724	8.973	-
0.2	7.5894	7.344	-
0.3	6.0523	5.885	-
0.4	4.5403	4.595	-
0.5	3.2766	3.467	3.260
0.6	2.2123	2.515	2.357
0.7	1.6047	1.725	1.661
0.8	1.1230	1.104	1.141
0.9	0.7863	0.647	0.771
1.0	0.5474	0.360	0.521
1.1	0.3796	0.240	0.364
1.3	0.1858	-	0.212
1.4	0.1279	-	0.162
1.5	0.07001	-	0.090
	(1s2s) ^{3}S state of He		
0.1	5.2629	4.749	-
0.2	5.0795	4.564	-
0.3	4.2004	4.112	-
0.4	3.4403	3.537	-
0.5	2.7189	2.912	-
0.6	2.1531	2.295	-
0.7	1.4564	1.733	-
0.8	1.3539	1.256	-
0.9	0.9728	0.885	-
1.0	0.6551	0.623	-
1.1	0.5577	0.463	-
1.2	0.3744	0.383	-
1.3	0.2898	0.347	-
1.5	0.2218	-	-

Table 4. Photoionization cross sections (Mb) of the ground state (1s1s) ^{1}S of Li$^+$.

k (Ry)	Hybrid Theory, Ref. [8]	Ref. [10]	Ref. [24]
1.6	1.1706	1.183	1.146
1.5	1.2768	1.297	1.248
1.4	1.3879	1.414	1.353
1.3	1.5035	1.533	1.459
1.2	1.6219	1.652	1.566
1.1	1.7396	1.770	1.674
1.0	1.8613	1.886	1.780
0.9	1.9792	1.998	1.885
0.8	2.0921	2.105	1.988
0.7	2.0005	2.206	2.087
0.6	2.2088	2.993	2.182
0.5	2.3870	2.384	2.271
0.4	2.4373	2.457	2.355
0.3	2.5231	2.520	2.432
0.2	2.5677	2.569	2.501

Photoionization cross sections of the metastable states of Li$^+$ ion have been calculated using the hybrid theory and 165 terms for the (1s2s) ^{1}S state and 120 terms for the (1s2s) ^{3}S state [9]. These results are shown in Table 5.

Table 5. Cross sections (Mb) for the metastable states of Li$^+$ ion.

K (Ry)	(1s2s) ^{1}S	(1s2s) ^{3}S
0.1	-	2.4456
0.2	2.5677	2.3780
0.3	2.5231	2.235
0.4	2.4622	2.0516
0.5	2.3869	1.8437
0.6	2.2998	1.6284
0.7	2.1999	1.4175
0.8	2.0925	1.2173
0.9	1.9789	1.0349
1.0	1.8605	0.8733
1.1	1.7414	0.7327
1.2	1.6219	0.6096
1.3	1.5037	0.5071
1.4	1.3886	0.4252
1.5	1.2777	0.3556
1.6	1.1716	0.2979
1.7	1.0712	0.2503
1.8	0.9770	0.2109
1.9	0.8892	0.1788
2.0	0.8079	0.1524
2.1	0.7332	0.1311
2.2	0.6649	0.1141
2.3	0.6034	0.1012
2.4	0.5488	0.0931
2.5	0.5025	0.0936

2. Radiative Attachment

Until this point, we have discussed the photodetachment process. However, the inverse process, namely, the radiative attachment is also possible. This process plays an important role in solar- and astrophysical problems. This is an important process, creating negative hydrogen ions which are important in understanding opacity of the solar system. The formation of the hydrogen molecule takes place through such processes:

$$e + H \rightarrow H^- + h\nu \quad (18)$$

$$H^- + H \rightarrow H_2 + e \quad (19)$$

Such recombination processes take place in the early Universe when the temperature of matter and radiation was close to a few thousand degrees. In Equations (18) and (19), H can be replaced by He$^+$ and Li^{2+} to form a He atom and Li$^+$ ion in the final state. The attachment cross section in terms of the photodetachment cross sections or photoionization cross section σ is given by

$$\sigma_a = \left(\frac{h\nu}{cp_e}\right)^2 \frac{g_f}{g_i} \sigma = \left(\frac{h\nu}{cp_e}\right)^2 \frac{1}{2mE} \frac{g_f}{g_i} \sigma \quad (20)$$

This relation follows from the principle of detailed balance. In Equation (20), $p_e = k$ is the incident electron momentum. The radiative-attachment cross sections are smaller than the photoabsorption cross sections. In Equation (20),

$$g_i = (2l_e + 1)(2S_e + 1)(2S_H + 1) = 3 \times 2 \times 2 = 12$$

and

$$g_f = (2 \times l_{h\nu} + 1)(2)(2S_{H^-} + 1) = 6(2S_{H^-} + 1)$$

The electron has an angular momentum = 1 = photon angular momentum, photon has two polarization directions, spin of the electron = 0.5 = spin of H, while the spin of the negative H^- = 0. Combining all these factors, we get g_i and g_f. These cross sections averaged over the Maxwellian velocity distribution $f(E)$ is given by

$$\alpha_R(T) = \langle \sigma_a v_e f(E) \rangle \tag{21}$$

The electron velocity is v_e, the recombination rate coefficient is given by

$$\alpha_R(T) = \left(\frac{2}{\pi}\right)^{0.5} \frac{c}{(mc^2 k_B T)^{1.5}} \frac{g_f}{g_i} \int dE (E+I)^2 \sigma e^{-E/k_B T} \tag{22}$$

$E = k^2$ is the energy of the electron in Equation (22), k_B is the Boltzmann constant, T is the electron temperature, and photon energy is $E = I + k^2$, where I is the threshold for photoabsorption. In Table 6, we give the recombination rates, averaged over the Maxwellian velocity distribution, at various temperature for the negative hydrogen ion, He, and Li$^+$. A comparison with R-matrix results is also given in Table 6.

Table 6. Recombination rate coefficients (cm^3/s) for (1s1s) state of H^-, He, and Li$^+$. A comparison with R-matrix results (interpolated) is also indicated.

T	$\alpha_R(T) \times 10^{15}$, H^-	$\alpha_R(T) \times 10^{13}$, He		$\alpha_R(T) \times 10^{11}$, Li$^+$
		Using Hybrid Theory Cross Sections Results	Using R-Matrix Cross Sections [25]	
-	-			-
1000	0.99	2.50	4.75	0.12
2000	1.28	2.30	3.43	1.04
5000	2.40	1.87	2.15	2.62
7000	2.82	1.66	1.79	2.92
10,000	3.20	1.45	1.48	3.03
12,000	3.37	1.35	1.36	3.02
15,000	3.56	1.23	1.22	2.95
17,000	3.65	1.17	1.19	2.89
20,000	3.75	1.10	1.08	2.79
22,000	3.79	1.05	1.04	2.73
25,000	3.83	0.99	0.99	2.63
30,000	3.83	0.92	0.93	2.49
35,000	3.77	0.87	0.89	2.36
40,000	3.67	0.82	0.86	2.25

The radiative rate coefficients for attachment to metastable states (1s2s) $^{1,3}S$ states of He and Li$^+$ are given in Table 7. A comparison of the results obtained using the hybrid theory with those obtained using the R-matrix formalism is also given in Table 7.

An extensive search to find the R-matrix calculations on recombination to Li ion failed to find any results. It seems such a calculation has not been carried out.

Nahar [25] has carried out R-matrix calculations of photoionization of the helium atom and recombination rate coefficients. Her photoionization results have been discussed above. The agreement between the cross sections obtained using the hybrid theory and R-matrix, along with the experimental results, is very good. The recombination rate coefficients to the ground state using the hybrid theory given in Table 6 agree with the results obtained using the R-matrix theory. The results for metastable states are indicated in Table 7. The agreement of the rate coefficients for the metastable states is quite good. This is surprising because the photoionization cross sections for metastable states obtained using the hybrid theory agree well with those obtained using the close-coupling approximation [22,23].

The reason that the method of polarized orbitals works well for atoms as well as for ions to provide accurate results for photoabsorption cross sections is the fact that the polarized target function depends on the nuclear charge Z only, as indicated in Equation (13).

Table 7. Recombination rate coefficients (cm^3/s) of the metastable states of He and Li$^+$, and comparison with the R-matrix results (interpolated).

	He				Li$^+$	
	^3S		^1S		^3S	^1S
	Hybrid	R-Matrix	Hybrid	R-Matrix	Hybrid	Hybrid
T	$\alpha_R(T) \times 10^{14}$		$\alpha_R(T) \times 10^{15}$		$\alpha_R(T) \times 10^{14}$	$\alpha_R(T) \times 10^{14}$
1000	2.13	4.33	8.27	17.13	4.68	2.99
2000	2.08	3.20	7.97	12.55	4.47	2.87
5000	1.71	2.05	7.30	7.85	3.48	2.37
7000	1.56	1.71	5.71	6.43	3.09	2.03
10,000	1.40	1.39	5.05	5.12	2.68	1.78
12,000	1.32	1.26	4.73	4.54	2.49	1.66
15,000	1.23	1.11	4.35	3.93	2.26	1.52
17,000	1.18	1.04	4.15	3.64	2.14	1.45
20,000	1.12	0.97	3.90	3.33	1.98	1.36
22,000	1.09	0.94	3.75	3.19	1.90	1.31
25,000	1.04	0.91	3.57	3.05	1.79	1.24
30,000	0.98	0.90	3.31	2.97	1.64	1.15
35,000	0.93	0.92	3.10	3.03	1.52	1.08
40,000	0.89	0.97	2.93	3.18	3.14	1.02

3. Photoejection with Excitation

Up to now we have considered photoabsorption when the remaining atom or ion is left in the ground state. However, it is possible to leave the remaining atom or ion in an excited state [26]. For example, in the photoionization process

$$h\nu + He \rightarrow He^+ + e \quad (23)$$

In the above equation, ionized helium can be in the excited 2^2S or 2^2P states. When the remaining atom or ion is in the 2^2P state, there is a possibility of emission of Lyman-α radiation of 304 Å, as in the photodetachment of the negative hydrogen ion or photoabsorption. The outgoing photoelectron can be in the angular momentum $l_f = 0$ or 2 when the resulting state is in ^2P state and $l_f = 1$ when the resultant state is ^2S state. Similar processes can take place when the targets are H^-, Li^+, Be^{2+}, and C^{4+}. The cross section in the dipole approximation is given by

$$= \frac{4\pi\alpha k\omega}{3(2l_i + 1)}(|M_0|^2 + |M_2|^2) \quad (24)$$

for ^2P states and

$$= \frac{4\pi\alpha\omega}{3(2l_i + 1)}|M_1|^2 \quad (25)$$

for ^2S states. The matrix M is defined as

$$M_{l_f} = (2l_f + 1)^{0.5}|\langle \Psi_f|z_1 + z_2|\Phi_i\rangle| \quad (26)$$

The continuum functions are calculated in the exchange approximation (cf. Equation (5)). Table 8 gives the ratios $R_1 = \sigma(2^2S)/\sigma(1^2S)$ and $R_2 = [\sigma(2^2S) + \sigma(2^2P)]/\sigma(1^2S)$ for the negative hydrogen ion. He and Li^{2+}. Komninos and Nicolaides [27] have calculated ratios of leaving the He ion in 2p and 2s states using the K-matrix theory. Jacobs and Burke [28] have also calculated these ratios using the close-coupling approximation. The agreement with the results of [27,28] is not good because the continuum functions are calculated in the exchange approximation, instead of using the hybrid theory. However, the calculations are much easier than using other approximations. The simpler approach might not give definitive results but is helpful in understanding various processes.

Table 8. Ratios R_1 and R_2.

k	R_1	R_2
-	H^-	
0.1	1.1760 (−4)	8.1873 (−2)
0.2	3.9946 (−4)	4.4460 (−2)
0.3	2.2783 (−3)	4.9332 (−2)
0.4	1.3944 (−2)	6.1779 (−2)
0.5	6.4958 (−2)	1.0498 (−1)
0.6	1.6743 (−1)	2.0920 (−1)
0.7	2.6673 (−1)	3.2210 (−1)
0.8	3.3490 (−1)	3.9742 (−1)
-	He	
0.1	8.0146 (−3)	1.2794 (−2)
0.2	8.6667 (−3)	1.3318 (−2)
0.3	9.6384 (−3)	1.4112 (−1)
0.4	1.0774 (−2)	1.5065 (−1)
0.5	1.1935 (−2)	1.6097 (−2)
0.6	1.3075 (−2)	1.7206 (−2)
0.7	1.4252 (−2)	1.8497 (−2)
0.8	1.5551 (−2)	2.0062 (−2)
-	Li^+	
0.2	4.9346 (−3)	6.4426 (−3)
0.3	4.9889 (−3)	6.5041 (−3)
0.4	5.0594 (−3)	6.5849 (−3)
0.5	5.1422 (−3)	6.6814 (−2)
0.6	4.0317 (−3)	5.2310 (−3)
0.7	5.3781 (−3)	6.9693 (−3)
0.8	5.5344 (−3)	7.1657 (−3)
0.9	5.7203 (−3)	7.4081 (−3)
1.0	5.9438 (−3)	7.6979 (−3)
1.1	6.2085 (−3)	8.0436 (−3)
1.2	6.5170 (−3)	8.4476 (−3)
1.3	6.8761 (−3)	8.9126 (−3)
1.4	7.2914 (−3)	9.4507 (−3)
1.5	7.7664 (−3)	1.0062 (−2)
1.6	8.3090 (−3)	1.0753 (−2)

The radiative rate coefficients averaged over the Maxwellian velocity distribution are given in Table 9.

Table 9. Recombination rate coefficients (cm^3/s) to (1s1s) ^1S state from ^2S and ^2P states.

T	$\alpha_R \times 10^{16}$, H^-	$\alpha_R \times 10^{16}$, He	$\alpha_R \times 10^{16}$, H^-	$\alpha_R \times 10^{16}$, He
-	Final State is ^2S		Final State is ^2P	
2000	4.36 (−3)	19.3	0.75	33.7
4000	2.86 (−2)	16.8	1.00	25.0
5000	5.83 (−2)	15.9	1.10	22.1
7000	1.73 (−1)	14.5	1.24	17.9
10,000	5.07 (−1)	13.2	1.35	14.0
12,000	8.30 (−1)	12.6	1.38	12.2
15,000	1.43	11.8	1.39	10.2
17,000	1.93	11.4	1.38	9.25
20,000	2.59	10.9	1.36	8.07

Table 9. Cont.

T	$\alpha_R \times 10^{16}$, H^-	$\alpha_R \times 10^{16}$, He	$\alpha_R \times 10^{16}$, H^-	$\alpha_R \times 10^{16}$, He
-	Final State is ^2S		Final State is ^2P	
22,000	3.17	10.7	1.34	7.43
25,000	3.89	10.3	1.30	6.64
30,000	4.97	9.98	1.24	5.64
35,000	5.86	9.55	1.18	4.89
40,000	6.55	9.28	1.12	4.31
45,000	7.06	9.06	1.06	3.86
50,000	7.43	8.88	1.00	3.48
60,000	7.81	8.62	0.904	2.91
70,000	7.88	8.42	0.818	2.50
80,000	7.76	8.26	0.742	2.18
90,000	7.53	8.11	0.676	1.93
100,000	7.25	7.97	0.619	1.73
200,000	4.48	6.38	0.307	0.82
300,000	2.96	4.94	0.189	0.511

4. Photoionization of Lithium and Sodium

Until this point, we discussed photoabsorption where the continuum functions were calculated using hybrid theory. However, we have some calculations mentioned above where the method of polarized orbitals was used. There are other calculations like photoionization of lithium [29] where the method of polarized orbitals was used. Cross sections for this process are given in Table 10.

Table 10. Photoionization cross sections (10^{-18} cm^2).

k	Cross Section	K	Cross Section
0.10	1.601	0.80	0.905
0.20	1.672	0.90	0.724
0.30	1.709	1.00	0.574
0.35	1.697	1.20	0.355
0.40	1.660	1.40	0.219
0.50	1.521	1.60	0.136
0.54	1.427	1.80	0.087
0.60	1.324	1.90	0.069
0.70	1.110	-	-

Similar calculations [30] have been carried out for the photoionization of sodium atoms. These results are given in Table 11 and are compared with the close-coupling results of Butler and Mendoza [31]. Cross sections at low energies agree fairly well with those obtained using the close-coupling approximation.

Table 11. Photoionization cross sections (10^{-20} cm^2) of Na atoms.

k	Pol. Orb. Approx. [30]	Close-Coupling Approx. [31]
0.0	8.419	8.496
0.1	5.748	5.390
0.2	1.228	1.142
0.3	0.319	0.264
0.4	3.536	3.513
0.5	7.832	7.773
0.6	11.111	11.416
0.7	12.627	13.411
0.8	12.815	13.740
0.9	12.157	12.957
1.0	11.047	12.102

Table 11. Cont.

k	Pol. Orb. Approx. [30]	Close-Coupling Approx. [31]
1.1	9.934	12.102
1.2	8.602	-
1.3	7.714	-
1.4	6.893	-
1.5	6.134	-
1.6	5.361	-
1.7	4.843	-
1.8	4.392	-
1.9	3.905	-
2.0	3.507	-

5. Photodetachment of $^3P^e$ State of Negative Hydrogen Ion

It is well known that the negative hydrogen ion has only one bound state. However, there is another bound triplet $(2p2p)P$ state of even parity, which is not well known. The bound state function has been calculated using Hylleraas type functions [32], where the energy of the state is given as $-0.2506536415\ Rm$, and Rm is the reduced Rydberg. This photodetachment process is similar to photoejection mentioned above in Equation (22). The continuum functions were calculated using the 1s-2s-2p close coupling approximation [33]. Cross sections are given in Table 12.

Table 12. Cross sections (cm^2) for the photodetachment of the $^3P^e$ state of a negative hydrogen ion.

Photon Energy (Ry)	Final State Is (2s)	Final State Is (2p)
0.002	8.74 (−17)	3.99 (−16)
0.004	6.67 (−17)	4.01 (−16)
0.006	4.61 (−17)	3.03 (−16)
0.008	3.52 (−17)	2.43 (−16)
0.010	2.90 (−17)	2.18 (−16)
0.014	2.16 (−17)	1.99 (−16)
0.018	1.69 (−17)	1.76 (−16)
0.022	1.38 (−17)	1.50 (−16)
0.026	1.16 (−17)	1.27 (−16)
0.030	1.01 (−17)	1.10 (−16)
0.040	7.93 (−18)	8.33 (−17)
0.050	6.81 (−18)	6.79 (−17)
0.150	2.64 (−18)	1.69 (−17)
0.250	8.63 (−18)	6.37 (−18)

Radiative attachment

$$e + H(2s) \rightarrow H(2\,^3P^e) + h\nu \qquad (27)$$

$$e + H(2p) \rightarrow H(2\,^3P^e) + h\nu \qquad (28)$$

These two processes are important sources of infrared emission. The radiative attachment cross sections are given in Ref. [30].

6. Photodetachment of Negative Positronum Ion

Photodetachment of the negative positronium ion is very much like the photodetachment of a negative hydrogen ion. It is indicated by

$$h\nu + Ps^- \rightarrow e^- + Ps \qquad (29)$$

This process also contributes to the opacity of the Sun and the stellar atmosphere. The binding energy of the positronium ion has been calculated by Bhatia and Drachman [34]. Following Ohmura and Ohmura [12], we write the wave function of the positronium ion as

$$\Phi(R_j, r_k) = C \frac{e^{-\gamma R_j}}{R_j} \phi(r_k) \qquad (30)$$

The constant C is determined using the exact wave function given in Ref. [34], where $\frac{3\gamma^2}{2}$ = binding energy = 0.024010113. We use plane waves for the scattering function [35]. The expression for the cross section is given in Equation (31). The cross section is the same for the process obtained by the charge conjugation of the process indicated in Equation (29). The cross section [35] is given by

$$\sigma = 1.32 \times 10^{-18} \frac{k^3}{(k^2 + \gamma^2)^3} \text{cm}^2 \qquad (31)$$

In Table 13, cross sections for the photodetachment are given, and they are also indicated in Figure 3.

Table 13. Photodetachment. Cross sections (cm^2).

Photon Energy (Ry)	Cross Section
0.26	1.58 (−17)
0.22	1.98 (−17)
0.20	2.41 (−17)
0.16	2.97 (−17)
0.13	3.81 (−17)
0.12	4.17 (−17)
0.11	4.59 (−17)
0.10	5.08 (−17)
0.08	6.27 (−17)
0.07	6.97 (−17)
0.065	7.33 (−17)
0.060	7.67 (−17)
0.05	8.13 (−17)
0.04	7.66 (−17)
0.03	4.16 (−17)

We can use the Thomas–Reiche–Kuhn sum rule to judge the accuracy of our calculation. The sum rule is given by

$$S_{-1} = \frac{1}{2\pi^2 \alpha a_0^2} \int_0^{\lambda_0} \frac{d\lambda}{\lambda} \sigma(\lambda) = \frac{8}{27} \langle (\vec{r}_1 + \vec{r}_2)^2 \rangle = \frac{8}{27} (4\langle r_1^2 \rangle - \langle r_{12}^2 \rangle) \qquad (32)$$

In the above equation, λ_0 is the threshold wave length for the photodetachment of the negative positronium ion. The expectation values of $\langle r_1^2 \rangle$ and $\langle r_{12}^2 \rangle$ have been calculated using the exact wave function of the positronium ion [34]. The left-hand side of Equation (32) is equal to 31.7 and the right-hand side is equal to 29.775. This shows that our cross section using the approximate wave functions exceed by 6.5%. This is confirmed by Ward et al. [36], who have carried out accurate calculations using accurate initial state wave function having 95 linear parameters of Ps$^-$ and continuum functions were obtained using the Kohn variational principle with 220 linear parameters. Their results for the cross sections are lower (cf. figure in their paper) than those obtained in Ref. [35].

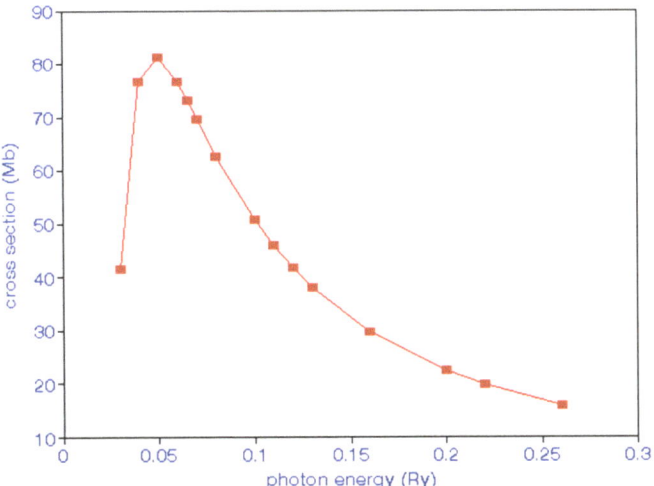

Figure 3. (Color online). Photodetachment cross sections (Mb) of Ps$^-$ vs. photon energy (Ry).

This calculation [35] has been extended to the photodetachment of the positronim ion when the positronium atom is left in nP states, n = 2, 3, 4, 5, 6, and 7 [37]. The ^2P state can decay into ^1S state which would correspond to Lyman-α Ps-radiation, just like 1216 Å radiation which has been observed from the center of galaxy [38], where it is due to a transition from ^2P to ^1S in a hydrogen atom. The photodetachment cross sections to various excited states are given below:

$$\sigma(2p) = 164.492 C(k)$$
$$\sigma(3p) = 26.3782 C(k)$$
$$\sigma(4p) = 9.1664 C(k)$$
$$\sigma(5p) = 4.3038 C(k)$$
$$\sigma(6p) = 2.3764 C(k)$$
$$\sigma(7p) = 0.2675 C(k)$$

where,

$$C(k) = \frac{10^{-20} k}{(\gamma^2 + k^2)} \text{cm}^2 \qquad (33)$$

Similar calculations can be carried out for leaving the positronium atom in ns states, n = 2 to 7. The transition from the 2s state to 1s state would be with the emission of 2 photons just like that in the case of a hydrogen atom.

We have discussed photoionization and photoabsorption for various systems using the expression for the cross section given in Equation (2). The derivation of this formula in Equation (2) is given in [39] (repeated in the Appendix A). In this article, photoionization cross sections of the (1s1s ^1S), (1s2s $^{1,3}S$), (1s3s ^3S) states of Be^{2+}, C^{4+}, and O^{6+}, along with radiative recombination rate coefficients at various electron temperatures, are given. Fitting formulae for photoionization cross sections are also given in [39].

Until this point, we have mostly mentioned two-electron systems and we have given cross sections using the exchange approximation, method of polarized orbitals, plane-wave approximation, R-matrix formulation, and hybrid theory. There are other calculations like coupled cluster study of photoionization by Tenoril et al. [40]. They use an asymptotic Lanczos algorithm to calculated photoionization and photodetachment cross sections of of He and give results in the form of a curve. It is difficult to get meaningful results for a comparison. However, they do give the sum rule S(0) = 1.999 for He, which is close to the exact value equal to 2, the number of electrons in the He atom, indicating the accuracy of their calculation. The exterior complex scaling has been used by Andric et al. [41] to calculate photoionization cross section of positive HCl ion. Measurements of photodetachment cross sections of Li$^-$, Be$^-$, and B$^-$ have been carried out using interacting beams by Pegg [42]. Photoionization cross sections of excited states of CO, N_2, and H_2O have been calculated by Ruberti et al. [43] using the many-electron Green's function approach. In a simple system like a hydrogen atom, Broad and Reinhardt [44] used L^2 basis to calculate photoionization of a hydrogen atom in the energy range 1.002 to 3.50 Rydberg; their results are given in Table 14. Their results appear close to those given by Joachain [45] in his book and also to the result 0.225 a_0^2 at the threshold. These cross sections obtained using L^2 basis are higher than those obtained using the R-matrix approach. Perhaps, there is possibility of improving the L^2 basis approach. It is very important to try other approximations in addition to the R-matrix approach.

Table 14. Photoionization cross sections (a_0^2) of a hydrogen atom.

Photon Energy (Ry)	Cross Section	R-Matrix Cross Section [a]
1.002	0.4478	-
1.500	0.1494	0.0747
2.000	0.0666	0.03327
2.500	0.0350	0.01438
3.000	0.0206	0.01045
3.500	0.0130	0.00654

[a] These cross sections calculated by S. N. Nahar are given in NORAD Atomic-Data.

Very recently, Nahar [46] has carried out very detailed and accurate calculations of photoionization cross sections and electron-ion recombination of n = 1 to very high n values of hydrogenic ions. Since hydrogen is very abundant in the universe, the results of this calculation are of immense importance in applications to solar- and astrophysics.

Paul and Ho [47] have calculated cross sections of H in the presence of Debye potential, while Kar and Ho [48] have calculated cross sections of the hydrogen negative ion in the Debye potential. Sahoo and Ho [49] have also calculated photoionization cross sections of Li and Na in the presence of the Debye potential. They find that in the presence of a Debye potential, the maximum of the photodetachment cross section of the negative ions moves to higher wave lengths as the Debye length decreases, making the plasma dense. The plasma is least dense when the Debye length is infinite.

7. Opacity

Opacity implies the loss of photons as in the photoabsorption indicated in Equations (1) and (29). We know the photodetachment cross sections of the negative hydrogen ion and of the positronium ion, we can compare their contributions to the opacity of the atmosphere of the Sun and the interstellar medium provided we include the free-free transitions:

$$h\nu + e + H \to e + H \tag{34}$$

$$h\nu + e + Ps \to e + Ps \tag{35}$$

In these processes the electron with energy k_0^2 absorbs an energy $h\nu$ of a photon in the initial state and the final energy of the electron is k^2, the change in energy is $h\nu = \Delta k^2 = |k_0^2 - k^2|$. The same kind of processes take place when the electrons in Equations (34) and (35) are replaced by positrons. The formula for free-free transitions has been given by Chandrasekhar and Breen [50]. Positrons had not been considered earlier and they contribute substantially, as indicated in Table 15, where a few values of cross sections are given at T = 6300 K.

Table 15. Comparison of bound-free (σ_{bf}) and free-free (σ_{ff}) cross sections (cm^2) for electrons and positrons, T = 6300 K.

$h\nu$ (Ry)	Electrons			Positrons		
-	σ_{bf}	σ_{ff}	$\sigma_{bf}+\sigma_{ff}$	σ_{bf}	σ_{ff}	$\sigma_{bf}+\sigma_{ff}$
0.26	2.26 (−17)	4.28 (−20)	2.27 (−17)	8.61 (−18)	4.14 (−21)	8.61 (−18)
0.24	2.47 (−17)	4.94 (−20)	2.47 (−17)	9.58 (−18)	4.74 (−21)	9.58 (−18)
0.22	2.70 (−17)	5.81 (−20)	2.70 (−17)	1.08 (−17)	3.49 (−21)	1.08 (−17)
0.20	2.96 (−17)	6.95 (−20)	2.96 (−17)	1.22 (−17)	6.45 (−21)	1.22 (−17)
0.18	3.25 (−17)	8.53 (−20)	3.25 (−17)	1.39 (−17)	7.71 (−2 1)	1.39 (−17)
0.16	3.56 (−17)	1.07 (−19)	3.57 (−17)	1.62 (−17)	9.49 (−21)	1.62 (−17)
0.14	3.87 (−17)	1.39 (−19)	3.88 (−17)	1.90 (−17)	1.19 (−20)	1.90 (−17)
0.12	4.15 (−17)	1.88 (−19)	4.17 (−17)	3.17 (−17)	1.56 (−20)	3.17 (−17)
0.10	4.13 (−17)	2.69 (−19)	4.16 (−17)	4.17 (−17)	2.15 (−20)	4.17 (−17)
0.08	3.50 (−17)	4.20 (−19)	3.35 (17)	3.42 (−17)	3.20 (−20)	3.42 (−17)
0.06	7.05 (−18)	7.45 (−19)	7.80 (−18)	8.96 (−17)	5.38 (−20)	8.97 (−17)
0.04	0.0 [a]	1.68 (−18)	1.68 (−18)	1.65 (−16)	1.13 (−19)	1.65 (−16)
0.03	0.00	2.99 (−18)	2.99 (−18)	2.53 (−16)	1.96 (−19)	2.53 (−16)
0.02	0.00	6.74 (−18)	6.74 (−18)	4.64 (−16)	4.30 (−19)	4.64 (−16)
0.01	0.00	2.70 (−17)	2.70 (−17)	1.30 (−15)	1.68 (−15)	1.30 (−15)
0.005	0.00	1.08 (−16)	1.08 (−16)	3.63 (−15)	6.72 (−15)	3.64 (−15)
0.003	0.00	3.00 (−16)	3.00 (−16)	7.69 (−15)	1.87 (−17)	7.71 (−15)
0.001	0.00	2.70 (−15)	2.70 (−15)	3.55 (−14)	1.68 (−16)	3.57 (−14)

[a] Photon energy less than 0.055 (Ry) is not sufficient for photodetachment.

In Figure 4, we give the total (detachment plus free-free) electron and positron cross sections. We find that the positron contribution is substantial and should be taken into account in the opacity calculations.

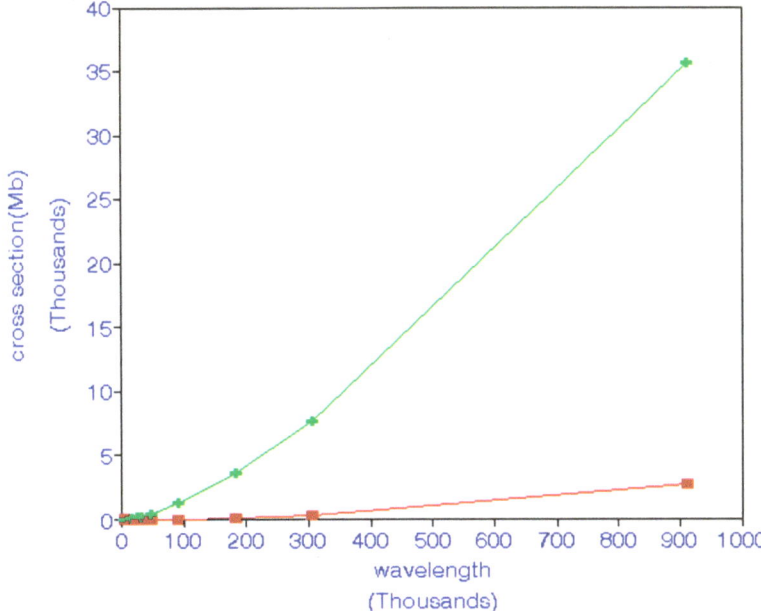

Figure 4. (Color online). The top curve refers to positron cross sections (Mb), while the lower curve refers to electron cross sections (Mb). Wavelengths are in units of Å.

Similarly, photoionization of other atoms and ions and free-free transitions contribute to opacity. However, hydrogen is the most abundant atom in the universe compared to other atoms and ions, whose concentrations decrease as Z, the nuclear charge, increases.

8. Conclusions

Here we described the photoabsorption process of two-electron systems for which scattering functions are required. There are various approaches to calculate continuum functions. We also described calculations in which we have used the exchange approximation, method of polarized orbitals, hybrid theory, close-coupling approximation, and R-matrix formalism. Further, we have mentioned photoabsorption cross sections of various molecules and sum rules. Cross sections were calculated using the coupled cluster formalism, which uses the asymptotic Lanczos algorithm, complex exterior scaling, and L^2 basis. These methods are briefly mentioned. Cross sections were compared with those obtained in other calculations and also with the experimental results. The photodetachment cross sections of the negative positronium ion, Ps^-, were calculated using the Ohmura and Ohmura approximation for the bound state and the plane-wave approximation for the final continuum state wave function. For a long time, it was thought that only processes involving electrons contribute to the opacity of the Sun and the interstellar medium. However, positrons do exist in many regions of the Sun and the interstellar medium [51]. Therefore, it is necessary to consider positrons in calculations of opacity. We have indicated in Table 15 that not only electrons but positrons also contribute substantially to the opacity of the Sun and of the interstellar medium.

Since the observation of the photoionization process in metals by Lenard [52] in 1902, there have been many experiments to observe photoionization in atoms, ions, and molecules and theoretical developments to calculate cross sections for this process. We have mentioned a few of the experiments and theoretical approaches.

Funding: This research work received no funding.

Institutional Review Board Statement: Not applicable.

Informed Consent Statement: Not applicable.

Data Availability Statement: There is no external data for this article.

Conflicts of Interest: The author declares no conflict of interest.

Appendix A

We briefly derive the photoionization formula for the hydrogen atom.
Photoionization is given by

$$h\nu + H \to H^+ + e \tag{A1}$$

The interaction Hamiltonian is given by

$$H' = -\frac{e}{2mc}(\vec{A} \cdot \vec{p} + \vec{p} \cdot \vec{A}) = -\frac{e}{mc}\vec{A} \cdot \vec{p} \tag{A2}$$

In the above equation, $\vec{A}$ is the vector potential and $\vec{p}$ is the electron momentum. The vector potential satisfies the condition $div\vec{A} = 0$ and it is represented by

$$\vec{A} = A_0 \vec{\varepsilon} e^{i\vec{k}\cdot\vec{r}} \tag{A3}$$

In the above equation $\vec{k}$ is the photon momentum which in magnitude is less than the radius of the atom, which implies that the exponential factor can be taken as equal to 1.0. This is called the dipole approximation and $\vec{\varepsilon}$ is the polarization direction, perpendicular to the z-axis, the incident photon direction. There are two polarization directions. However, for the derivation we need to consider only one of them and we can consider the polarization in the direction of the x-axis. The density of states is given by

$$\rho(k) = \frac{mk}{(2\pi)^3 \hbar^2} \sin(\theta) d\theta d\varphi \tag{A4}$$

Here θ is the angle between $\vec{k}$ and the photon direction. The incident flux is given by

$$\frac{\omega^2 A_0^2}{2\pi c(\hbar\omega)} = \frac{\omega A_0^2}{2\pi \hbar c} \tag{A5}$$

The electron momentum satisfies the commutation relation

$$\langle \vec{p} \rangle = m\frac{d\vec{r}}{dt} = m\langle \frac{1}{i\hbar}[H, \vec{r}] \rangle = \frac{mE}{i\hbar}\hbar\omega\langle \vec{r} \rangle \tag{A6}$$

Therefore,

$$\langle H\prime \rangle = -\frac{e}{mc}\langle \vec{A} \cdot \vec{p} \rangle = -\frac{e\omega A_0}{ic}\langle \vec{\varepsilon} \cdot \vec{r} \rangle \tag{A7}$$

The transition probability is given by

$$\omega_p = \frac{2\pi}{\hbar}\rho(k)|\langle H\prime \rangle|^2 = \frac{2\pi}{\hbar}\rho(k)\frac{e^2\omega^2 A_0^2}{c^2}|\langle \vec{\varepsilon} \cdot \vec{r} \rangle|^2 \tag{A8}$$

Therefore, the differential cross section is given by

$$\sigma(\theta, \varphi)\sin(\theta)d\theta d\varphi = \frac{\omega_p}{Flux} \tag{A9}$$

This gives

$$\sigma(\theta, \varphi) = \frac{2\pi}{\hbar} \frac{mk}{(2\pi)^3 \hbar^2} \frac{e^2\omega^2}{c^2} \frac{A_0^2 |\langle \vec{\varepsilon} \cdot \vec{r} \rangle|^2 (2\pi\hbar c)}{\omega A_0^2} \tag{A10}$$

Since

$$\vec{\varepsilon} \cdot \vec{r} = (\vec{\varepsilon} \cdot \hat{k})(\hat{k} \cdot \vec{r}) = \sin(\theta)\cos(\varphi)(\hat{k} \cdot \vec{r}) \tag{A11}$$

$$\sigma(\theta, \varphi)\sin(\theta)d\theta d\varphi = \frac{mke^2\hbar\omega}{2\pi\hbar^3 c}|\langle \hat{k} \cdot \vec{r} \rangle|^2 \sin^3(\theta)\cos^2(\varphi)d\theta d\varphi \tag{A12}$$

Using $\int \sin^3(\theta)\cos(\varphi)^2 d\theta d\varphi = \frac{4\pi}{3}$, we get the total cross section

$$\sigma = \frac{2mk}{\hbar^2}\frac{e^2}{\hbar c}\hbar\omega\frac{|\langle \hat{k} \cdot \vec{r} \rangle|^2}{3} \tag{A13}$$

where, $\frac{2m}{\hbar^2} = \frac{1}{1Ry \cdot a_0^2}$ and $\frac{e^2}{\hbar c} = \alpha$, the fine-structure constant, we get, using $\hbar\omega = I + k^2$

$$\sigma = k\alpha(I + k^2)|\langle \hat{k} \cdot \vec{r} \rangle|^2/3 \tag{A14}$$

The cross section given above is in the units of a_0^2, I and k^2 are in Ry units.

Since the normalization is a plane-wave normalization, the scattering function has a normalization

$$(4\pi(2l_f + 1))^{0.5} \tag{A15}$$

Here $l_f = 1$ and

$$\hat{k} \cdot \vec{r} = r\cos(\theta_1) = z \tag{A16}$$

We get

$$\sigma(a_0^2) = 4\pi\alpha k(I+k^2)\langle z\rangle^2 \quad \text{(A17)}$$

Since there is no φ dependence because the scattering functions and bound state functions are functions of angle θ_1, x and y components do not contribute because, in Equation (4), only the magnetic quantum $m = 0$ is being considered, x corresponds to $m = 1$ and y corresponds to $m = -1$. We get

$$\sigma(a_0^2) = 4\pi\alpha k(I+k^2)\langle z\rangle^2 \quad \text{(A18)}$$

This formula can be generalized to more than one electron, as indicated in Equation (2). We have considered polarization direction in x-axis only. We could also add to (A11) the polarization in the y-direction but then we would have to average the result.

References

1. Hoffman, B. *The Strange Story of the Quantum*; Dover Publications: Mineola, NY, USA, 1947; p. 13.
2. Compton, A.H. A Quantum Theory of the Scattering of X-rays by Light Elements. *Phys. Rev.* **1923**, *21*, 483–502. [CrossRef]
3. Bothe, W.; Geiger, H. Uber das Wesen des Compton effects ein experimenteller Breitrag Theorie der Strahlung. *Z. Phys.* **1923**, *21*, 152.
4. Wildt, R. Electron Affinity in Astrophysics. *Astrophys. J.* **1939**, *89*, 295–301. [CrossRef]
5. Morse, P.M.; Allis, W.P. The Effect of Exchange on the Scattering of Slow Electrons from Atoms. *Phys. Rev.* **1933**, *44*, 269–276. [CrossRef]
6. Temkin, A. A Note on the Scattering of Electrons from Atomic Hydrogen. *Phys. Rev.* **1959**, *116*, 358–363. [CrossRef]
7. Sloan, I.H. The method of polarized orbitals for the elastic scattering of slow electrons by ionized helium and atomic hydrogen. *Proc. R. Soc. Lond. Ser. A Math. Phys. Sci.* **1961**, *281*, 151–163.
8. Bhatia, A.K. Hybrid theory of electron-hydrogen elastic scattering. *Phys. Rev. A* **2007**, *75*, 032713. [CrossRef]
9. Bhatia, A.K. Hybrid theory of P-wave electron-Li^{2+} elastic scattering and photoabsorption in two-electron systems. *Phys. Rev. A* **2013**, *87*, 042705. [CrossRef]
10. Bell, K.L.; Kingston, A.E. Photoionization of Li^+. *Proc. Phys. Soc.* **1967**, *90*, 337–342. [CrossRef]
11. Wishart, A.W. The bound-free photodetachment of H^-. *J. Phys. B* **1979**, *12*, 3511. [CrossRef]
12. Ohmura, T.; Ohmura, H. Electron-Hydrogen Scattering at Low Energies. *Phys. Rev.* **1960**, *118*, 154–157. [CrossRef]
13. Miyake, S.; Stancil, P.C.; Sadeghpour, H.R.; Dalgarno, A.; McLaughlin, B.M.; Forrey, R.C. Resonant H^- Photodetachment Enhanced Photodestruction and Consequences for Radiative Feedback. *Astrophys. J. Lett.* **2010**, *709*, L168–L171. [CrossRef]
14. Branscomb, L.M.; Smith, S.J. Experimental Cross Sections for Photodetachment of Electrons from H^- and D^-. *Phys. Rev.* **1955**, *98*, 1028. [CrossRef]
15. Smith, S.J.; Burch, D.S. Relative Measurements of the Photodetachment for H^-. *Phys. Rev.* **1959**, *116*, 1125. [CrossRef]
16. Nahar, S.N. The Ultraviolet Properties of Evolved Stellar Population. In *New Quests in Stellar Astrophysics II*; Chavez, M., Bertone, E., Rosa-Gonzalez, D., Rodriguez-Merino, L.H., Eds.; Springer: New York, NY, USA, 2011; p. 245.
17. West, J.B.; Marr, G.V. The absolute photoionization of helium, neon, argon and krypton I the extreme vacuum ultraviolet region of the spectrum. *Proc. R. Soc. Lond. Ser. A Math. Phys. Sci.* **1976**, *349*, 397–421.
18. Samson, J.A.R.; He, Z.X.; Yin, L.; Haddad, G.N. Precision measurements of the absolute photoionization cross of He. *J. Phys. B* **1994**, *27*, 887. [CrossRef]
19. Madden, R.P.; Codling, K. New Autoionizing Atomic Energy Levels in He, Ne, and Ar. *Phys. Rev. Lett.* **1963**, *10*, 516–518. [CrossRef]
20. Bhatia, A.K.; Temkin, A. Calculation of autoionization of He and H^- using the projection-operator formalism. *Phys. Rev. A* **1975**, *11*, 2018–2024. [CrossRef]
21. Yan, M.; Sadeghpour, H.R.; Dalgarno, A. Photionization Cross Sections of He and H_2. *Astrophys. J.* **1998**, *496*, 1044. [CrossRef]
22. Norcross, D.W. Photoionization of the He metastable states. *J. Phys. B* **1971**, *4*, 652–657. [CrossRef]
23. Jacobs, V. Low-Energy Photoionization of the ^{11}S and ^{21}S states of He. *Phys. Rev. A* **1971**, *3*, 289–298. [CrossRef]
24. Daskhan, M.; Ghosh, A.S. Photoionization of He and Li^+. *Phys. Rev. A* **1984**, *29*, 2251–2254. [CrossRef]
25. Nahar, S.N. Photoionization and electron-ion recombination of He I. *New Astron.* **2010**, *15*, 417–426. [CrossRef]
26. Bhatia, A.K.; Drachman, R.J. Photoejection with excitation in H^- and other systems. *Phys. Rev. A* **2015**, *91*, 012702. [CrossRef]
27. Bhatia, A.K.; Temkin, A.; Silver, A. Photoionization of lithium. *Phys. Rev. A* **1975**, *12*, 2044–2051. [CrossRef]
28. Komninos, Y.; Nicolaides, C.A. Many-electron approach to atomic photoionization: Rydberg series of resonances and partial photoionization cross sections in helium, around the $n = 2$ threshold. *Phys. Rev. A* **1986**, *34*, 1995–2000. [CrossRef] [PubMed]
29. Jacobs, V.L.; Burke, P.G. Photoionization of He above the $n = 2$ threshold. *J. Phys. B* **1972**, *5*, L67. [CrossRef]
30. Dasgupta, A.; Bhatia, A.K. Photoionization of sodium atoms and electron scattering from ionized sodium. *Phys. Rev. A* **1985**, *31*, 759–771. [CrossRef]

31. Butler, K.; Mendoza, C. Accuracy of excitation and ionization cross sections. *J. Phys. B* **1983**, *16*, L707. [CrossRef]
32. Bhatia, A.K. Transitions (1s2p) $^3P^o$–$(2p^2)$ $^3P^e$ in He and (2s2p) $^3P^o$–$(2p^2)$ $^3P^e$ in H^-. *Phys. Rev. A* **1970**, *2*, 1667. [CrossRef]
33. Jacobs, V.L.; Bhatia, A.K.; Temkin, A. Photodetachment and Radiative Attachment involving the $2p^2\,^3P^e$ state of H^-. *Astrophys. J.* **1980**, *242*, 1278. [CrossRef]
34. Bhatia, A.K.; Drachman, R.J. New calculation of the properties of the positronium ion. *Phys. Rev. A* **1983**, *28*, 2523–2525. [CrossRef]
35. Bhatia, A.K.; Drachman, R.J. Photodetachment of the positronium negative ion. *Phys. Rev. A* **1985**, *32*, 3745–3747. [CrossRef] [PubMed]
36. Ward, S.J.; Humberston, J.W.; McDowell, M.R.C. Elastic scattering of electrons from positronium and photodetachment of positronium negative ion. *J. Phys. B* **1987**, *20*, 127. [CrossRef]
37. Bhatia, A.K. Photodetachment of the Positronium Negative Ion with Excitation in the Positronium Atom. *Atoms* **2019**, *7*, 2. [CrossRef]
38. Iallement, R.; Quemerais, E.; Bartex, J.-L.; Sandel, B.R.; Izamodenov, V. Voyager measurements of Hydrogen Lyman-α Diffuse Emission from the milky Way. *Science* **2011**, *334*, 1665–1669. [CrossRef] [PubMed]
39. Bhatia, A.K. P-wave electron-Be^{3+}, C^{5+}, and O^{7+} elastic scattering and photoabsorption in two-electron systems. *J. At. Mol. Condens. Nano Phys.* **2014**, *1*, 71–86. [CrossRef]
40. Tenorio, B.N.C.; Nascimento, M.A.C.; Coriani, S.; Rocha, A.B. Coupled Cluster Study of Photoionization and Photodetachment Cross Sections. *J. Chem. Theory Comput.* **2016**, *12*, 4440–4459. [CrossRef]
41. Andric, L.; Baccarelli, I.; Grozdanov, T.; McCarroll, R. Calculations of photodissociation cross sections by the smooth exterior complex scaling method. *Phys. Lett. A* **2002**, *298*, 41–48. [CrossRef]
42. Pegg, D.J. Interacting beam measurements of photodetachment. *Nucl. Instrum. Methods Phys. Res. B Beam Interact. Mater. Atoms* **1995**, *99*, 140. [CrossRef]
43. Ruberti, M.; Yun, R.; Gokhberg, K.; Kopelke, S.; Cederbaum, L.S.; Tarantelli, F.; Averbukh, V. Total photoionization cross-sections of excited electronic states by the algebraic diagrammatic construction-Stieltjes-Lanczos method. *J. Chem. Phys.* **2014**, *140*, 184107. [CrossRef] [PubMed]
44. Broad, J.T.; Reinhardt, W.P. Calculation of photoionization cross sections using L^2 basis sets. *J. Chem. Phys.* **1974**, *60*, 2182–2183. [CrossRef]
45. Joachain, C.J. *Quantum Scattering Theory*; North-Holland Publishing Company: New York, NY, USA, 1975; p. 654.
46. Nahar, S.N. Photoionization and Electron-Ion Recombination of $n = 1$ to Very High n-Values of Hydrogenic Ions. *Atoms* **2021**, *9*, 73. [CrossRef]
47. Paul, S.; Ho, Y.K. Hydrogen Atom in Debye Plasma. *Phys. Plasmas* **2009**, *16*, 06302. [CrossRef]
48. Kar, S.; Ho, Y.K. Photodetachment of the hydrogen negative ion in weakly coupled plasmas. *Phys. Plasmas* **2008**, *15*, 13301. [CrossRef]
49. Sahoo, S.; Ho, Y.K. Photoionization of Li and Na in Debye plasma environments. *Phys. Plasmas* **2006**, *13*, 63301. [CrossRef]
50. Chandrasekhar, S.; Breen, S. On the continuous absorption coefficient of the negative hydrogen ion. *Astrophys. J.* **1946**, *104*, 430. [CrossRef]
51. Gopalswamy, N. Positron Processes in the Sun. *Atoms* **2020**, *8*, 14. [CrossRef]
52. Lenard, P. Uber die lichtelektrische Wirkung. *Ann. Phys.* **1902**, *8*, 149. [CrossRef]

Article

Theoretical Spectra of Lanthanides for Kilonovae Events: Ho I-III, Er I-IV, Tm I-V, Yb I-VI, Lu I-VII

Sultana N. Nahar

Department of Astronomy, The Ohio State University, Columbus, OH 43210, USA; nahar.1@osu.edu; Tel.: +1-614-292-1888

Abstract: The broad emission bump in the electromagnetic spectra observed following the detection of gravitational waves created during the kilonova event of the merging of two neutron stars in August 2017, named GW170817, has been linked to the heavy elements of lanthanides (Z = 57–71) and a new understanding of the creation of heavy elements in the r-process. The initial spectral emission bump has a wavelength range of 3000–7000 Å, thus covering the region of ultraviolet (UV) to optical (O) wavelengths, and is similar to those seen for lanthanides. Most lanthanides have a large number of closely lying energy levels, which introduce extensive sets of radiative transitions that often form broad regions of lines of significant strength. The current study explores these broad features through the photoabsorption spectroscopy of 25 lanthanide ions, Ho I-III, Er I-IV, Tm I-V, Yb I-VI, and Lu I-VII. With excitation only to a few orbitals beyond the ground configurations, we find that most of these ions cover a large number of bound levels with open $4f$ orbitals and produce tens to hundreds of thousands of lines that may form one or multiple broad features in the X-ray to UV, O, and infrared (IR) regions. The spectra of 25 ions are presented, indicating the presence, shapes, and wavelength regions of these features. The accuracy of the atomic data used to interpret the merger spectra is an ongoing problem. The present study aims at providing improved atomic data for the energies and transition parameters obtained using relativistic Breit–Pauli approximation implemented in the atomic structure code SUPERSTRUCTURE and predicting possible features. The present data have been benchmarked with available experimental data for the energies, transition parameters, and Ho II spectrum. The study finds that a number of ions under the present study are possible contributors to the emission bump of GW170817. All atomic data will be made available online in the NORAD-Atomic-Data database.

Keywords: atomic data; energies; transition parameters; photo-excitation cross-sections; photoabsorption spectra; lanthanide ions - Ho I-III; Er I-IV; Tm I-V; Yb I-VI; Lu I-VII; broad emission bumps

Citation: Nahar, S.N. Theoretical Spectra of Lanthanides for Kilonovae Events: Ho I-III, Er I-IV, Tm I-V, Yb I-VI, Lu I-VII. *Atoms* **2024**, *12*, 24. https://doi.org/10.3390/atoms12040024

Academic Editor: Eugene T. Kennedy

Received: 22 November 2023
Revised: 13 March 2024
Accepted: 18 March 2024
Published: 17 April 2024

Copyright: © 2024 by the author. Licensee MDPI, Basel, Switzerland. This article is an open access article distributed under the terms and conditions of the Creative Commons Attribution (CC BY) license (https://creativecommons.org/licenses/by/4.0/).

1. Introduction

In 2017, the LIGO and VIRGO collaborations detected the first gravitational waves generated by the merging of two neutron stars, GW170817. This was followed by the detection of electromagnetic waves, the spectrum of which showed similarity to those of the heavy elements of lanthanides (Z = 57–71) [1,2]. The observed spectrum exhibited a broad feature or an emission bump at a wavelength range of 3000–7000 Å, covering the ultraviolet (UV) to optical (O) wavelength region. The feature moved towards the infrared (IR) region during the 10 days of observation, 18–27 August 2017, indicating the effect of opacity, i.e., the absorption of the traveling radiation by the plasma medium and re-emission with some loss of energy. The detection of the electromagnetic waves has provided new scope for an understanding of how heavy elements are formed through the r-process. Heavy elements are known to be formed by neutron capture in the s(slow)-process inside a star or by a r-(rapid)-process during a supernova explosion. The spectrum gives evidence of a new means of the creation of heavy elements during the merging of two neutron stars or two black holes, or a combination of the two. The interpretation of the emission bump of GW170817 will have a considerable impact in broadening our knowledge for a more complete picture of the characteristic atomic features and of the creation of elements. Since this finding, the need for atomic data for lanthanides and other heavy elements that can be used to interpret and provide information has increased. Over the next decade, it is expected that hundreds of mergers will be detected with the full network of current and upcoming gravitational wave detectors and electromagnetic telescopes.

Lanthanides are heavy elements with a large number of electrons ($Z = 57$–71). They have the core ion configuration of Xe, $[1s^2 2s^2 2p^6 3s^2 3p^6 4s^2 3d^{10} 4p^2 5s^2 5p^6 4d^{10}]$. In the ground state, the outer electron of a lanthanide can be in 4f, 6s, or 5d orbitals. The configurations of lanthanides can be described as $[Xe] 4f^i 6s^j 6p^k 5d^l$, where i, j, k, l are various occupancy numbers. These configurations introduce extensive numbers of radiative transitions that can form broad absorption features. Although lanthanides have been under study for a long time, the focus has been both academic and industrial, largely because of the luminescence properties and intense narrow-band emission, which have a large range of applications, such as in optical amplifiers, active waveguides, and fluorescent tubes.

Since the detection of the electromagnetic spectra, studies have been carried out to interpret the broad feature and identify the heavy elements that created it. Such study includes other elements, in addition to lanthanides, that may fall within the wavelengths of the bump. However, the accuracy in the large sets of atomic data for these elements has been a longstanding problem. Theoretically, the computation of lanthanide opacities is a formidable atomic physics problem, since these atoms and ions have a large number of electrons, causing complex electron–electron correlations and relativistic effects, and open 4d- and 4f-orbitals introduce large numbers of fine structure energy levels resulting from a large Hamiltonian matrix. Among the existing past and recent calculations carried out for the atomic data, we can note the work of Kasen et al. [3], who used the Breit–Pauli intermediate coupling code AUTOSTRUCTURE, which was created from and hence has the same atomic structure methodology as SUPERSTRUCTURE. Tanaka and Hotokezaka [4] and Tanaka et al. [5] used the relativistic HULLAC code with parametric potential, and Tanaka et al. [6] and Radžiūtė et al. [7] used the Dirac–Fock code GRASP2. Fontes et al. [8] used the semi-empirical Dirac–Fock–Slater code of Cowan [9]. Using the atomic data as well as the observed spectrum, a new periodic table with the origin of the creation of elements was produced by Johnson [10]. Kobayashi et al. [11] produced another table using theoretical and observational models, which differed somewhat from that of Johnson [10]. Kobayashi et al. also stated, "Although our calculations provide opacities of a wide range of r-process elements, the detailed spectral features in the model cannot be compared with the observed spectra because our atomic data . . . do not have enough accuracy in the transition wavelengths".

Among the experimental work, the energy levels of lanthanides were measured by Martin et al. [12]. These values are listed in the NIST [13] compilation table. Carlson et al. [14] computed ionization threshold energies for the ground configuration of lanthanides, which are quoted in the NIST [13] table and are particularly useful when measured data are not available.

Individual transition probabilities, theoretical and experimental, are available for a limited number of transitions. They have been evaluated and compiled by NIST [13]. For the five lanthanides studied here, these values were obtained mainly by Meggers et al. [15], Morton [16], Komarovski [17], Wickliffe and Lawler [18], Sugar et al. [19], Penkin and Komarovski [20], and Fedchak et al. [21]. Recently, Irvine et al. [22] used calibration-free laser-induced breakdown spectroscopy (CF-LIBS) to calculate 967 transition probabilities of 13 earth elements in a plasma environment. Obaid et al. [23] measured the spectrum of photo-fragmented Ho and found a broad feature.

We present very large sets of atomic data for the 25 ions of five lanthanides, Ho, Er, Tm, Yb, and Lu, and compare them with the measured energies, transition probabilities, and Ho spectra. We present the spectral features of all 25 lanthanide ions plotted over a wide range of wavelengths.

2. Theory

Distinct lines in a spectrum are generated mainly by radiative transitions for photo-excitation by absorption or de-excitation by the emission of photons. The present study is carried out for these transitions in lanthanides. For an element X with charge Z, the process is expressed as

$$X^{+z} + h\nu \rightleftharpoons X^{+z*} \qquad (1)$$

The transitions can be of several types depending on the selection rules, such as dipole allowed (E1), with the same and different spins of the initial and final states, or forbidden for lower magnitudes than those of E1; these follow different selection rules. The selection rules are determined by the angular part of the probability integral of the transition per unit time, P_{ij}, between two levels, i and j (e.g., [24]).

$$P_{ij} = 2\pi \frac{c^2}{h^2 \nu_{ji}^2} |<j| \frac{e}{mc} \hat{\mathbf{e}} \cdot \mathbf{p} e^{i\mathbf{k}\cdot\mathbf{r}} |i>|^2 \rho(\nu_{ji}). \qquad (2)$$

where $\mathbf{p}$ and $\mathbf{k}$ are the momenta of the electron and the photon, respectively; ν_{ij} is the frequency of the photon; and ρ is the density of the radiation field. Various terms in $e^{i\mathbf{k}\cdot\mathbf{r}}$ yield various multipole transitions. The first term gives the electric dipole transitions E1, the second term the electric quadrupole E2 and magnetic dipole M1, the third term the electric octupole E3 and magnetic quadrupole M2 transition, etc. Various transition parameters, such as the line strength (S), oscillator strength (f), and radiative decay rate (A), can be derived from the probability P_{ij}.

The transition matrix element with the first term of the expansion of Equation (2), for the E1 transition, can be written as $<\Psi_{B_j}||\mathbf{D}||\Psi_{B_i}>$, where $\mathbf{D} = \sum_i \mathbf{r}_i$ is the dipole operator and i indicates the number of electrons in the ion; Ψ_{B_i} and Ψ_{B_j} are the initial and final bound levels. The generalized line strength can be reduced as

$$S = \left| \left\langle \Psi_f | \sum_{j=1}^{N+1} r_j | \Psi_i \right\rangle \right|^2 \qquad (3)$$

The oscillator strength (f_{ij}) and radiative decay rate (A_{ji}) for the bound-bound transition are obtained from S as

$$f_{ij} = \left[\frac{E_{ji}}{3g_i} \right] S, \; A_{ji}(sec^{-1}) = \left[0.8032 \times 10^{10} \frac{E_{ji}^3}{3g_j} \right] S \qquad (4)$$

where g_i and g_j are the statistical weight factors; E_{ij} is the transition energy in Ry. The photoabsorption cross-section can be obtained as

$$\sigma(\nu) = 8.064 \frac{E_{ij}}{3g_i} S \; [Mb] \qquad (5)$$

Equation (5) is similar to that of the photoionization cross-section. While the photoionization and photoabsorption cross-sections are basically the same as both correspond to the absorption of a photon and can be seen in photoabsorption spectra, the photoionization cross-section is continuous as it depends on any photon energy beyond the ionization threshold, and the photoabsorption cross-section depends on the excitation energies of the bound and autoionizing states. Both can be plotted for spectral features. Researchers who do not use close-coupling approximation for the automatic generation of resonances in photoionization typically use distorted wave approximation for the background cross-section of photoionization and use the photoabsorption cross-sections as resonance lines over a smooth background.

A Lorentzian or Gaussian function is often multiplied with a photoabsorption line to broaden it in to order to simulate the width of a resonance in photoionization cross-sections. When an experiment is carried out, the lines are broadened by the bandwidth of the experimental beam and the detector. In such cases, the calculated lines can be convoluted with the bandwidth of the beam to simulate the observed spectrum. We present photoabsorption spectra without the broadening of lines. Ions, such as lanthanide ions, with a large number of quantum states generate many lines due to transitions among these states. They form almost a continuous curve in the cross-sections, as will be seen for most of the lanthanide ions studied here.

The present work includes E1 transitions to produce the synthetic spectrum of the ion and illustrate the spectral features. The magnitudes of the decay rates for forbidden transitions are typically several orders of magnitude lower than those of E1. However, E2, E3, M1, and M2 results are available for distribution in the NORAD-Atomic-Data database [25]. The present study considers both the bound-bound and bound-free (continuum) excitation as computed by the program SUPERSTRUCTURE (SS) [26].

Computations of the energies and transition parameters have been carried out in the relativistic Breit–Pauli approximation, where the Hamiltonian is given by (e.g., [24,26])

$$\mathbf{H}_{BP} = \mathbf{H}_{NR} + \mathbf{H}_{mass} + \mathbf{H}_{Dar} + \mathbf{H}_{so} +$$

$$\frac{1}{2}\sum_{i \neq j}^{N} [g_{ij}(so + so') + g_{ij}(ss') + g_{ij}(css') + g_{ij}(d) + g_{ij}(oo')] \quad (6)$$

where the non-relativistic Hamiltonian is given by

$$H_{NR} = \left[\sum_{i=1}^{N}\left\{-\nabla_i^2 - \frac{2Z}{r_i} + \sum_{j>i}^{N}\frac{2}{r_{ij}}\right\}\right] \quad (7)$$

and the one-body mass correction, Darwin, and spin-orbit interaction terms are, respectively,

$$H_{mass} = -\frac{\alpha^2}{4}\sum_i p_i^4, \ H_{Dar} = \frac{\alpha^2}{4}\sum_i \nabla^2\left(\frac{Z}{r_i}\right), \ H_{so} = \frac{Ze^2\hbar^2}{2m^2c^2r^3}\mathbf{L}.\mathbf{S} \quad (8)$$

while the two-body Breit interaction term is

$$H_B = \sum_{i>j}[g_{ij}(so + so') + g_{ij}(ss')] \quad (9)$$

The present approximation includes the contributions of all these terms and part of the last three terms of H_{BP} incorporated in the atomic structure code SS [26,27]. The prime notation for spin and orbital angular momenta in the two-body interaction terms indicates the quantity belonging to the other electron.

The electron–electron interaction, as implemented in the atomic structure program SUPERSTRUCTURE [24,26,27], is represented by the Thomas–Fermi–Dirac–Amaldi (TFDA) potential, which includes the electron exchange effect and configuration interactions. Electrons are treated as a Fermi sea of electrons, constrained by the Pauli exclusion principle, filling in cells up to the highest Fermi level of momentum $p = p_F$ at temperature T = 0. As T rises, electrons are excited out of the Fermi sea close to the 'surface' level and approach a Maxwellian distribution. Solutions of the Schrodinger equation provide the wavefunctions and energies of the fine structure levels. Based on quantum statistics, the TFDA model gives a continuous function $\phi(x)$ such that the potential is represented by [27,28]

$$V(r) = \frac{\mathcal{Z}_{\text{eff}}(\lambda_{nl}, r)}{r} = -\frac{Z}{r}\phi(x), \tag{10}$$

where

$$\phi(x) = e^{-Zr/2} + \lambda_{nl}(1 - e^{-Zr/2}), \; x = \frac{r}{\mu}, \; \mu = 0.8853\left(\frac{N}{N-1}\right)^{2/3} Z^{-1/3} = constant. \tag{11}$$

λ_{nl} is the Thomas–Fermi scaling parameter of the nl orbital wavefunction. Depending on its value, typically around 1 and less than 2, the orbital wavefunction can be compressed towards the nucleus or extended outwards. The function $\phi(x)$ is a solution of the potential equation

$$\frac{d^2\phi(x)}{dx^2} = \frac{1}{\sqrt{x}}\phi(x)^{\frac{3}{2}} \tag{12}$$

The boundary conditions on $\phi(x)$ are such that

$$\phi(0) = 1, \; \phi(\infty) = -\frac{Z-N+1}{Z}. \tag{13}$$

The atomic wavefunction may be obtained by using an exponentially decaying function appropriate for a bound state, e.g., the Whittaker function.

3. Computation

As mentioned above in the Theory section, the transition parameters f, A, and photoabsorption cross-sections σ were obtained using the atomic structure program SUPERSTRUCTURE (SS [26,27]). The wavefunction of each atomic species of the 25 lanthanide ions was optimized with a set of configurations and a set of Thomas–Fermi scaling parameters λ_{nl} for the orbitals. Both sets for each ion are presented in Table 1.

Table 1. Sets of optimized configurations with identifying number within parentheses and Thomas–Fermi orbital scaling parameters (λ_{nl}) used in SS to compute the energies and transition parameters. All listed configurations correspond to nine inner closed or filled orbitals, $[1s^2 2s^2 2p^6 3s^2 3p^6 4s^2 3d^{10} 4p^2 5s^2]$, plus additional ones depending on the ion being studied. N_T is the total number of transitions, allowed and forbidden, computed for each ion.

	Ho I (11 orbitals filled), N_T = 1,019,566
Configurations:	$4f^{11}6s^2(1)$, $4f^{10}6s^25d(2)$, $4f^{10}6s^26p(3)$
λ_{nl}	1.30 (1s), 1.25 (2s), 1.12 (2p), 1.07 (3s), 1.05 (3p), 1.0 (3d), 1.0 (4s), 1.0 (4p), 1.0 (5s), 1.60 (5p), 0.90 (4d), 0.99 (4f), 1.1 (6s), 1.1 (6p), 1.2 (5d)
	Ho II (10 orbitals filled), N_T = 408,070
Configurations:	$4d^{10}4f^{11}6s(1)$, $4d^94f^{12}6s(2)$, $4d^{10}4f^{11}6p(3)$, $4d^{10}4f^{11}5d(4)$
λ_{nl}	1.30 (1s), 1.25 (2s), 1.12 (2p), 1.07 (3s), 1.05 (3p), 1.00 (3d), 1.0 (4s), 1.0 (4p), 1.0 (5s), 1.00 (5p), 1.0 (4d), 1.0 (4f), 1.0 (6s), 1.0 (6p), 1.0 (5d)
	Ho III (11 orbitals filled), N_T = 1,309,895
Configurations:	$4f^{11}(1)$, $4f^{10}5d(2)$, $4f^{10}6s(3)$, $4f^{10}6p(4)$
λ_{nl}	1.30 (1s), 1.25 (2s), 1.12 (2p), 1.07 (3s), 1.05 (3p), 1.00 (3d), 1.0 (4s), 1.0 (4p), 1.2 (5s), 0.925 (4d), 1.50 (5p), 1.00 (4f), 0.95 (6s), 1.20 (6p), 1.25 (5d)
	Er I (11 orbitals filled), N_T = 206,202
Configurations:	$4f^{12}6s^2(1)$, $4f^{11}6s^25d(2)$, $4f^{11}6s^26p(3)$, $4f^{12}6s6p(4)$, $4f^{12}6s5d(5)$
λ_{nl}	1.30 (1s), 1.25 (2s), 1.12 (2p), 1.07 (3s), 1.05 (3p), 1.00 (3d), 1.0 (4s), 1.05 (4p), 1.0 (5s), 1.0 (5p), 1.0 (4d), 1.0 (4f), 1.05 (6s), 1.09 (6p), 1.0 (5d)

Table 1. Cont.

Er II (11 orbitals filled), N_T = 897,374	
Configurations:	$4f^{12}6s(1), 4f^{11}6s^2(2), 4f^{11}6s5d(3), 4f^{12}6p(4), 4f^{12}5d(5), 4f^{11}6s6p(6)$
λ_{nl}	1.30 (1s), 1.25 (2s), 1.12 (2p), 1.07 (3s), 1.05 (3p), 1.00 (3d), 1.0 (4s), 1.0 (4p), 1.2 (5s), 0.955 (5p), 1.02 (4d), 1.0 (4f), 1.0 (6s), 1.0 (6p), 1.0 (5d)
Er III (9 orbitals filled), N_T = 409,161	
Configurations:	$4d^{10}5p^6 4f^{12}(1), 4d^{10}5p^6 4f^{11}5d(2), 4d^{10}5p^6 4f^{11}6s(3), 4d^{10}5p^6 4f^{11}6p(4),$ $4d^{10}5p^5 4f^{12}6s(5), 4d^{10}5p^5 4f^{13}(6), 4d^{10}5p^6 4f^{11}6s^2(7), 4d^9 5p^6 4f^{13}(8)$
λ_{nl}	1.30 (1s), 1.25 (2s), 1.12 (2p), 1.07 (3s), 1.05 (3p), 1.00 (3d), 1.0 (4s), 1.0 (4p), 1.8 (5s), 1.17 (4d), 1.01 (5p), 1.0 (4f), 1.2 (6s), 1.0 (6p), 1.0 (5d)
Er IV (11 orbitals filled), N_T = 1,309,955	
Configurations:	$4f^{11}(1), 4f^{10}5d(2), 4f^{10}6s(3), 4f^{10}6p(4)$
λ_{nl}	1.30 (1s), 1.25 (2s), 1.12 (2p), 1.07 (3s), 1.05 (3p), 1.0 (3d), 1.0 (4s), 1.0 (4p), 1.0 (5s), 1.07 (4d), 1.0 (5p), 1.0 (4f), 1.0 (5d), 1.0 (6s), 1.0 (6p)
Tm I (11 orbitals filled), N_T = 118,759	
Configurations:	$4f^{13}6s^2(1), 4f^{12}6s^2 5d(2), 4f^{13}6s5d(3), 4f^{13}6s6p(4), 4f^{12}6s^2 6p(5), 4f^{13}6p5d(6),$ $4f^{13}6p^2(7), 4f^{13}5d^2(8), 4f^{14}6s(9), 4f^{14}6p(10), 4f^{14}5d(11)$
λ_{nl}	1.30 (1s), 1.25 (2s), 1.22 (2p), 1.1 (3s), 1.12 (3p), 1.1262 (3d), 1.002 (4s), 1.0606 (4p), 0.9 (5s), 1.05436 (5p), 0.97512 (4d), 1.0 (4f), 0.712 (6s), 1.16173 (6p), 1.096 (5d)
Tm II (11 orbitals filled), N_T = 34,184	
Configurations:	$4f^{13}6s(1), 4f^{12}6s^2(2), 4f^{12}6s5d(3), 4f^{13}5d(4), 4f^{13}6p(5), 4f^{12}5d^2(6),$ $4f^{12}6s6p(7), 4f^{12}6s5d(8), 4f^{14}(9)$
λ_{nl}	1.30 (1s), 1.25 (2s), 1.12 (2p), 1.07 (3s), 1.05 (3p), 1.01 (3d), 0.92 (4s), 0.80 (4p), 1.53 (5s), 0.9 (5p), 1.004 (4d), 1.02 (4f), 1.014 (6s), 0.9 (6p), 0.95 (5d)
Tm III (11 orbitals filled), N_T = 849,878	
Configurations:	$4f^{13}(1), 4f^{12}5d(2), 4f^{12}6s(3), 4f^{12}6p(4), 4f^{11}6s5d(5), 4f^{11}6s6p(6)$
λ_{nl}	1.30 (1s), 1.25 (2s), 1.12 (2p), 1.07 (3s), 1.05 (3p), 1.0 (3d), 1.0 (4s), 1.0 (4p), 1.0 (5s), 1.12 (5p), 1.0 (4d), 0.97 (4f), 0.98 (6s), 1.0 (6p), 0.98 (5d)
Tm IV (10 orbitals filled), N_T = 1,096,164	
Configurations:	$5p^6 4f^{12}(1), 5p^6 4f^{11}5d(2), 5p^6 4f^{11}6s(3), 5p^6 4f^{11}6p(4), 5p^5 4f^{13}(5), 5p^5 4f^{12}5d(6),$ $5p^5 4f^{12}6s(7), 5p^6 4f^{10}6s^2(8)$
λ_{nl}	1.30 (1s), 1.25 (2s), 1.12 (2p), 1.07 (3s), 1.05 (3p), 1.0 (3d), 1.0 (4s), 1.0 (4p), 1.0 (5s), 1.0 (4d), 1.03 (5p), 1.0 (4f), 1.0 (6s), 1.0 (6p), 1.0 (5d)
Tm V (11 orbitals filled), N_T = 801,717	
Configurations:	$4f^{11}(1), 4f^{10}5d(2), 4f^{10}6s(3), 4f^{10}6p(4)$
λ_{nl}	1.3 (1s), 1.25 (2s), 1.12 (2p), 1.07 (3s), 1.05 (3p), 1.0 (3d), 1.0 (4s), 1.0 (4p), 1.0 (5s), 1.0 (4d), 1.0 (5p), 1.0 (4f), 1.0 (6s), 1.0 (6p), 1.0 (5d)
Yb I (11 orbitals filled), N_T = 109,127	
Configurations:	$4f^{14}6s^2(1), 4f^{14}6s6p(2), 4f^{13}6s^2 5d(3), 4f^{14}6s5d(4), 4f^{13}6s^2 6p(5), 4f^{13}6s5d^2(6),$ $4f^{13}6s6p5d(7), 4f^{14}6p^2(8), 4f^{14}6p5d(9), 4f^{14}5d^2(10)$
λ_{nl}	1.30 (1s), 1.25 (2s), 1.12 (2p), 1.07 (3s), 1.05 (3p), 1.03 (3d), 1.05 (4s), 1.02 (4p), 0.935 (5s), 0.937 (5p), 1.0 (4d), 1.0 (4f), 1.0 (6s), 1.0 (6p), 1.0 (5d)
Yb II (11 orbitals filled), N_T = 39,009	
Configurations:	$4f^{14}6s(1), 4f^{13}6s^2(2), 4f^{14}5d(3), 4f^{13}6s5d(4), 4f^{14}6p(5), 4f^{13}5d^2(6), 4f^{13}6s6p(7),$ $4f^{13}6p5d(8)$
λ_{nl}	1.30 (1s), 1.45 (2s), 1.20 (2p), 1.10 (3s), 1.07 (3p), 1.0 (3d), 1.0 (4s), 1.05 (4p), 0.918 (5s), 0.90 (5p), 1.025 (4d), 1.0 (4f), 1.1007 (6s), 0.97 (6p), 0.97 (5d)
Yb III (11 orbitals filled), N_T = 925,575	
Configurations:	$4f^{14}(1), 4f^{13}5d(2), 4f^{13}6s(3), 4f^{13}6p(4), 4f^{12}5d^2(5), 4f^{12}6s5d(6), 4f^{12}6p5d(7),$ $4f^{12}6s6p(8), 4f^{12}6s^2(9)$
λ_{nl}	1.30 (1s), 1.25 (2s), 1.12 (2p), 1.07 (3s), 1.05 (3p), 1.15 (3d), 1.05 (4s), 0.819 (4p), 1.24 (5s), 0.887 (5p), 0.98 (4d), 1.02 (4f), 1.05 (6s), 0.95 (6p), 1.0 (5d)
Yb IV (10 orbitals filled), N_T = 400,325	
Configurations:	$4d^{10}4f^{13}(1), 4d^{10}4f^{12}5d(2), 4d^{10}4f^{12}6s(3), 4d^{10}4f^{12}6p(4), 4d^9 4f^{14}(5), 4d^{10}4f^{11}6s5d(6)$
λ_{nl}	1.3 (1s), 1.25 (2s), 1.12 (2p), 1.07 (3s), 1.05 (3p), 1.0 (3d), 1.0 (4s), 1.0 (4p), 1.0 (5s), 0.995 (5p), 1.0 (4d), 1.0 (4f), 1.0 (6s), 1.0 (6p), 1.0 (5d)

Table 1. Cont.

	Yb V (10 orbitals filled), N_T = 208,128
Configurations:	$4f^{12}5p^6(1)$, $4f^{13}5p^5(2)$, $4f^{11}5p^65d(3)$, $4f^{12}5p^56s(4)$, $4f^{11}5p^66s(5)$, $4f^{11}5p^66p(6)$
λ_{nl}	1.30 (1s), 1.25 (2s), 1.12 (2p), 1.07 (3s), 1.05 (3p), 1.0 (3d), 1.0 (4s), 1.0 (4p), 1.0 (5s), 1.0 (4d), 1.03 (5p), 1.0 (4f), 1.0 (6s), 1.15 (6p), 1.2 (5d)
	Yb VI (10 orbitals filled), N_T = 486,262
Configurations:	$45p^54f^{12}(1)$, $5p^64f^{10}5d(2)$, $5p^44f^{12}6s(3)$, $5p^44f^{13}(4)$
λ_{nl}	1.30 (1s), 1.25 (2s), 1.12 (2p), 1.07 (3s), 1.05 (3p), 1.0 (3d), 1.0 (4s), 1.0 (4p), 1.1 (5s), 1.0 (4d), 1.0 (5p), 1.0 (4f), 1.0 (6s), 1.0 (6p), 1.0 5d
	Lu-I (12 orbitals filled), N_T = 13,936
Configurations:	$5d6s^2(1)$, $6s^26p(2)$, $5d6s6p(3)$, $6s5d^2(4)$, $6s6p^2(5)$, $6p5d^2(6)$, $5d6p^2(7)$, $5p^3(8)$, $5d^3(9)$
λ_{nl}	1.30 (1s), 1.25 (2s), 1.12 (2p), 1.07 (3s), 1.05 (3p), 1.0 (3d), 1.0 (4s), 1.0 (4p), 0.95 (5s), 1.135 (4d), 0.94 (5p), 1.0 (4f), 1.0 (6s), 0.98 (6p), 1.0 5d
	Lu-II (11 orbitals filled), N_T = 109,566
Configurations:	$4f^{14}6s^2(1)$, $4f^{14}6s5d(2)$, $4f^{14}6s6p(3)$, $4f^{14}5d^2(4)$, $4f^{14}5d6p(5)$, $4f^{14}6p^2(6)$, $4f^{13}6s5d^2(7)$, $4f^{13}6s^25d(8)$, $4f^{13}6s5d6p(9)$, $4f^{13}6s^26p(10)$
λ_{nl}	1.30 (1s), 1.25 (2s), 1.12 (2p), 1.07 (3s), 1.05 (3p), 1.0 (3d), 1.0 (4s), 1.0 (4p), 0.95 (5s), 0.937 (5p), 1.0 (4d), 1.0 (4f), 1.0 (6s), 1.0 (6p), 0.99 5d
	Lu-III (11 orbitals filled), N_T = 8564
Configurations:	$4f^{14}6s(1)$, $4f^{14}5d(2)$, $4f^{14}6p(3)$, $4f^{13}5d^2(4)$, $4f^{13}5d6s(5)$, $4f^{13}6s^2(6)$, $4f^{13}6p5d(7)$, $4f^{13}6s6p(8)$
λ_{nl}	1.30 (1s), 1.25 (2s), 1.12 (2p), 1.07 (3s), 1.05 (3p), 1.00 (3d), 1.0 (4s), 1.05 (4p), 1.0 (5s), 0.95 (5p), 0.98 (4d), 1.0 (4f), 1.03 (6s), 0.97 (6p), 0.98 5d
	Lu-IV (11 orbitals filled), N_T = 926,436
Configurations:	$4f^{14}(1)$, $4f^{13}5d(2)$, $4f^{13}6s(3)$, $4f^{13}6p(4)$, $4f^{12}5d^2(5)$, $4f^{12}5d6s(6)$, $4f^{12}6s^2(7)$, $4f^{12}6s6p(8)$, $4f^{12}6p5d(9)$
λ_{nl}	1.30 (1s), 1.25 (2s), 1.12 (2p), 1.07 (3s), 1.05 (3p), 1.00 (3d), 1.0 (4s), 0.80 (4p), 1.40 (5s), 0.90 (5p), 0.98 (4d), 1.02 (4f), 1.05 (6s), 0.92 (6p), 0.97 5d
	Lu-V (11 orbitals filled), N_T = 850,668
Configurations:	$4f^{13}(1)$, $4f^{12}5d(2)$, $4f^{12}6s(3)$, $4f^{12}6p(4)$, $4f^{11}5d6s(5)$, $4f^{11}6s6p(6)$
λ_{nl}	1.30 (1s), 1.25 (2s), 1.12 (2p), 1.07 (3s), 1.05 (3p), 1.0 (3d), 1.0 (4s), 1.0 (4p), 1.0 (5s), 0.999 (5p), 0.993 (4d), 1.01 (4f), 0.98 (6s), 1.02 (6p), 1.0 5d
	Lu-VI (10 orbitals filled), N_T = 317,817
Configurations:	$4f^{12}5p^6(1)$, $4f^{13}5p^5(2)$, $4f^{11}5p^65d(3)$, $4f^{12}5p^56s(4)$, $4f^{11}5p^66s(5)$, $4f^{11}5p^66p(6)$
λ_{nl}	1.30 (1s), 1.25 (2s), 1.12 (2p), 1.07 (3s), 1.05 (3p), 0.80 (3d), 1.0 (4s), 1.0 (4p), 1.4 (5s), 0.98 (4d), 1.0 (5p), 1.0 (4f), 0.95 (6s), 1.3 (6p), 1.3 5d
	Lu-VII (10 orbitals filled), N_T = 304,178
Configurations:	$4f^{13}5p^4(1)$, $4f^{12}5p^5(2)$, $4f^{14}5p^3(3)$, $4f^{13}5p^35d(4)$, $4f^{13}5p^36s(5)$, $4f^{13}5p^36p(6)$
λ_{nl}	1.30 (1s), 1.25 (2s), 1.12 (2p), 1.07 (3s), 1.05 (3p), 1.0 (3d), 1.0 (4s), 1.0 (4p), 1.0 (5s), 0.98 (4d), 0.97 (5p), 1.0 (4f), 1.0 (6s), 1.0 (6p), 1.12 5d

The optimization process for the energies was considerably complex due to the sensitivity of the potential with a large number of electrons. A large number of angular quantum numbers due to a large number of electrons with open orbitals 4f, 5p, 5d, 6s, 6p, particularly 4f, introduces a very large number of energy levels. Hence, a slight variation in the Thomas–Fermi scaling parameters λ_{nl} for the orbital wavefunctions would perturb the electron–electron interaction and change the energy values and the order in the fine structure levels. A numerical challenge arose when exceeding the dimension of the Hamiltonian matrix that SS can accommodate. The number of digital spaces for the dimension surpassed those allotted to SS. Hence, the optimization of the set of configurations was carried out carefully such that the order of the calculated energy levels, particularly the ground and low-lying energies, could match to those of the measured energies available in the NIST [13] table.

Table 1 presents the optimized set of configurations, specifying the occupancy of the outer orbitals that can vary while specifying the number of inner orbitals that remain closed, as well as the Thomas–Fermi scaling parameters of the orbitals for each ion. The top line of each set gives the total number of radiative transitions (N_T), which includes both the allowed E1 and forbidden E2, E3, M1, M2 transitions, produced by a set of configurations.

All atomic data from SS were processed using the program PRCSS [29] to obtain clean tables and easy applications of them. They were further processed to compute absorption cross-sections, sum them if the transition energies were the same, and display the spectral features using the program SPECTRUM [30].

4. Results and Discussion

The present study reports results on the energies and transitions and corresponding photoabsorption spectra of 25 ions of lanthanides, Ho I-III, Er I-IV, Tm I-V, Yb I-VI, and Lu I-VII. We investigate the spectral features of these ions. Most of the ions have produced close to hundreds of thousands lines, with excitation to a few orbitals upwards. All atomic data for all 25 lanthanides ions considered in the present study are available in the NORAD-Atomic-Data database [25].

The present results correspond to a limited number of configurations for each ion. For most ions, configurations with orbitals up to 6p, 5d are only considered; for some ions, such as Ho I, no additional configurations beyond those already included could be considered due to exceeding the limit on the dimensions of various arrays, such as that of the Hamiltonian matrix, in the program. These issues are mentioned in the Computation section. For some other ions, it was possible to add more configurations, but they were omitted at the end, as they produced energies that were much higher than and had an energy gap with the lower levels. It was not possible to verify the accuracy of these high-lying energies due to the lack of availability of observed energies. These configurations also introduced additional electron–electron correlation interactions, which impacted the order and numerical values of the lower energies. This implies that significantly more configurations would be needed for a converged and larger set of accurate energies. Thus, we selected the set of configurations that provided the overall best set of energies when compared to those available at NIST for each ion.

It was observed that these large ions do not show the isoelectronic behavior that is often seen with lower Z elements. Hence, the isoelectronic set of large ions does not necessarily have the same set of optimized configurations. The energy tables of the present lanthanide ions available in the NIST table confirm this behavior, namely that these ions may not have the same symmetries for the ground and low-lying levels. The photoabsorption features of the isoelectronic series, such as Ho I, Er II, Tm III, Yb IV, and Lu V, as will be seen later, do not have similar features.

We discuss some general information and data files for the 25 ions, before giving examples of the characteristics of individual ions.

4.1. General Information on the Atomic Data

Each lanthanide ion produced a significantly large number of energy levels. A file containing all energies for each ion, as mentioned above, is available electronically in the NORAD-Atomic-Data database [25]. A sample set of energies for Ho I is presented in Table 2 to demonstrate the format of the complete energy table of the ions. In Table 2, the number of energy levels obtained from the set of configurations (Table 1) is quoted at the top of Table 2. The total number may include both the bound levels below the ionization threshold and the Rydberg levels in continuum above it.

Table 2 contains, for each energy level, the running index (ie), the symmetry of the level $SL\pi(C\#)$ with the configuration number ($C\#$) from Table 1 within parentheses next to it, the total angular momenta J, and the energy in Ry relative to the ground level. In an atomic structure calculation, such as the present case with SS, all energies are computed relative to the ground level, which is set to zero. Hence, all energies presented are positive. The program SS does not distinguish between the bound and continuum levels. This is the same format that is followed by NIST [13], which also presents relative energies to the ground level. However, NIST provides the ionization threshold energy, which is obtained separately. The present energies are compared in Table 3 with the measured values that are available on the NIST website.

Table 2. Sample table of energies for Ho I demonstrating the format of the complete energy table for each lanthanide ion. The total number of levels obtained is given at the top. The column headings are as follows: ie is the running index, $SL\pi(C\#)$ is the symmetry (total spin S, total orbital angular momentum L, and parity π), $cf\#$ is the configuration number as given in Table 1, J is the total angular momentum, and E is the relative energy in Ry.

Number of Fine Structure Levels = 1629ie	$SL\pi$(cf#)	J	E(Ry)
1	4Io(1)	15/2	0.00000E+00
2	4Io(1)	13/2	3.60033E-02
3	4Io(1)	11/2	5.85101E-02
4	4Io(1)	9/2	7.43454E-02
5	4Me(2)	15/2	9.79376E-02
6	6Le(2)	13/2	1.01223E-01
7	4Fo(1)	9/2	1.09269E-01
8	6Le(2)	11/2	1.22576E-01
9	2Ho(1)	11/2	1.28064E-01
10	4So(1)	3/2	1.35965E-01

Table 3. Ground and low-lying excited energy levels of Ho, Er, Tm, Yb, Lu ions obtained from SUPER-STRUCTURE (SS) are presented and compared with measured values, largely obtained by Martin et al. [12], available in the compiled table of NIST [13]. Each configuration below corresponds to electrons outside the core ion configuration of Xe and filled 4d orbitals unless 4d has vacancies. N_E is the number of energy levels and N_{E1} is the number of corresponding E1 transitions obtained from the configuration set of each ion.

	Config	$SL\pi$	J	E(SS, Ry)	E(NIST [12], Ry)
		Ho I, N_E = 1629, N_{E1} = 210,522			
1	$4f^{11}6s^2$	$^4I^o$	15/2	0.0	0.0
2	$4f^{11}6s^2$	$^4I^o$	13/2	0.03603	0.0493879
3	$4f^{11}6s^2$	$^4I^o$	11/2	0.05851	0.0784160
4	$4f^{11}6s^2$	$^4I^o$	9/2	0.07434	0.0974668
5	$4f^{11}6s^2$	^4M	17/2	0.13881	0.0763542
6	$4f^{10}5d6s^2$	^4M	15/2	0.09794	0.0767935
7	$4f^{10}5d6s^2$	^6L	13/2	0.10122	0.0833543
8	$4f^{10}5d6s^2$	^6L	19/2	0.19610	0.0887711
9	$4f^{11}6s^2$	$^4F^o$	9/2	0.10927	0.1193261
10	$4f^{10}5d6s^2$	^6L	11/2	0.12257	0.1543452
		Ho II, N_E = 924, N_{E1} = 81,623			
1	$4f^{11}6s$	$^5I^o$	8	0.0	0.0
2	$4f^{11}6s$	$^5I^o$	7	0.00724	0.005808
3	$4f^{11}6s$	$^3I^o$	7	0.05528	0.051186
4	$4f^{11}6s$	$^5I^o$	6	0.05793	0.053306
5	$4f^{11}6s$	$^5I^o$	5	0.08928	0.080652
6	$4f^{11}6s$	$^3I^o$	6	0.09113	0.082029
7	$4f^{11}5d$	$^5G^o$	6	0.11003	0.098771
8	$4f^{11}6s$	$^5I^o$	4	0.11108	0.102111
9	$4f^{11}6s$	$^3I^o$	9	0.11702	0.148388
10	$4f^{11}5d$	$^5H^o$	7	0.11832	0.152628
		Ho III, N_E = 1837, N_{E1} = 258,124			
1	$4f^{11}$	$^4I^o$	15/2	0.0	0.0
2	$4f^{11}$	$^4I^o$	13/2	0.03114	0.04956
3	$4f^{11}$	$^4I^o$	11/2	0.04969	0.07877
4	$4f^{11}$	$^4I^o$	9/2	0.06203	0.09815
6	$4f^{11}$	$^4F^o$	9/2	0.08576	0.12147
7	$4f^{11}$	$^4F^o$	7/2	0.1149	0.16282
8	$4f^{11}$	$^4F^o$	5/2	0.1245	0.17656
9	$4f^{11}$	$^4F^o$	3/2	0.1261	
10	$4f^{11}$	$^2H^o$	11/2	0.10275	0.15392

Table 3. Cont.

	Config	SLπ	J	E(SS, Ry)	E(NIST [12], Ry)
		Er I, N_E = 993, N_{E1} = 88,827			
1	$4f^{12}6s^2$	^3H	6	0.0	0.0
2	$4f^{12}6s^2$	^3H	5	0.06787	0.063408
3	$4f^{12}6s^2$	^3H	4	0.06158	0.097970
4	$4f^{12}6s^2$	^3F	4	0.11386	0.045884
5	$4f^{12}6s^2$	^3F	3	0.13971	0.112792
6	$4f^{12}6s^2$	^3F	2	0.15311	0.119357
7	$4f^{11}5d6s^2$	$^5G^o$	6	0.089544	0.065397
8	$4f^{11}5d6s^2$	$^5I^o$	7	0.097936	0.070140
9	$4f^{11}5d6s^2$	$^5K^o$	9	0.104784	0.078556
10	$4f^{11}5d6s^2$	$^5I^o$	8	0.114163	0.085204
		Er II, N_E = 1476, N_{E1} = 189,738			
1	$4f^{12}6s$	^4H	13/2	0.0	0.0
2	$4f^{12}6s$	^4H	11/2	0.06837	0.0661522
3	$4f^{12}6s$	^2H	9/2	0.06211	0.046772
4	$4f^{12}6s$	^4H	7/2	0.06260	0.049242
5	$4f^{11}6s^2$	$^4I^o$	15/2	0.18768	0.062192
6	$4f^{11}6s^2$	$^4I^o$	13/2	0.24412	0.121552
7	$4f^{11}6s^2$	$^4I^o$	11/2	0.28125	0.154443
8	$4f^{12}6s$	^2I	11/2	0.36487	0.065153
9	$4f^{12}6s$	^4F	9/2	0.11218	0.065567
10	$4f^{11}5d6s$	$^6L^o$	13/2	0.09742	0.097206
		Er III, N_E = 1000, N_{E1} = 82,286			
1	$4f^{12}$	^3H	6	0.0	0.0
2	$4f^{12}$	^3H	5	0.045163	0.063513
3	$4f^{12}$	^3H	4	0.035074	0.098284
4	$4f^{12}$	^3F	2	0.09552	
5	$4f^{12}$	^3F	3	0.08826	
6	$4f^{12}$	^3F	4	0.07184	0.04631
7	$4f^{11}5d$	$^5G^o$	6	0.10297	0.15469
8	$4f^{11}5d$	$^5H^o$	7	0.10991	0.160818
9	$4f^{11}5d$	$^5K^o$	9	0.14872	0.172929
10	$4f^{11}5d$	$^5L^o$	8	0.13645	0.181508
		Er IV, N_E = 1837, N_{E1} = 257,713			
1	$4f^{11}$	$^4I^o$	15/2	0.0	0.0
2	$4f^{11}$	$^4I^o$	13/2	0.0598	0.0591
3	$4f^{11}$	$^4I^o$	11/2	0.0970	0.0921
4	$4f^{11}$	$^4I^o$	9/2	0.1230	0.1125
5	$4f^{11}$	$^4F^o$	9/2	0.1770	0.1383
6	$4f^{11}$	$^4F^o$	7/2	0.2333	0.1863
7	$4f^{11}$	$^4F^o$	5/2	0.2525	0.2011
8	$4f^{11}$	$^4F^o$	3/2	0.2555	0.2042
9	$4f^{11}$	$^2H^o$	11/2	0.2103	
10	$4f^{11}$	$^4S^o$	3/2	0.2193	0.1667
		Tm I, N_E = 470, N_{E1} = 23,804			
1	$4f^{13}6s^2$	$^2F^o$	7/2	0.0	0.0
2	$4f^{11}6s^2$	$^2F^o$	5/2	0.0891	0.0799
3	$4f^{12}6s^25d$	^4F	9/2	0.1343	0.11956
4	$4f^{12}6s^25d$	^2K	15/2	0.2030	0.1716
5	$4f^{12}6s^25d$	^4G	11/2	0.21430	0.14205
6	$4f^{12}6s^25d$	^4K	13/2	0.25119	0.15906
7	$4f^{12}6s^25d$	^4K	17/2	0.31023	0.14997
8	$4f^{12}6s^25d$	^4F	7/2	0.20683	0.15452
9	$4f^{12}6s^25d$	^4I	15/2	0.38814	0.17034
10	$4f^{12}6s^25d$	^4G	9/2	0.28376	0.17165

Table 3. Cont.

	Config	SLπ	J	E(SS, Ry)	E(NIST [12], Ry)
		Tm II, $N_E = 1129$, $N_{E1} = 2467$			
1	$4f^{13}6s$	$^3F^o$	4	0.0	0.0
2	$4f^{13}6s$	$^1F^o$	3	0.000309	0.00216
3	$4f^{13}6s$	$^3F^o$	2	0.1050	0.0799
4	$4f^{13}6s$	$^3F^o$	3	0.1053	0.0816
5	$4f^{12}6s^2$	^3H	6	0.1449	0.1135
6	$4f^{12}6s^2$	^3H	5	0.2272	0.1879
7	$4f^{12}6s^2$	^3H	4	0.1812	0.2272
8	$4f^{12}5d6s$	^5F	5	0.1362	0.1510
9	$4f^{12}5d6s$	^3K	7	0.14733	0.17878
10	$4f^{12}5d6s$	^5G	6	0.15250	0.2032
		Tm III, $N_E = 1437$, $N_{E1} = 181{,}768$			
1	$4f^{13}$	$^2F^o$	7/2	0.0	0.0
2	$4f^{13}$	$^2F^o$	5/2	0.06669	0.07995
3	$4f^{12}5d$	^4F	9/2	0.24034	0.20866
4	$4f^{12}5d$	^4H	15/2	0.27574	0.235825
5	$4f^{12}5d$	^4H	11/2	0.27975	0.239230
6	$4f^{12}5d$	^4K	13/2	0.30738	0.261724
7	$4f^{12}6s$	^2H	13/2	0.21097	0.230575
8	$4f^{12}6s$	^2H	11/2	0.21716	0.236207
9	$4f^{12}5d$	^4H	7/2	0.27029	0.251029
10	$4f^{12}5d$	^4I	17/2	0.28719	0.251127
		Tm IV, $N_E = 1606$, $N_{E1} = 160{,}013$			
1	$4f^{12}$	^3H	6	0.0	0.0
2	$4f^{12}$	^3H	5	0.08038	0.0737
3	$4f^{12}$	^3H	4	0.06591	0.114
4	$4f^{12}$	^3F	4	0.12865	0.0514
5	$4f^{12}$	^3F	3	0.15760	0.1308
6	$4f^{12}$	^3F	2	0.17087	0.1353
7	$4f^{12}$	^1G	4	0.21265	0.1943
8	$4f^{12}$	^1D	2	0.33334	
9	$4f^{12}$	^1I	6	0.40132	
10	$4f^{12}$	^3P	0	0.43598	
		Tm V, $N_E = 1837$, $N_{E1} = 259{,}539$			
1	$4f^{11}$	$^4I^o$	15/2	0.0	0.0
2	$4f^{11}$	$^4I^o$	13/2	0.04067	
3	$4f^{11}$	$^4I^o$	11/2	0.06393	
4	$4f^{11}$	$^4I^o$	9/2	0.07922	
5	$4f^{11}$	$^4F^o$	9/2	0.10784	
6	$4f^{11}$	$^2H^o$	11/2	0.12898	
7	$4f^{11}$	$^4S^o$	3/2	0.13396	
8	$4f^{11}$	$^4F^o$	7/2	0.14556	
9	$4f^{11}$	$^4F^o$	5/2	0.15756	
10	$4f^{11}$	$^4F^o$	3/2	0.15875	
		Yb I, $N_E = 455$, $N_{E1} = 22{,}002$			
1	$4f^{14}6s^2$	^1S	0	0.0	0.0
2	$4f^{14}6s6p$	$^3P^o$	0	0.11722	0.157544
3	$4f^{14}6s6p$	$^3P^o$	1	0.12353	0.163955
4	$4f^{14}6s6p$	$^3P^o$	2	0.13864	0.179614
5	$4f^{13}5d6s^2$	$^3H^o$	2	0.17604	0.211309
6	$4f^{13}5d6s^2$	$^3H^o$	5	0.21229	0.235651
7	$4f^{13}5d6s^2$	$^3D^o$	3	0.23746	0.250103
8	$4f^{13}5d6s^2$	$^3H^o$	4	0.25282	0.256836
9	$4f^{14}5d6s$	^3D	1	0.37221	0.223161
10	$4f^{14}5d6s$	^3D	2	0.37380	0.225556

Table 3. *Cont.*

	Config	SLπ	J	E(SS, Ry)	E(NIST [12], Ry)
		Yb II, N_E = 264, N_{E1} = 8033			
1	$4f^{14}6s$	^2S	1/2	0.0	0.0
2	$4f^{13}6s^2$	$^2F^o$	7/2	0.12512	0.195182
3	$4f^{13}6s^2$	$^2F^o$	5/2	0.22718	0.287669
4	$4f^{14}5d$	^2D	3/2	0.13701	0.209234
5	$4f^{14}5d$	^2D	5/2	0.15567	0.221736
6	$4f^{13}6s5d$	$^4P^o$	5/2	0.20627	0.243846
7	$4f^{13}6s5d$	$^4P^o$	3/2	0.23529	0.262062
8	$4f^{13}6s5d$	$^4P^o$	1/2	0.28170	0.306676
9	$4f^{14}6s6p$	$^2P^o$	1/2	0.25680	0.246605
10	$4f^{14}6s6p$	$^2P^o$	3/2	0.27636	0.276954
		Yb III, N_E = 1485, N_{E1} = 203,904			
1	$4f^{14}$	^1S	0	0.0	0.0
2	$4f^{13}5d$	$^3P^o$	2	0.31312	0.304234
3	$4f^{13}5d$	$^3H^o$	5	0.35368	0.337353
4	$4f^{13}5d$	$^3D^o$	3	0.381433	0.356681
5	$4f^{13}5d$	$^3H^o$	4	0.397868	0.365965
6	$4f^{13}6s$	$^3F^o$	4	0.313955	0.315810
7	$4f^{13}6s$	$^1F^o$	3	0.314706	0.318858
8	$4f^{13}5d$	$^3H^o$	6	0.387298	0.365965
9	$4f^{13}5d$	$^3P^o$	1	0.393126	0.361962
10	$4f^{13}5d$	$^3D^o$	2	0.406379	0.367132
		Yb IV, N_E = 963, N_{E1} = 40,767			
1	$4f^{13}$	$^2F^o$	7/2	0.0	0.0
2	$4f^{13}$	$^2F^o$	5/2	0.10849	0.093077
3	$4f^{12}5d$	^4F	9/2	0.68093	0.715611
4	$4f^{12}5d$	^2K	15/2	0.71781	0.748930
5	$4f^{12}5d$	^4G	11/2	0.72798	0.753373
6	$4f^{12}5d$	^4K	13/2	0.75809	0.779549
7	$4f^{12}5d$	^4F	7/2	0.74697	0.768627
8	$4f^{12}5d$	^4K	17/2	0.75311	0.775715
9	$4f^{12}5d$	^4G	9/2	0.78667	0.802991
10	$4f^{12}5d$	^4I	11/2	0.79105	0.803512
		Yb V, N_E = 873, N_{E1} = 59,027			
1	$4f^{12}5p^6$	^3H	6	0.0	0.0
2	$4f^{12}5p^6$	^3H	4	0.031148	
3	$4f^{12}5p^6$	^3H	5	0.056967	
4	$4f^{12}5p^6$	^3F	4	0.085510	
5	$4f^{12}5p^6$	^3F	3	0.10029	
6	$4f^{12}5p^6$	^3F	2	0.10611	
7	$4f^{12}5p^6$	^1G	4	0.13819	
8	$4f^{12}5p^6$	^1D	2	0.21423	
9	$4f^{12}5p^6$	^1I	6	0.26570	
10	$4f^{12}5p^6$	^3P	0	0.28940	
		Yb VI, N_E = 1407, N_{E1} = 13,807			
1	$4f^{12}5p^5$	$^4I^o$	15/2	0.0	0.0
2	$4f^{12}5p^5$	$^4H^o$	11/2	0.020645	
3	$4f^{12}5p^5$	$^4I^o$	13/2	0.021470	
4	$4f^{12}5p^5$	$^2H^o$	9/2	0.024775	
5	$4f^{12}5p^5$	$^2G^o$	11/2	0.036494	
6	$4f^{12}5p^5$	$^2G^o$	5/2	0.072901	
7	$4f^{12}5p^5$	$^2H^o$	11/2	0.074037	
8	$4f^{12}5p^5$	$^4H^o$	11/2	0.082992	
9	$4f^{12}5p^5$	$^4F^o$	7/2	0.093051	
10	$4f^{12}5p^5$	$^2G^o$	9/2	0.103729	

Table 3. Cont.

	Config	SLπ	J	E(SS, Ry)	E(NIST [12], Ry)
			Lu I, N_E = 148, N_{E1} = 3220		
1	$5d6s^2$	^2D	3/2	0.0	0.0
2	$5d6s^2$	^2D	5/2	0.020658	0.018170
3	$6s^26p$	$^2P^o$	1/2	0.083090	0.037691
4	$6s^26p$	$^2P^o$	3/2	0.097783	0.068130
5	$5d6s6p$	$^4F^o$	3/2	0.116829	0.158809
6	$5d6s6p$	$^4F^o$	5/2	0.123975	0.168626
7	$5d6s6p$	$^4F^o$	7/2	0.137319	0.186195
8	$5d6s6p$	$^4F^o$	9/2	0.152215	0.206033
9	$5d^26s$	^4F	3/2	0.251670	0.171785
10	$5d^26s$	^4F	5/2	0.258495	0.176816
			Lu II, N_E = 455, N_{E1} = 22,154		
1	$6s^2$	^1S	0	0.0	0.0
2	$5d6s$	^3D	1	0.088774	0.107495
3	$5d6s$	^3D	2	0.098492	0.113319
4	$5d6s$	^3D	3	0.132393	0.129392
5	$5d6s$	^1D	2	0.162822	0.157946
6	$6s6p$	$^3P^o$	0	0.235055	0.248452
7	$6s6p$	$^3P^o$	1	0.248583	0.259740
8	$6s6p$	$^3P^o$	2	0.288652	0.295736
9	$5d^2$	^3F	2	0.286572	0.267974
10	$5d^2$	^3F	3	0.318543	0.281482
			Lu III, N_E = 145, N_{E1} = 159		
1	$4f^{14}6s$	^2S	1/2	0.0	0.0
2	$4f^{14}5d$	^2D	3/2	0.0510604	0.052011
3	$4f^{14}5d$	^2D	5/2	0.0956229	0.078805
4	$4f^{14}6p$	$^2P^o$	1/2	0.342017	0.349932
5	$4f^{14}6p$	$^2P^o$	3/2	0.379465	0.407384
6	$4f^{13}5d^2$	$^4G^o$	5/2	0.741428	
7	$4f^{13}5d^2$	$^4F^o$	7/2	0.743474	
8	$4f^{13}5d^2$	$^4D^o$	3/2	0.745181	
9	$4f^{13}5d^2$	$^4I^o$	11/2	0.758697	
10	$4f^{13}5d^2$	$^4F^o$	7/2	0.766942	
			Lu IV, N_E = 1485, N_{E1} = 204,567		
1	$4f^{14}$	^1S	0	0.0	0.0
2	$4f^{13}5d$	$^3P^o$	2	0.860650	0.824085
3	$4f^{13}5d$	$^3H^o$	5	0.894914	0.863590
4	$4f^{13}5d$	$^3D^o$	3	0.919319	0.886969
5	$4f^{13}5d$	$^3H^o$	4	0.934682	0.898131
6	$4f^{13}5d$	$^3H^o$	6	0.927311	0.895222
7	$4f^{13}5d$	$^3P^o$	1	0.939756	0.897644
8	$4f^{13}5d$	$^3P^o$	2	0.943373	0.908169
9	$4f^{13}5d$	$^3F^o$	4	0.971074	0.931466
10	$4f^{13}5d$	$^3F^o$	3	0.971264	0.939247
			Lu V, N_E = 1437, N_{E1} = 182,086		
1	$4f^{13}$	$^2F^o$	7/2	0.0	0.0
2	$4f^{13}$	$^2F^o$	5/2	0.12751	0.107464
3	$4f^{12}5d$	^4F	7/2	1.45797	1.373872
4	$4f^{12}5d$	^4F	9/2	1.39228	1.412341
5	$4f^{12}5d$	^4G	9/2	1.48568	1.429373
6	$4f^{12}5d$	^4G	7/2	1.50702	1.448762
7	$4f^{12}5d$	^4G	5/2	1.50048	1.449241
8	$4f^{12}5d$	^4F	5/2	1.53991	1.476007
9	$4f^{12}5d$	^4H	7/2	1.53119	1.481346
10	$4f^{12}5d$	^4G	9/2	1.51306	1.483601

Table 3. Cont.

	Config	$SL\pi$	J	E(SS, Ry)	E(NIST [12], Ry)
		Lu VI, N_E = 873, N_{E1} = 59,028			
1	$4f^{12}5p^6$	^3H	6	0.0	0.0
2	$4f^{12}5p^6$	^3H	4	0.030443	
3	$4f^{12}5p^6$	^3H	5	0.055132	
4	$4f^{12}5p^6$	^3F	4	0.083617	
5	$4f^{12}5p^6$	^3F	3	0.099283	
6	$4f^{12}5p^6$	^3F	2	0.106151	
7	$4f^{12}5p^6$	^1G	4	0.133692	
8	$4f^{12}5p^6$	^1D	2	0.219180	
9	$4f^{12}5p^6$	^1I	6	0.280176	
10	$4f^{12}5p^6$	^3P	0	0.305074	
		Lu VII, N_E = 777, N_{E1} = 68,947			
1	$4f^{13}5p^4$	$^2H^o$	11/2	0.0	0.0
2	$4f^{13}5p^4$	$^2H^o$	9/2	0.0125831	
3	$4f^{13}5p^4$	$^4F^o$	3/2	0.0421028	
4	$4f^{13}5p^4$	$^4G^o$	7/2	0.0666561	
5	$4f^{13}5p^4$	$^4G^o$	5/2	0.104410	
6	$4f^{13}5p^4$	$^4F^o$	5/2	0.104410	
7	$4f^{13}5p^4$	$^4D^o$	3/2	0.123075	
8	$4f^{13}5p^4$	$^4D^o$	1/2	0.127465	
9	$4f^{13}5p^4$	$^2G^o$	9/2	0.147603	
10	$4f^{13}5p^4$	$^4F^o$	7/2	0.224123	

Each lanthanide ion has produced an extensive set of transitions, including both allowed and forbidden types, among its large number of energy levels. The allowed transitions are strong. The forbidden transitions are much weaker compared to E1 transitions. We have obtained a very large set of forbidden transitions of types E2, E3, M1, and M2 for each ion. The atomic data file for each ion contains four tables of transitions: (i) a table of dipole allowed E1 transitions where the spin remains the same as the transition; (ii) a table of dipole allowed E1 transitions where the spin changes (these transitions are also known as intercombination transitions); (iii) a table of forbidden E3 and M2 transitions (they follow the same selection rules); and (iv) a table of E2 and M1 transitions (they follow the same selection rules). At the end of each table, the total number of transitions is given. The file containing the complete sets of transition parameters, i.e., the line strengths, oscillator strengths, and radiative decay rates, is available electronically from the NORAD-Atomic-Data database [25]. It is the same file that contains the energy table.

Table 4 presents an example set of E1 transitions with unchanged spin belonging to Ho I, to demonstrate the format of the complete table. As explained in the caption, i and j are the transitional level numbers; $SLpCi$ is the symmetry with total spin S, total orbital angular momentum L, and parity p; Ci is the configuration number of the level that it belongs to; g_i is the statistical weight factor; f_{ij} is the oscillator strength; and a_{ji} is the radiative decay rate for the transition. The other tables for forbidden transitions are self-explanatory for the transitional quantities, as explained for Table 4. Hence, sample tables for these are not presented. The A-values from the present work are compared with the available published values in Table 5.

We calculated the photoabsorption cross-sections σ of the dipole allowed (E1) transitions and plotted the spectrum for each lanthanide ion to display their characteristic features. A comparison of partial features is presented in Figure 1 and the full characteristic features are shown in Figures 2–26. All points in the figures correspond to actual line strengths. A single strong line can be a sum of lines. There are many lines at or very close to those where other transitions occur. These overlapped and almost equal wavelength transitions have been added together to obtain the total intensity. No broadening is considered. Collisional, Doppler, and Stark broadening depend on the physical conditions of the plasma. Hence, these lines can be broadened in a model depending on the plasma condition. Due to the large number of points, almost all spectra appear continuous. However, some sparse points are also visible in some energy regions for some ions. A separate file containing the photoabsorption cross-sections for each ion is available at the NORAD-Atomic-Data database [25].

We found one experimental spectrum of a lanthanide, Ho II, measured by the group at the University of Connecticut [23], who presented the results (Figure 1, top panel) at a Division of Atomic, Molecular, and Optical Physics (DAMOP) meeting of the American Physical Society in 2016. The red curve in Figure 1 corresponds to the Ho II yield from fragmentation and the black dotted curve to solid Ho photoabsorption. The measured energy range, 150–180 eV, is in the soft X-ray region and is much smaller than that covered for Ho II in the present work. In Figure 1, the lower panel compares the present Ho II photoabsorption spectrum with the experimental one obtained by Obaid et al. [23]. The comparison indicates that the black curve is due to the photoabsorption of the spectrum of Ho II following photo-fragmentation. Although there is an energy shift of 10 eV in the predicted spectrum, we find very good agreement in the features between the two spectra. Similar to the observed feature, the predicted spectrum shows a rise in line strength that increases with the energy and remains strong over an energy range before dropping off at about 190 eV. The observed spectrum is a smooth curve since the spectral lines have been averaged out by the bandwidth of the experimental set-up.

Table 4. Sample table of dipole allowed E1 transitions with same spin for Ho I to demonstrate the format of the complete table of transitions. i and j are the energy level numbers of transitional levels i and j, SLpiCi and SLpCj are the transitional level symmetries with their configuration numbers, gi and gj are the statistical weight factors (as given in the energy table), wl(A) is the wavelength of the transition in Å, fij is the oscillator strength, and aji(s-1) is the radiative decay rate in sec^{-1}.

i-j	SLp Ci-SLp Cj	gi-gj	wl(A)	fij	aji(s-1)
29-3	4Ie 2-4Io 1	12-10	6454.84	4.24E-06	8.15E+02
29-4	4Ie 2-4Io 1	10-10	7270.34	4.48E-06	5.65E+02
42-2	4Ke 2-4Io 1	14-12	4678.36	9.76E-06	3.47E+03
42-3	4Ke 2-4Io 1	12-12	5289.56	1.11E-05	2.65E+03
42-4	4Ke 2-4Io 1	10-12	5824.98	8.06E-07	1.32E+02
...					
361-455	2Po 1-2Pe 2	4-2	946.19	7.21E-02	1.07E+09

Table 5. Comparison of A-values between the present calculated results and those available in the NIST [13] compilation table. fij is the oscillator strength and aji(s-1) is the radiative decay rate in sec^{-1} for transitions from level i to j.

$A_{ji}(s^{-1})$		Transition
NIST	SS	
Ho I		
3.73×10^7	2.77×10^7	$4f^{11}6s^2(^4I^o_{15/2}) - 4f^{10}5d6s^2(^4K_{13/2})$
1.62×10^8	1.00×10^8	$4f^{11}6s^2(^4I^o_{15/2}) - 4f^{10}5d6s^2(^4I^o)_{15/2}$
Ho II		
6.35×10^7	3.09×10^7	$4f^{11}6s(^4I^o_8) - 4f^{11}6p(^4I_8)$
4.87×10^7	4.64×10^7	$4f^{11}6s(^4I^o_7) - 4f^{11}6p(^4I_8)$
Ho III: No A-value is available		
Er I		
1.16×10^8	2.49×10^8	$4f^{12}6s^2(^3H_6) - 4f^{12}6s6p(^3H^o_6)$
7.28×10^7	7.26×10^7	$4f^{12}6s^2(^3H_6) - 4f^{11}5d6s^2(^3I^o_5)$
Er II		
2.0×10^7	4.67×10^7	$4f^{12}6s(^4H_{13/2}) - 4f^{11}5d6s(^4I^o_{11/2})$
1.4×10^7	1.01×10^7	$4f^{12}6s(^4H_{13/2}) - 4f^{12}6p(^4I^o_{13/2})$

Table 5. Cont.

NIST	$A_{ji}(s^{-1})$ SS	Transition
	Er III, Er IV: No A-value is available	
	Tm I	
5.3×10^6	3.33×10^6	$4f^{13}6s^2(^2F^o_{7/2}) - 4f^{12}5d6s^2(^2G_{9/2})$
1.47×10^7	1.81×10^7	$4f^{13}6s^2(^2F^o_{7/2}) - 4f^{12}5d6s^2(^2G_{7/2})$
	Tm II	
1.06×10^8	7.29×10^7	$4f^{13}6s(^3F^o_4) - 4f^{12}5d6s(^3G_5)$
1.57×10^7	2.19×10^7	$4f^{13}6s(^3F^o_4) - 4f^{12}5d6s(^3F_4)$
	Tm III, IV, V: No A-value is available	
	Yb I	
1.00×10^8	1.66×10^8	$4f^{14}6s^2(^1S_0) - 4f^{13}5d6s^2(^1P^o_1)$
6.83×10^7	9.12×10^7	$4f^{14}6s^2(^1S_0) - 4f^{13}5d6s^2(^3P^o_1)$
	Yb II	
6.83×10^7	9.21×10^7	$4f^{14}6s^2(^1S_0) - 4f^{13}5d6s^2(^3P^o_1)$
1.92×10^8	1.66×10^8	$4f^{14}6s^2(^1S_0) - 4f^{14}6s6p(^3P^o_1)$
	Yb III, IV, V, VI: No A-value is available	
	Lu I	
7.90×10^6	4.53×10^6	$5d6s^2(^2D_{3/2}) - 5d6s6p(^2P^o_{3/2})$
1.85×10^8	3.10×10^8	$5d6s^2(^2D_{3/2}) - 5d6s6p(^2F^o_{5/2})$
	Lu II	
4.53×10^8	3.97×10^8	$6s^2(^1S_0) - 6s6p(^1P^o_1)$
7.14×10^7	5.43×10^7	$6s^2(^1S_0) - 6s6p(^1P^o_1)$
	Lu III, IV, V, VI, VII: No A-value is available	

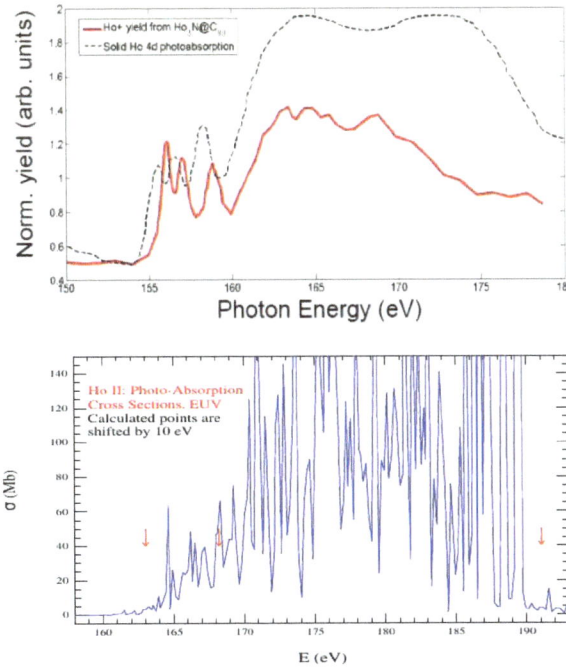

Figure 1. Photoabsorption cross-sections (σ) of Ho II. Top: Experimental photoabsorption spectrum (black dashed curve) of Ho [23]. Bottom: Predicted spectrum of Ho II from the present work. The predicted

energy is shifted by about 10 eV. Arrows point to energies E = 155, 160, and 180 eV, around which a change in feature in the measured spectrum is noticeable. The similarities in the features indicate that the black curve in the top panel corresponds to the photoabsorption features of Ho II following the fragmentation of Ho.

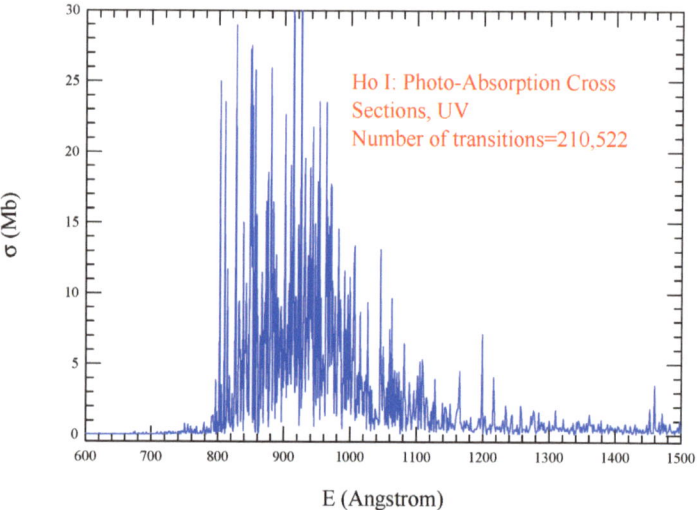

Figure 2. Photoabsorption cross-sections (σ) of Ho I demonstrating broad spectral feature in the UV wavelength region of 800–1150 Å.

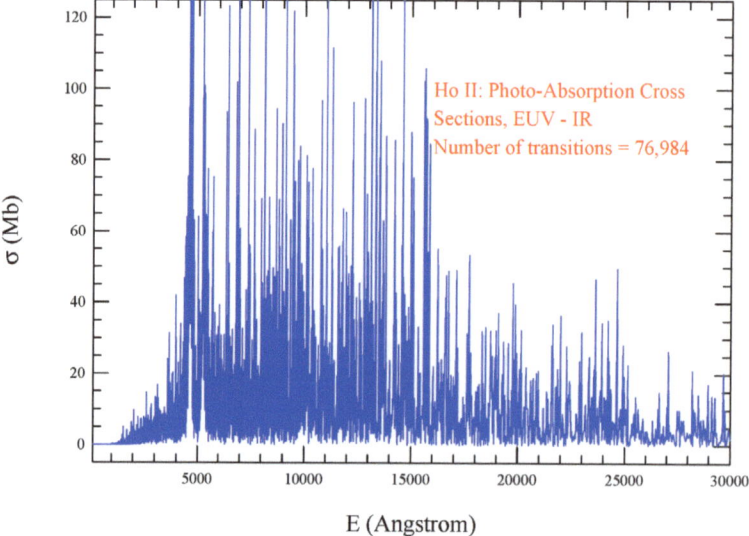

Figure 3. Photoabsorption cross-sections (σ) of Ho II demonstrating broad spectral feature in the UV wavelength region.

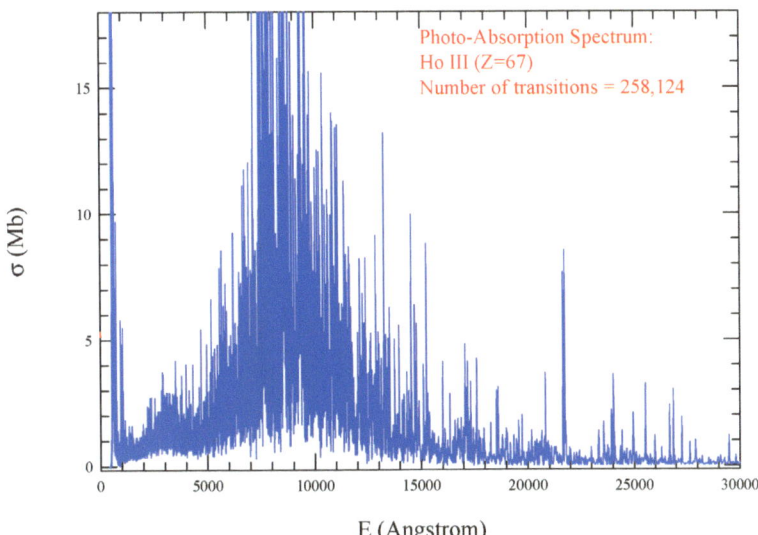

Figure 4. Photoabsorption cross-sections (σ) of Ho III demonstrating a very broad spectral feature in the O–IR wavelength region, particularly ranging from 4000 to 15,500 Å.

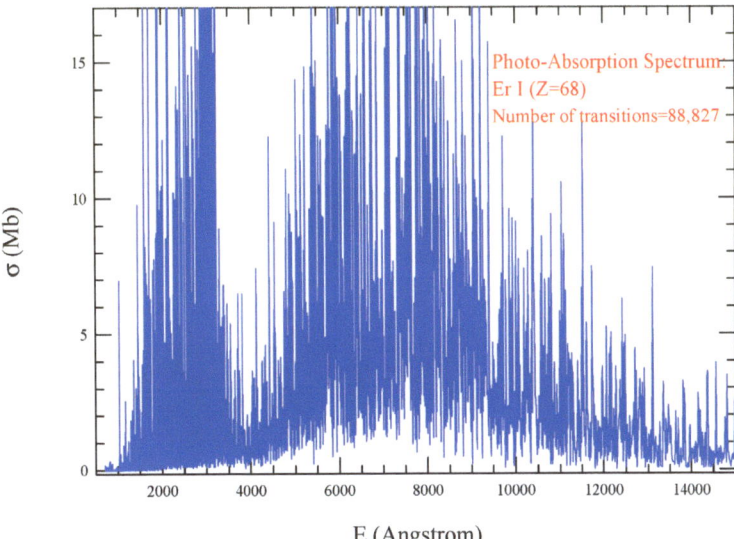

Figure 5. Photoabsorption cross-sections (σ) of Er I demonstrating one broad, 1000–3500 Å, and one very broad, 4000–13,200 Å, spectral feature in the O–IR wavelength region.

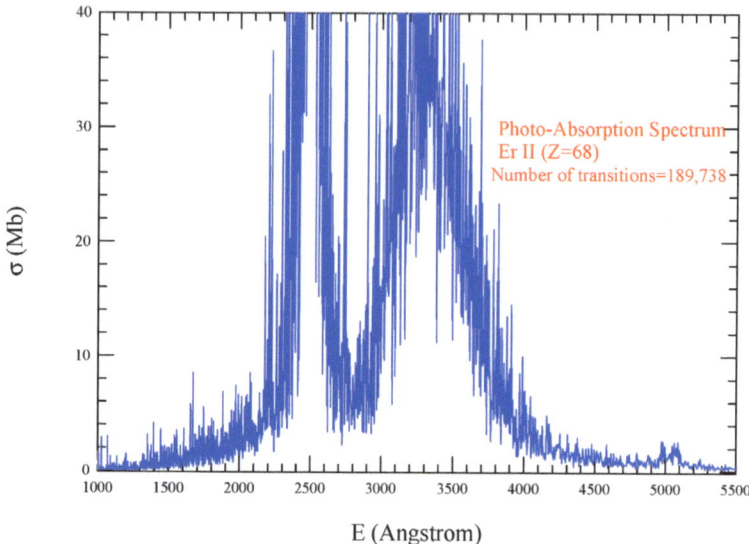

Figure 6. Photoabsorption cross-sections (σ) of Er II demonstrating two broad spectral features in the UV (2200 Å)–near-optical (4000 Å) region.

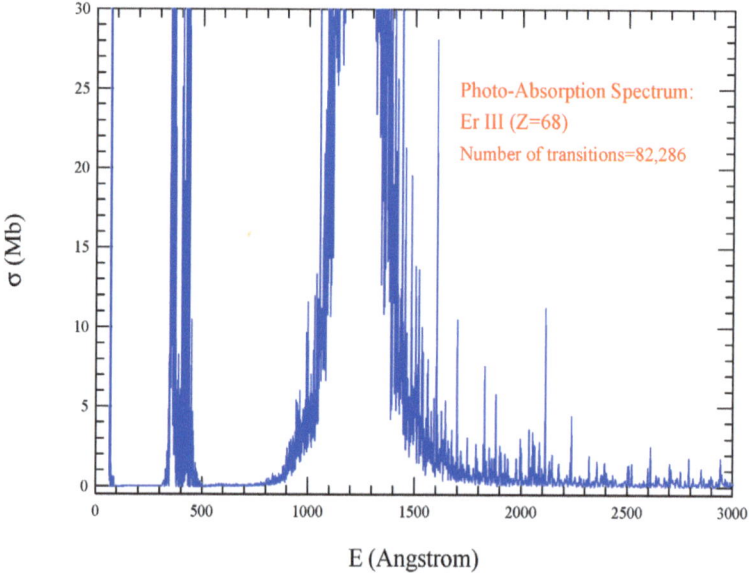

Figure 7. Photoabsorption cross-sections (σ) of Er III demonstrating three regions of high peak strong lines from X-ray to UV regions: the first one is in the narrow X-ray region (around 80 Å), one is a relatively narrow region in the EUV range (300–500 Å), and one is a relatively large broad spectral region in the UV range (900–1700 Å).

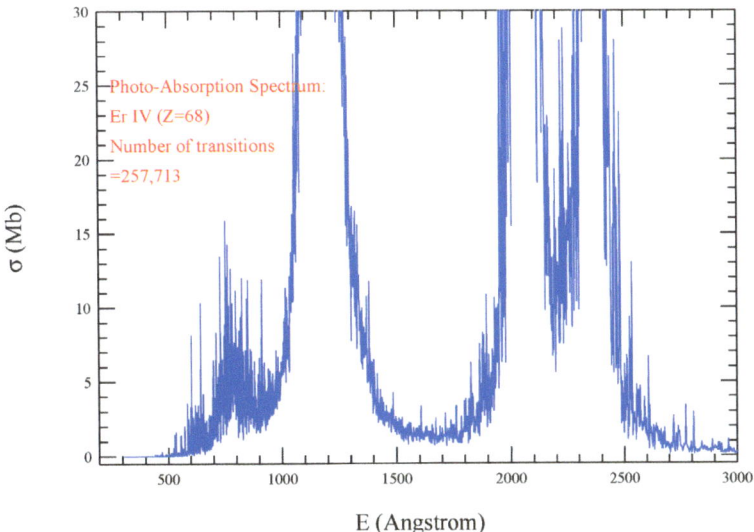

Figure 8. Photoabsorption cross-sections (σ) of Er IV demonstrating multiple regions of high peak strong lines in the UV region.

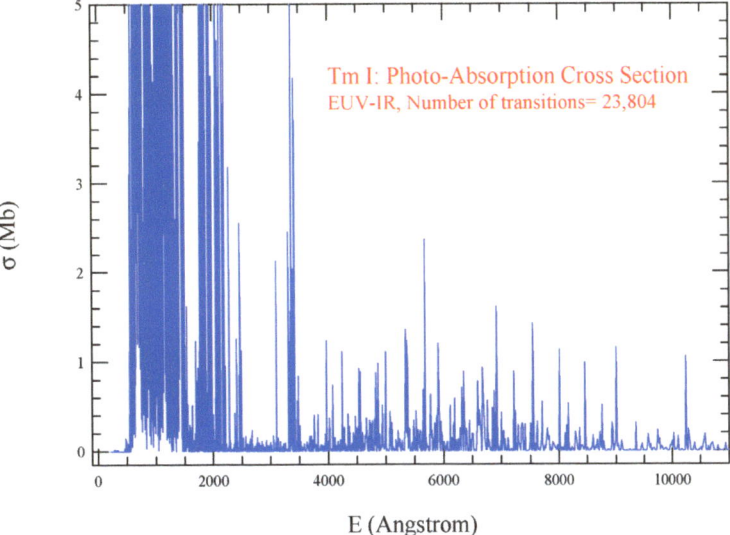

Figure 9. Photoabsorption cross-sections (σ) of Tm I demonstrating three regions of high peak strong lines: one in the EUV region of 500–1500 Å, the second one in 1700–2500 Å, and the third one in the narrow band around 3400 Å. Noticeable lines become more sparse with larger wavelengths.

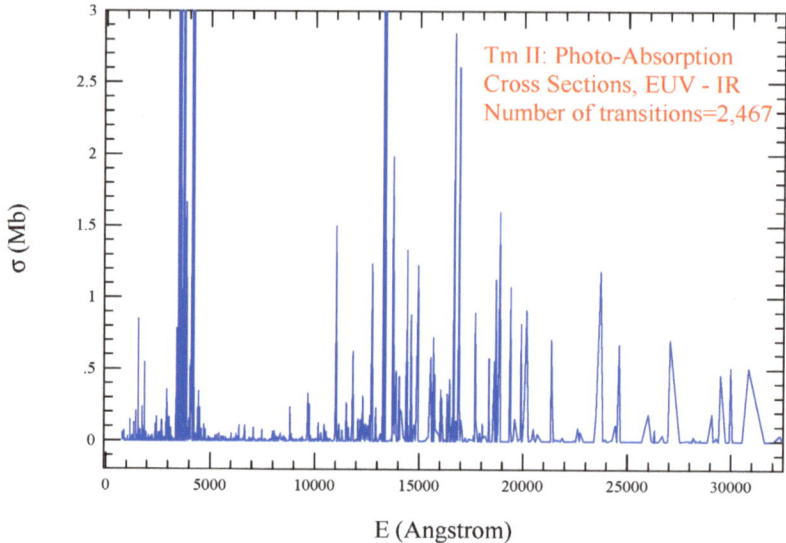

Figure 10. Photoabsorption cross-sections (σ) of Tm II demonstrating the visible presence of strong lines in the UV and IR regions and almost no strong lines in the optical (4000–7000 Å) region. Compared to other lanthanides discussed here, this ion has a smaller number of transitions and a relatively wider broad feature exists in the IR region.

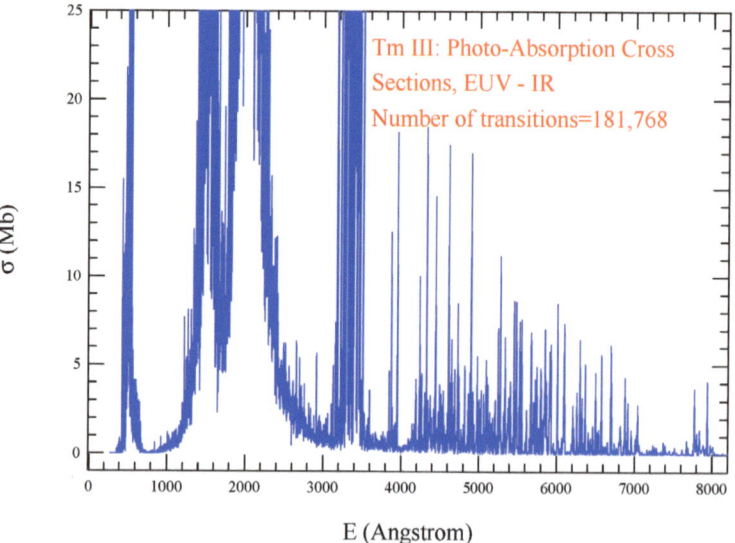

Figure 11. Photoabsorption cross-sections (σ) of Tm III demonstrating multiple broad structures from EUV to O wavelength range, with the widest one being in the range of 1200–2500 Å.

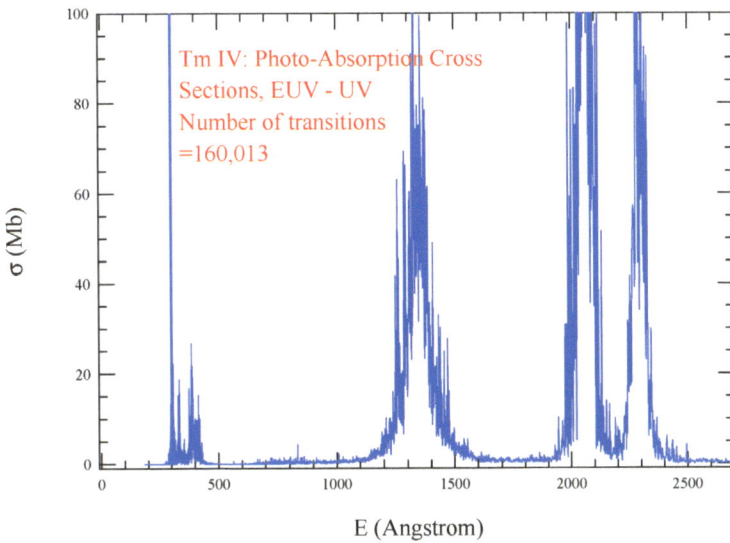

Figure 12. Photoabsorption cross-sections (σ) of Tm IV. There are four distinct broad regions, with strong lines, in the wavelength regions from EUV up to UV.

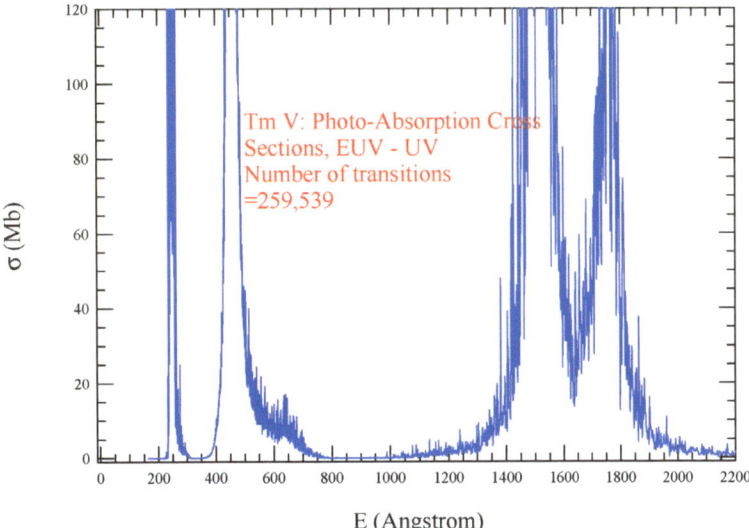

Figure 13. Photoabsorption cross-sections (σ) of Tm V in the energy range of EUV to UV. It demonstrates multiple broad structures from the EUV to the UV wavelength range, with the widest one being in the range of 1400–1900 Å with a dip at around 1650 Å.

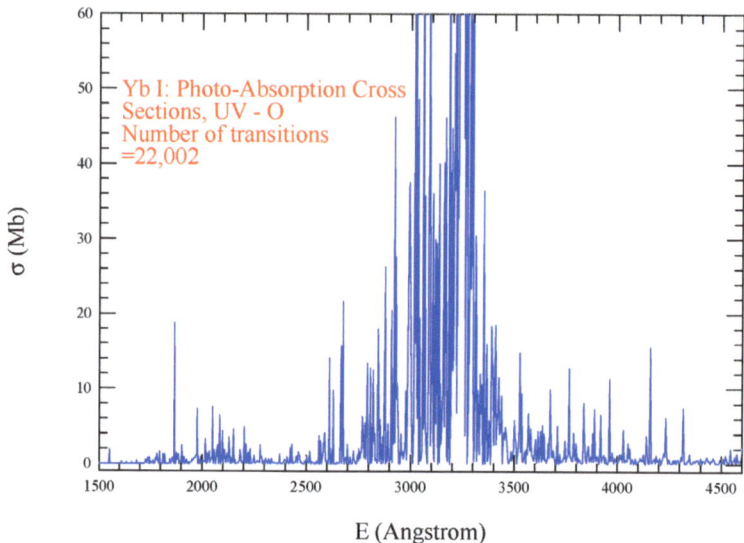

Figure 14. Photoabsorption cross-sections (σ) of Yb I in the energy range of UV to O. A region of strong lines appears in the wavelength range of about 2600–3500 Å.

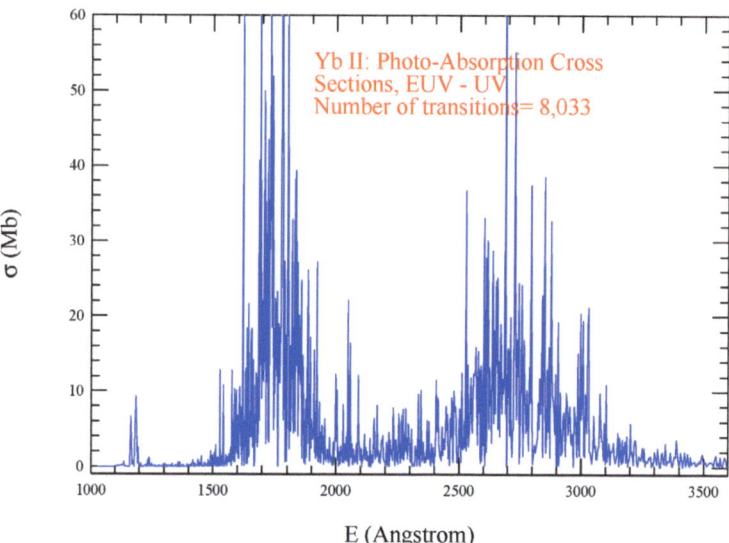

Figure 15. Photoabsorption cross-sections (σ) of Yb II with dominating strong lines in the energy range of EUV–UV. The spectrum shows two broad features, one in the wavelength range of 1500–2100 Å and another one in 2500–3100 Å.

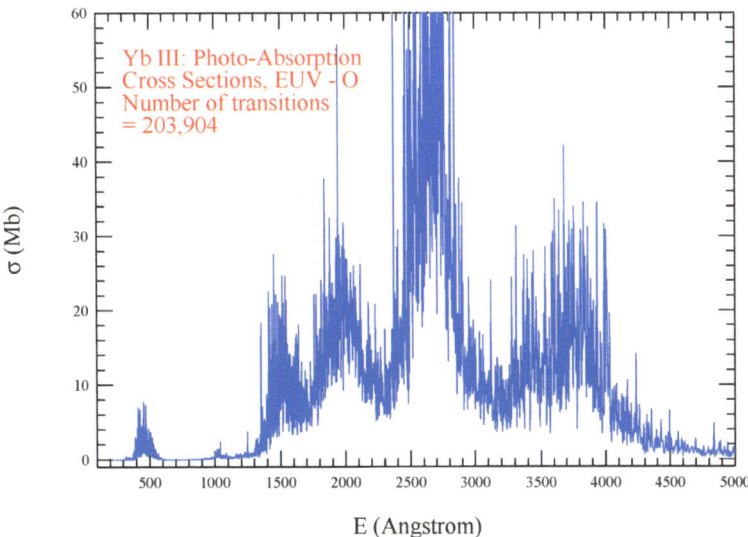

Figure 16. Photoabsorption cross-sections (σ) of Yb III in the energy range of EUV to O. The spectrum has multiple broad features dominated by strong lines in the energy range of EUV–O. The strongest absorption bump is in the wavelength range of 2200–3000 Å.

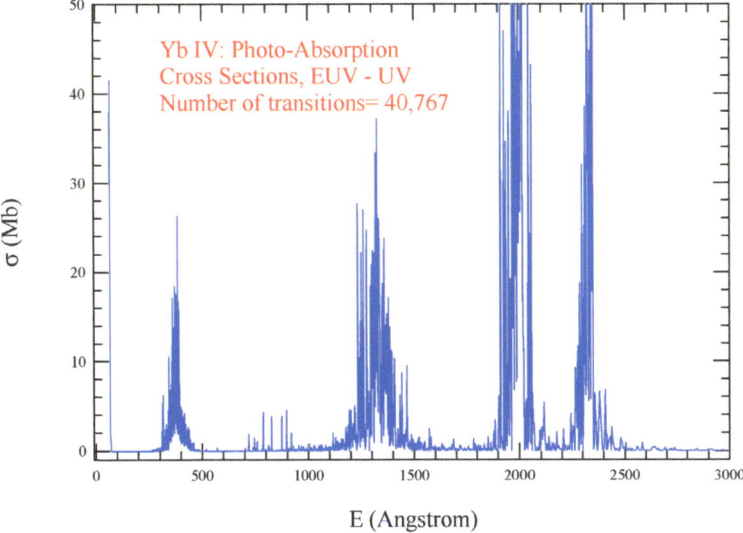

Figure 17. Photoabsorption cross-sections (σ) of Yb IV in the energy range of E to EUV to UV, exhibiting multiple regions of strong lines.

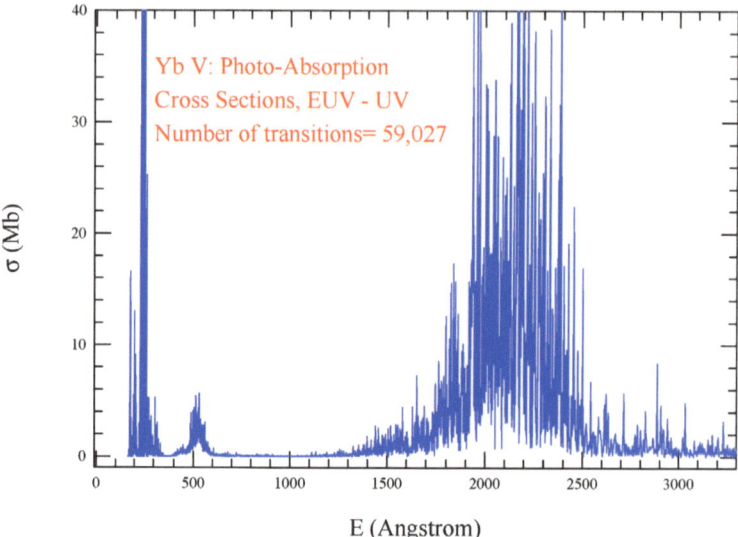

Figure 18. Photoabsorption cross-sections (σ) of Yb V in the energy range of E to EUV to UV. It has two broad features dominated by strong lines: one in the EUV region followed by a lower peak structure and a broader absorption bump in the wavelength region of 1600–2500 Å.

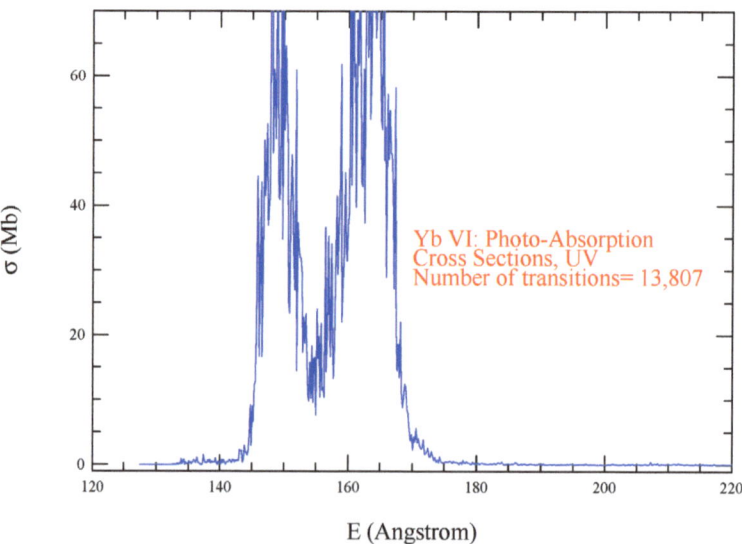

Figure 19. Photoabsorption cross-sections (σ) of Yb VI in the energy range of UV. It has two absorption bumps next to each other in the energy range of 145–170 Å.

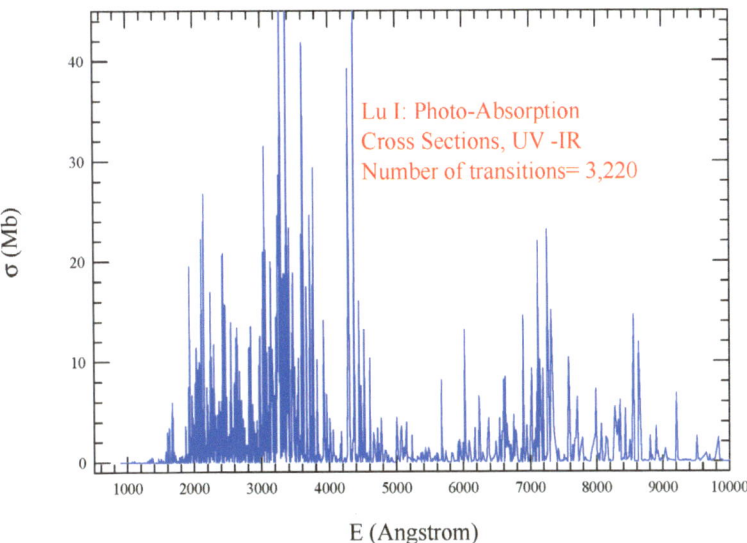

Figure 20. Photoabsorption cross-sections (σ) of Lu I in the energy range of UV–IR, showing several energy regions of strong lines.

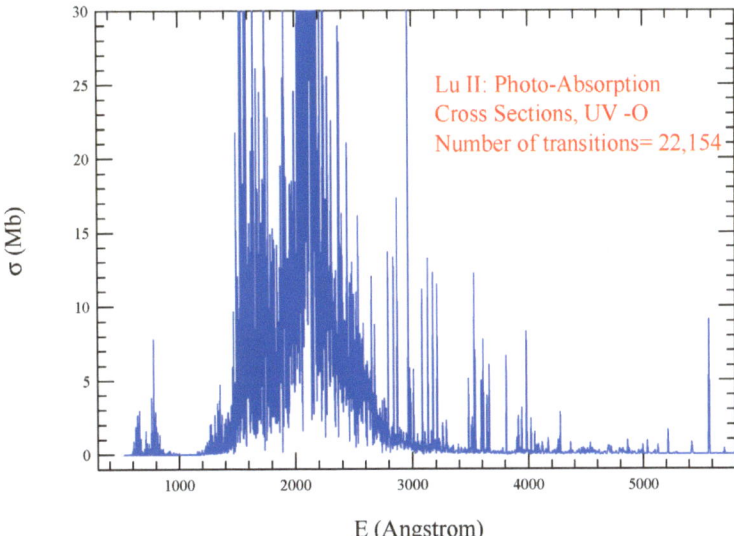

Figure 21. Photoabsorption cross-sections (σ) of Lu II in the energy range of UV–O. Prominent lines are seen in the energy region from UV to O. The spectrum has a wide broad region of strong photoabsorption lines in UV ranging from 1400 to 3300 Å.

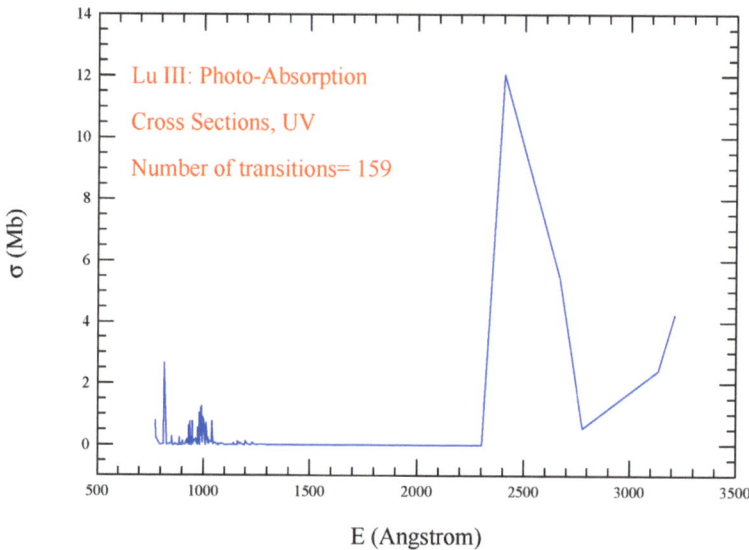

Figure 22. Photoabsorption cross-sections (σ) of Lu III in the energy range of UV. The spectrum has only a few strong lines.

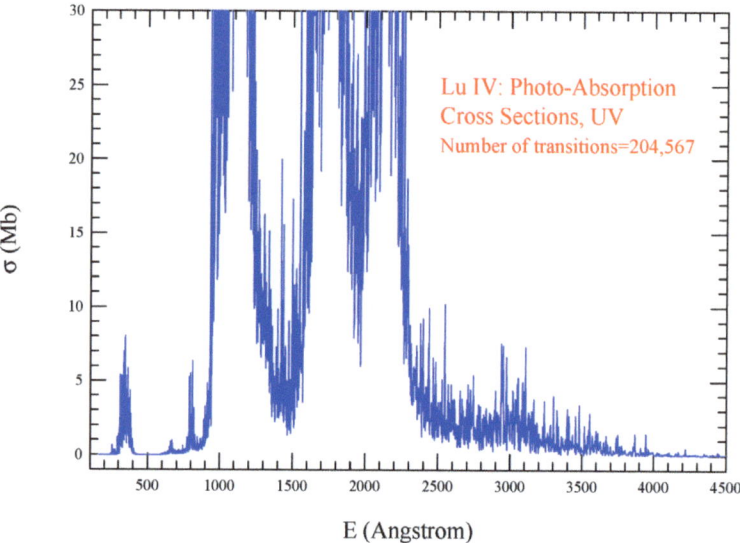

Figure 23. Photoabsorption cross-sections (σ) of Lu IV in the energy range of UV, showing multiple absorption bumps in the UV region.

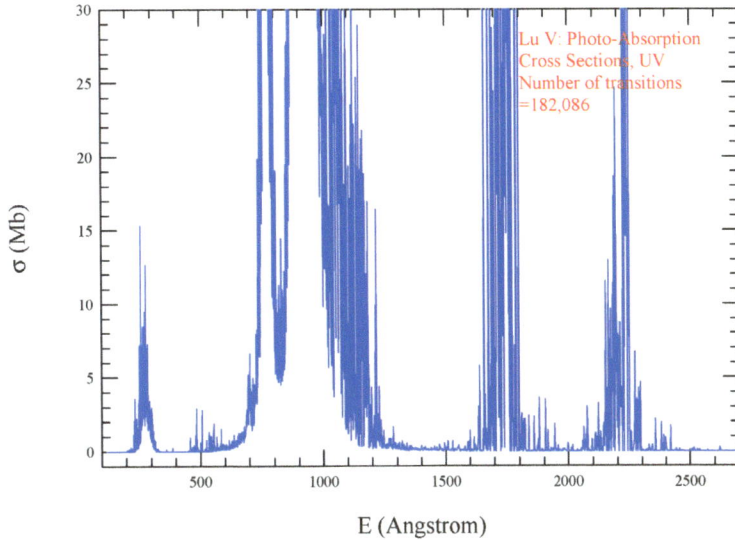

Figure 24. Photoabsorption cross-sections (σ) of Lu V with prominent lines in the UV energy region. The spectrum shows the presence of multiple broad photoabsorption bumps. The broadest one is in the EUV range of 700–1200 Å with a dip around 800 Å.

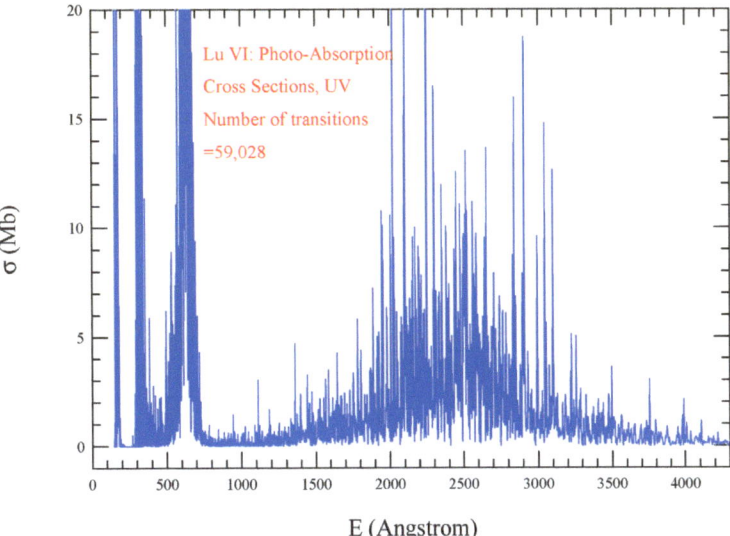

Figure 25. Photoabsorption cross-sections (σ) of Lu VI showing the presence of strong lines in the energy region from EUV to near O. The spectrum shows 3 energy regions of very strong lines in the EUV range and a relatively broad feature at about 1700–3500 Å.

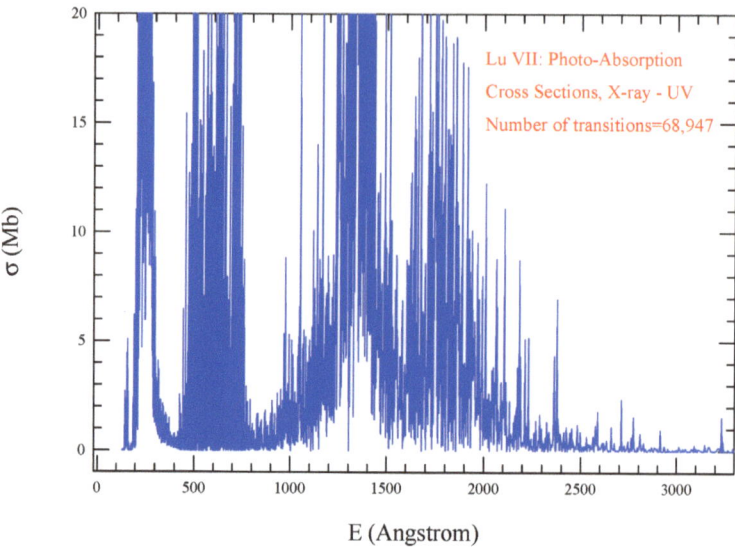

Figure 26. Photoabsorption cross-sections (σ) of Lu VII with prominent absorption lines in the energy range of EUV–UV. The spectrum shows multiple broad absorption energy bumps in the energy region. The prominence of strong lines in the absorption bumps, one from 150 Å in the X-ray range to 350 Å in the EUV range and one from 400–800 Å in UV, can be seen. Two absorption bumps exist next to each, covering a energy width of about 1000 to 2200 Å.

4.2. Benchmarking of Energies

A small set of energies, with 10 levels for brevity, of each of the 25 lanthanide ions is presented in Table 3. The total number of energy levels, N_E, obtained from the set of configurations, and the total number of dipole allowed (E1) transitions, N_{E1}, obtained among them are specified at the top. N_{E1} is the subset of the total number of combined allowed and forbidden transitions, N_T, specified in Table 1. N_{E1} also corresponds to the number of transitions included in producing the photoabsorption spectrum for the ion.

Table 3 also benchmarks the present energies with the measured values, largely from Martin et al. [12], which are available in the NIST [13] table. There are two reasons for the comparison of a small set. With a larger number of levels, the comparison table will be very long with 25 ions. The other reason is the difficulty in the correct matching of levels to compare. Although only 10 energy levels have been used for the comparison with measured values for each ion, some of the ions have a relatively large number of levels available in the NIST table and some do not have any except for the ground angular momentum. A significant number of levels are not assigned full spectroscopic designation. The comparison of energies in Table 3 reveals the complex issues in the spectroscopic identification of levels. The NIST tables show the large mixing of the levels from different LS states as well as configurations, indicating possible variations in the spectroscopic designations for the level from various theoretical approaches. In most cases of the present lanthanide ions, NIST provides only partial identification for a level, the *J*-value, and the parity. Similar to those in NIST, SS provides the final spectroscopic identification based on the leading percentage contributions of the configurations and states. The leading percentage contribution depends on the method being used and the wavefunctions describing the atomic states. Some differences in potential or wavefunction representation in various approaches usually introduce differences in identification. However, general agreement on the spectroscopic identification can often be found among most of them. Nonetheless, these differences in identification are found to be more noticeable for lanthanides.

Identification becomes sensitive to perturbations from the mixing of levels and configurations. This is not unusual for large atomic systems with many electrons, such as lanthanides, which have highly sensitive electron–electron interactions. We attempted to match the levels with the exact identifications, such as the J-values and parity and any specified configuration, for the energy comparisons. In some cases, it appears that better agreement exists if the sequential calculated energies of the same parity are compared with the observed values, which brings the question of the possibility of higher accuracy for the calculated values than the assigned J-values, which are affected by the percentage contributions of other levels. Parity is not affected by percentage contributions.

We adopted a matching scheme for the calculated optimized set of energies to compare them with the measured energy levels. We first ensured that the calculated ground level agreed exactly with those in the NIST table, and that the other fine structure levels of the ground state matched. Next, we compared the level energies with the exact identifications for both the calculated and experimental levels. However, if there were missing L and S numbers in the NIST table or an unusual difference in the values or order of levels was noticed, we compared the levels with the same J-values and parities, and the configuration. If the NIST table did not provide L and S values, we used those designated by SS.

We found that, for higher excited levels, the calculated values tend to diverge towards larger values than those of the measured ones. During optimization, we made an attempt to reduce this divergence even when the energy order was shifted. The energy levels of the 25 ions are discussed below.

Ho I-III:

For Ho I, the comparison encountered problems due to differences in the spectroscopic identification of levels. The NIST table presents the full spectroscopic identification of the ground state $^4I^o$ and its four fine structure levels. The present results agree with them, with about 20% uncertainty in the values. Given the difficulty encountered due to the sensitivity of electron–electron correlation, this agreement can be considered good. The NIST table does not give complete identifications for the next four sets of levels. Thus, in Table 3, we list them with the L and S values of the calculated levels from SS that have the same J and parity. In the next four levels of Ho I in Table 3, we see that the differences in energy values vary significantly. The last two levels in the table have been selected following their energy positions in the calculated energies. The comparison with the measured values is good in regard to the sensitivities of Ho I.

The problem in the proper matching of levels is also one of the reasons for the mismatched order of the calculated energy levels with those of the measured levels. To match for comparison, we have used largely the NIST energy order and SS identifications for the designation of levels. However, even with such identifications, the comparison shows poor to good agreement. Hence, the energies of Ho I may require a larger configuration set for better optimization in future work.

The calculated energies and order of the low-lying Ho II levels have good agreement with those in the NIST [13] table. However, NIST does not provide spectroscopic values of the total spin S and orbital angular momentum L. Hence, the identifications produced by SS are used to designate these values in the comparison table.

The calculated energies of Ho III are comparable but are lower than those of [12]. The calculated energy order also differs after the first five levels and thus introduces a discrepancy. Further optimization in the wavefunction expansions increased the difference between the calculated and measured values. Hence, the present set of values is chosen for overall agreement. It is possible that the differences in energies between the calculated and measured values are due to uncertainties introduced by configuration mixing and hence the lack of proper spectroscopic identifications of levels.

Er I-IV:

The energy levels of Er I show, in general, good to poor agreement with the measured values. These ions were challenging computationally and in terms of spectroscopic identification in similar ways to the Ho ions. The sensitivity of the electron–electron interaction caused the repetition of the computation many times to optimize the wavefunctions to match the ground and excited levels. When the energy levels are similar to those of the observed values, the identifications and order of levels would be different from those of NIST. We adopted the comparison strategy mentioned in the above section for the best matching of the levels and the comparison of energies, which showed good to fair agreement.

The calculated energies of Er II show agreement similar to that for Er I. Except for one odd parity state, $^4I^o$, none of the Er II levels, including the ground state, have full spectroscopic designation in the NIST table. NIST provides J-values and parities. Hence, the levels have been assigned with the L and S values given by SS. For the Er II comparison, with the preservation of the parities and J-values, we find that the $^4I^o$ levels show the largest discrepancy. This discrepancy would have been smaller if we had considered configuration $4f^{11}5d6s$ instead of the NIST-assigned configuration $4f^{11}6s^2$. The agreement is quite good with the $4f^{11}5d6s$ odd parity level of J = 13/2. Hence, the comparison will need verification with other accurate calculations or experiments.

The optimization produced calculated energies for Er III that were somewhat lower than the measured values, but they remained in fair to good agreement. The NIST table assigns spectroscopic designation only for three levels. Hence, while the parity and J-values are matched for comparison, the L and S values are assigned following those from SS.

The calculated energies of Er IV can be discussed in terms of the same points as for Er I-III. For Er IV, the calculated values are somewhat higher than the values quoted in NIST. The agreement is fair to good.

Tm I-V:

Table 3 shows good to poor agreement between the calculated and measured energies of Tm I–Tm-V. The reasons for the differences are the same as those for the Ho and Er ions. These lanthanide ions have strong electron–electron interactions that can be perturbed easily by slight changes in wavefunction, similarly to other lanthanides. The problem of the matching identification of levels and partial identification introduced uncertainty in the comparison of the levels and hence caused a discrepancy between the calculated and measured energies.

The Tm I energies are very sensitive to the configurations and Thomas–Fermi scaling parameters. A slight change in the scaling parameters, which expands or contracts the wavefunction, would affect the order and values of the levels and the energy values. It is also difficult to compare them as NIST gives the configurations but is missing the L and S values. The energies from SS are closer to the energies of the NIST values if the quantum states designated to them are ignored. This again highlights the need for high-accuracy calculations and experimental measurements.

Tm II and Tm III show similar identification problems. NIST provides a number for the total angular momentum, instead of an alphabetic character, and no spin information. Thus, the present comparison is made largely based on the order of matching of the parity and J-values. For Tm II, levels 6 and 7 appear to have reverse identification. The agreement between the present and the compiled set of NIST is good to fair. The comparison of Tm III can be considered good on average.

For Tm IV, the present values in general are in good agreement with the measured energy values of Martin et al. [12]. However, two levels, ^3H and ^3F, both with j = 4, appear to be misidentified, with a change in energy values of 0.114 Ry and 0.0514 Ry. The identification of these levels can be adopted easily in the reverse order as their leading percentages are very close, 60 and 63, respectively. The NIST table list only seven levels. Hence, the three additional levels in Table 3 are calculated ones.

For Tm V, there is no energy level available except for the ground level $4f^{11}(^3I^o_{15/2})$, which we confirm in the present work.

Yb I-VI:

Yb I identification for the four levels above $4f^{14}6s6p(^3P^o)$ is not defined in the NIST table and hence the SS identifications that match the parity and j-values are used. Yb I also perturbs easily with a slight change in the spatial extension of the orbital wavefunctions.

They were optimized to match the observed low-lying energies by Martin et al. [12], but the order of the spectroscopic identifications has less agreement. The present comparison shows large differences, but they can be reduced by the matching of the configurations, which was not considered.

Yb II also had very a sensitive electron–electron interaction potential, which would change the order of the energy levels or the energy values with a slight change in the wavefunction. Since the NIST table reports an over 90% leading percentage for the lowest-lying level designations, the present optimization focused more on the energy order than the energy values to achieve it. The comparison shows fair to good agreement between the calculated and observed values.

The calculated energies for Yb III are seen to be in good agreement with the measured values. SS identifications were used for the levels whenever NIST did not have them.

The calculated Yb IV energies agree well with the measured values. The identifications of levels given in Table 3 correspond to those predicted by SS as most levels are not identified in the NIST table.

There is no measured energy for Yb V available on the NIST website, except for the ground level designation of $^3H_6$. Our calculated ground level agrees with the level designation. Hence, all energy levels of Yb V in Table 3 are calculated values.

Yb VI has 68 levels belonging to the ground configuration $4f^{12}5p^5$. This means that there is no low-lying allowed transition. The transition energy for the first dipole allowed transition of the ground level $^4I^o_{15/2}$ is at about 1 Ry. The NIST table does not contain any observed energy of Yb VI besides giving the J-value of the ground level, which was predicted by [14]. The present ground level agrees with this J-value. We noted that a different set of configurations can also produce energies for Yb VI that are different in values and energy order from the present set. It also gives a different spectroscopic designation for the ground level. We choose the present set as it has the same ground level configuration as that given in the NIST table. There is a need for observed values as guidance to determine the configuration set for the ion.

Lu I-VII:

The optimization of wavefunctions for the ordering of the Lu I energies was found to be sensitive to the presence of excited configuration $5d^26s$, whose levels would raise the ground level to a higher excited state. A ground level that has even parity needs odd parity levels for dipole allowed transitions. Hence, we attempted to perform the optimization of the energies of odd parity levels such that lower energies were achieved for the levels through the optimization of the wavefunctions. The resultant set of energies, as seen in Table 3, shows good to poor agreement with the set of NIST.

The energies of Lu II show good agreement with the measured values.

For Lu III, we are able compare only a limited number of energy levels since most of the energies in the NIST table belong to orbitals that are not accessible to SS. However, the comparison shows good agreement. The NIST table does not list the energies of configuration $4f^{13}5d^2$, which is included in the present calculations and found to produce a large number of bound levels. Being a $^2S_{1/2}$ level of configuration $4f^{14}6s$, the ground level can have dipole allowed transitions only to $^2P^o_{1/2,3/2}$ and hence only a limited number of transitions is possible. We obtained only 159 dipole allowed transitions, including those to levels of $^2S_{1/2}$, of the ground level to be included in producing the spectrum. We included a few other configurations that generated bound levels but have not been observed.

For Lu IV, overall good agreement is found between the calculated and measured values. The levels have been identified following SS, as NIST does not provide the full spectroscopic designation of any excited level.

For Lu V, the calculated values of the ground level and lower energies agree well with the measured values listed by NIST [13]. The full spectroscopic designation of any excited level of Lu V is not available in the NIST table. The leading percentage values in the NIST table indicate highly mixed states. Hence, in Table 3, these levels have been designated with the *LS* term obtained from SS when the *J*-value and parity match for both the calculated and observed levels.

For Lu VI, there is no fine structure level, except for the ground level, available in the NIST table. We carried out the optimization of the levels such that the absolute values were the lowest and the ground level matched $4f^{12}5p^6(^3H_6)$ given by NIST. All nine excited levels of Lu VI in Table 3 are calculated values.

There are no fine structure energy levels for Lu VII in the NIST table, except for the ground configuration $4f^{13}5p^4$, with which the present work agrees. Similar to Lu VI, the optimization of energies was carried out by lowering the energy values.

4.3. Benchmarking of Transitions in Lanthanide Ions

We perform a comparison of a number of A-values of the present lanthanide ions in Table 5 with the compiled values available at NIST [13]. A limited number of A-values for some ions of Ho, Er, Tm, Yb, and Lu is available in the NIST [13] compilation table. The NIST references for the A-values of lanthanides are Meggers et al. [15], Morton [16], Komarovski [17], Wickliffe and Lawler [18], Sugar et al. [19], Penkin and Komarovski [20], and Fedchak et al. [21]. The comparisons show variable agreement between theory and computation. The order of magnitude agrees, and the absolute values agree to different degrees. Typically, the A-values calculated from two different approaches or programs show general agreement in the transitions, but not for all transitions. Hence, in the present case, the overall agreement should be good given the sensitivity of electron–electron interactions and the impact of slight changes in the wavefunction in lanthanides.

We compare only a couple of transitions. The reason for this is that the comparisons are expected to have a certain amount of uncertainty due to the lack of proper spectroscopic identification, particularly of the excited level to which the ion is excited. This problem is due to the high mixing of levels. The other issue is the identical set of quantum numbers. A large number of possible angular momenta resulting from the vector addition of the individual angular momenta of a large number of electrons introduce multiple sets of similar quantum numbers that can be assigned to a level. A single configuration can produce a number of levels with different energies but with the same J-values and transitions that occur between the same set of two J-values of the transitional levels belonging to the same set of configurations, although with different energies. The small differences in energies, such as for lanthanides, do not resolve these issues since the calculated energies are not as precise as the measured values.

The present work aims at the overall improvement of the accuracy for collective features of transitions, such as those observed in lanthanide ions.

4.4. Spectral Features of Lanthanide Ions

Ho I–Ho III:

The photoabsorption spectrum of Ho I is presented in Figure 2. A total of 210,522 transitions were included to plot the spectrum. However, those with very low cross-sections were beyond the scale of the plot. Figure 2 shows that the dominant strength of lines lies in the UV range of about 800 to 1200 Å. The range could deviate by some Å due to the differences between the calculated and measured energies.

Figure 3 presents the photoabsorption spectrum of Ho II, which includes 76,984 transitions. However, very weak transitions lay outside the range of the plot. The figure shows the visible presence of lines from 4000 Å in UV to 18,000 Å in the IR region.

Figure 4 presents the photoabsorption spectrum of Ho III, which shows the dominance of lines from X-ray to FIR 22,000 Å. A large absorption bump is found in the wavelength range of 4000 to 15,500 Å. The number of transitions included in the figure is 258,124.

Er I–Er IV:

The spectrum of Er I presented in Figure 5 shows two broad features of strong lines, one in the UV region of 1000–3500 Å and the other in the O–IR region of wavelengths 4500–15,000 Å. The number of E1 transitions included in the spectrum is 88,827.

The spectrum of Er II presented in Figure 6 shows two broad features of strong lines, both in the UV region: one in the wavelength range of 2200–2700 Å and the other one in 2800–4000 Å. The number of E1 transitions included in the spectrum is 189,738.

The spectrum of Er III presented in Figure 7 shows two broad features of strong lines, both in the EUV region: one in the wavelength range of 300–500 Å and the next one in 900–1700 Å. The number of E1 transitions included in the spectrum is 82,286.

The spectrum of Er IV presented in Figure 8 shows multiple broad features of strong lines in various wavelength ranges in the EUV–UV region. The number of E1 transitions included in the spectrum is 247,713.

Tm I–Tm V:

The spectrum of Tm I, presented in Figure 9, shows a few regions of strong lines; the broadest region is in the EUV wavelength range of 500–1500 Å and the next one is at about 1700–2500 Å. Beyond this, at about 3400 Å, a narrow region of strong spectral lines is noticeable. The number of transitions included is 23,804.

The spectrum of Tm II, presented in Figure 10, shows the visible presence of strong lines in the UV and IR regions, and almost no strong lines in the optical (4000–7000 Å) region. Compared to other lanthanides discussed here, this ion has a smaller number of transitions and a relatively wider broad feature exists in the IR region.

The spectrum of Tm III, presented in Figure 11, shows multiple broad regions with strong lines, from the EUV to the optical wavelength range. The widest one is in the range of 1200–2500 Å. The number of transitions included is 181,768.

The spectrum of Tm IV, presented in Figure 12, shows four distinct broad regions, with strong lines, in the wavelength regions of EUV up to UV. The number of transitions included is 160,013.

The photoabsorption spectrum of Tm V, presented in Figure 13, demonstrates the dominance of strong lines in the energy range of EUV to UV with a gap of about 700–1300 Å. It shows multiple broad structures from the EUV to the UV wavelength range, with the widest one being in the range of 1400–1900, Å with a dip at around 1650 Å. The number of transitions included is 259,539.

Yb I–Yb VI:

The photoabsorption spectrum of Yb I, presented in Figure 14, demonstrates the dominance of strong lines in the energy range of EUV to UV, with a broad feature in the wavelength range of about 2600–3500 Å. The number of transitions included is 22,002.

The photoabsorption cross-sections (σ) of Yb II, presented in Figure 15, show dominating strong lines in the energy region of EUV–UV. The spectrum also shows two broad features, one in the wavelength range of 1500–2100 Å and another one at 2500–3100 Å. The number of transitions included is 8083.

The photoabsorption cross-sections (σ) of Yb III, presented in Figure 16, has multiple broad features dominated by strong lines in the energy range of EUV–O. The strongest absorption bump is in the wavelength range of 2200–3000 Å. The number of transitions included is 203,904.

The photoabsorption cross-sections (σ) of Yb IV, presented in Figure 17, have multiple broad features dominated by strong lines in the energy range of EUV–UV. The number of transitions included is 40,767.

The photoabsorption cross-sections (σ) of Yb V, presented in Figure 18, show noticeable features in the energy range of EUV–UV. It has two broad features dominated by strong lines: one in the EUV region, followed by a structure of low peaks and a broader absorption bump in the wavelength region of 1600–2500 Å. The number of transitions included is 59,027.

The photoabsorption spectrum (σ) of Yb VI, presented in Figure 19, shows visible features in the EUV region, particularly two absorption bumps next to each other in the energy range of 145–170 Å. The number of transitions included is 13,807.

Lu I–Lu VIII:

The photoabsorption spectrum (σ) of Lu I, presented in Figure 20, shows prominent strong lines in the energy region of UV to IR. It has a few broader regions of strong photoabsorption lines in the energy regions of 1800 to 3000 Å, 3000–4000 Å, and 4300–4600 Å. The higher-energy region also demonstrates the presence of some strong lines. The number of transitions included is 3220.

The photoabsorption spectrum (σ) of Lu II, presented in Figure 21, shows prominent lines in the energy region of UV to O. It has a wide broad region of strong photoabsorption lines in UV from 1400 to 2700 Å, with a dip around 1900 Å in the UV range. The number of transitions included is 22,142.

The photoabsorption cross-sections (σ) of Lu III, presented in Figure 22, show only a few strong lines. The number of transitions included is 159.

The photoabsorption cross-sections (σ) of Lu IV, presented in Figure 23, show the prominence of lines in the UV region. There are multiple broad absorption bumps in the spectrum. The number of transitions included is 204,567.

The photoabsorption cross-sections (σ) of the Lu V, presented in Figure 24, show the prominence of lines in the UV region. There are multiple broad absorption bumps in the spectrum. The broadest one is in the EUV range of 700–1200 Å with a dip at around 800Å. The number of transitions included is 182,086.

The photoabsorption cross sections (σ) of Lu VI, presented in Figure 25, show the prominence of strong lines from EUV to the near-O region. The spectrum shows three energy regions of very strong lines in the EUV range and a relatively broad feature at about 1700–3500 Å. The number of transitions included is 59,028.

The photoabsorption cross-sections (σ) of Lu VII, presented in Figure 26, show the prominence of lines from EUV to UV. The spectrum shows multiple broad absorption energy bumps in the energy region. Two large absorption bumps are seen next to each other, covering an energy width of about 1000–2200 Å. The number of transitions included is 68,947.

The lanthanide ions described in the present report, with strong lines forming spectral features in the wavelength range of about 3000 to about 7000 Å, are the possible ion contributors of the broad feature of GW170817. It is also possible that ions with features in the UV range near 3000 Å and in the IR range beyond 7000 Å also make contributions but are not noticeable for reasons such as weaker transitions, the Doppler shift of the wavelengths, and shifts due to energy loss due to the opacity, which indicates absorption in the medium that the radiation passes through.

5. Conclusions

We summarize the present report as follows.

1. We present atomic data for the energy levels and radiative transitions of 25 ions of lanthanides, Ho I-II, Er I-IV, Tu I-V, Yb I-VI, and Lu I-VII. Compared to the available datasets, these are probably the largest sets of atomic data for these lanthanide ions and can be applied for broad features, such as those from kilonovae events.
2. These data, as extensive sets, are expected to be much more accurate than those available and hence should enable higher-precision astrophysical applications in broad features and fill the gaps in data needed for modeling. It should also be noted that the improved accuracy varies according to how the ion has been represented in the present study.
3. The calculated energies have been benchmarked with the measured values, largely from Martin et al. [12], available on the NIST webpage [13]. The comparison shows overall good agreement, within a few percent, to fair to poor agreement, where the difference can be a factor close to 2 for the energies. This difference increases with higher energies.
4. The radiative transition probabilities have been compared with those available at NIST [13], compiled from a number of sources. The agreement is fair to good. One factor in the

differences is the proper identification of the levels. Much greater improvements will be needed over the present work for line diagnostic applications, using programs such as GRASP, which can provide limited but more accurate energies and transition parameters.
5. We present the spectral features of these 25 lanthanide ions that illustrate the dominance of lines in various regions from X-ray to infrared.
6. Very good agreement with the observed features in Figure 1 is found when compared with the calculated spectral features of Ho II. The observed features were generated by the photoabsorption of Ho II as the Ho compound was fragmented. The agreement is very good given given the strong electron–electron correlation interaction of a large atomic system like Ho II.
7. Lanthanides have highly mixed levels and are very sensitive to slight changes in the representation of the potential and wavefunctions. These characteristics can lead easily to different sets of levels. Hence, guidance through experimentally determined levels is of great need and importance.
8. All atomic data will be available online at the NORAD-Atomic-Data database (https://norad.astronomy.osu.edu, accessed on 7 July 2007) [25].

Funding: All computations of this project were carried out on the high-performance computers of the Ohio Supercomputer Center (OSC).

Data Availability Statement: All data are available online at NORAD-Atomic-Data database at the Ohio State University: https://norad.astronomy.osu.edu accessed since 7 July 2007.

Acknowledgments: SNN acknowledges the many useful discussions with and guidance of Werner Eissner, who passed away in 2021. SNN acknowledges the support of the Ohio State University to maintain the NORAD-Atomic-Data database and the Ohio Supercomputer Center (OSC) for the provision of the computational time.

Conflicts of Interest: The authors declare no conflicts of interest.

References

1. Abbott, B. P.; Abbott, R; Abbott, T. D.; Acernese, F.; Ackley, K.; Adams, C.; Adams, T.; Addesso, P.; Adhikari, R. X.; Adya, V. B.; et al. GW170817: Observation of Gravitational Waves from a Binary Neutron Star Inspiral. *Phys. Rev. Lett.* **2017**, *119*, 161101. [CrossRef]
2. Pian, E.; D'Avanzo, P.; Benetti, S.; Branchesi, M.; Brocato, E.; Campana, S.; Cappellaro, E.; Covino, S.; D'Elia, V.; Fynbo, J. P. U.; et al. Spectroscopic identification of r-process nucleosynthesis in a double neutron-star merger. *Nature* **2017**, *551*, 67. [CrossRef]
3. Kasen, D.; Badnell, N.R.; Barnes, J. Opacities and Spectra of the R-Process Ejecta from Neutron Star Mergers. *Astrophys. J.* **2013**, *774*, 25. [CrossRef]
4. Tanaka, M.; Hotokezaka, K. Radiative Transfer Simulations of Neutron Star Merger Ejecta. *Astrophys. J.* **2013**, *775*, 113. [CrossRef]
5. Tanaka, M.; Kato, D.; Gaigalas, G.; Kawaguchi, K. Systematic opacity calculations for kilonovae. *Mon. Not. R. Astron. Soc.* **2020**, *496*, 1369–1392. [CrossRef]
6. Tanaka, M.; Kato, D.; Gaigalas, G.; Rynkun, P.; Radžiūtė, L.; Wanajo, S.; Sekiguchi, Y.; Nakamura, N.; Tanuma, H.; Murakami, I.; et al. Properties of Kilonovae from Dynamical and Post-merger Ejecta of Neutron Star Mergers. *Astrophys. J.* **2018**, *852*, 109
7. Radžiūtė, L.; Gaigalas, G.; Kato, D.; Rynkun, P.; Tanaka, M. Extended Calculations of Energy Levels and Transition Rates for Singly Ionized Lanthanide Elements. I. Pr–Gd. *Astrophys. J. Suppl. Ser.* **2020**, *248*, 17. [CrossRef]
8. Fontes, C.J.; Fryer, C.L.; Hungerford, A.L.; Wollaeger, R.T.; Korobkin, O. A line-binned treatment of opacities for the spectra and light curves from neutron star mergers. *Mon. Not. R. Astron. Soc.* **2020**, *493*, 4143. [CrossRef]
9. Cowan, R. The Theory of Atomic Structure and Spectra; University of California Press: Berkeley, CA, USA, 1981.
10. Johnson, J.A. Populating the periodic table: Nucleosynthesis of the elements. *Science* **2019**, *363*, 474–478.
11. Kobayashi, C.; Karakas, A.I.; Lugaro, M. The Origin of Elements from Carbon to Uranium. *Astrophys. J.* **2020**, *900*, 179. [CrossRef]
12. Martin, W.C.; Zalubas, R.; Hagan, L. *NSRDS-NBS 60*; National Standard Reference Data Series. National Bureau of Standards: Washington, DC, USA, 1978; 422p. . [CrossRef]
13. Kramida, A.; Ralchenko, Y.; Reader, J.; NIST ASD Team. NIST Atomic Spectra Database (Ver. 5.8). 2020. Available online: https://physics.nist.gov/PhysRefData/ASD/levels_form.htm (accessed on 1 January 1996).
14. Carlson, T.A.; Nestor, C.W., Jr.; Wasserman, N.; McDowell, J.D. Calculated Ionization Potentials for Multiply Charged Ions. *At. Data Nucl. Data Tables* **1970**, *2*, 63–99. [CrossRef]
15. Meggers, W.F.; Corliss, C.H.; Scribner, B.F. *National Bureau of Standards Monograph 145*; National Bureau of Standards: Washington, DC, USA, 1975; 600p. . [CrossRef]

16. Morton, D.C. Atomic Data for Resonance Absorption Lines. Ii. Wavelengths Longward Of the Lyman Limit for Heavy Elements. *Astrophys. J. Suppl. Ser.* **2000**, *130*, 403–436; Erratum in **2001**, *132*, 411. [CrossRef]
17. Komarovskii, V.A. Oscillator strength for spectral lines and probabilities of electron transitions in atoms and single charged ions of lantanides. *Opt. Spektrosk.* **1991**, *71*, 559–592.
18. Wickliffe, M.E.; Lawler, J.E. Atomic transition probabilities for Tm I and Tm II. *J. Opt. Soc. Am. B* **1997**, *14*, 737–753. [CrossRef]
19. Sugar, J.; Meggers, W.F.; Camus, P. Spectrum and energy levels of neutral thulium. *J. Res. Natl. Bur. Stand. Sect. A* **1973**, *77*, 1–43. [CrossRef] [PubMed]
20. Penkin, N.P.; Komarovskii, V.A. J. Oscillator strengths of spectral lines and lifetimes of atoms of rare earth elements with unfilled 4f-shell. *Quant. Spectrosc. Radiat. Transf.* **1976**, *16*, 217–252. (In Russian)
21. Fedchak, J.A.; Den Hartog, E.A.; Lawler, J.E.; Palmeri, P.; Quinet, P.; Biémont, E. Experimental and theoretical radiative lifetimes, branching fractions, and oscillator strengths for Lu I and experimental lifetimes for Lu II and Lu III. *Astrophys. J.* **2000**, *542*, 1109–1118.
22. Irvine, S.; Andrews, H.; Myhre, K.; Coble, J. Radiative transition probabilities of neutral and singly ionized rare earth elements (La, Ce, Pr, Nd, Sm, Gd, Tb, Dy, Ho, Er, Tm, Yb, Lu) estimated by laser-induced breakdown spectroscopy. *J. Quant. Spectrosc. Radiat. Transf.* **2023**, *297*, 108486. [CrossRef]
23. Obaid, R.; Xiong, H.; Ablikim, U.; Augustin, S.; Schnorr, K.; Battistoni, A.; Wolf, T.; Carroll, A.M.; Bilodeau, R.; Osipov, T.; et al. In Proceedings of the 47th Annual Meeting of the APS Division of Atomic, Molecular and Optical Physics (DAMOP) Annual Meeting, Providence, RI, USA, 23–27 May 2016. Abstract: B9.00007.
24. Pradhan, A.K.; Nahar, S.N. *Atomic Astrophysics and Spectroscopy*; Cambridge University Press: Cambridge, UK, 2011.
25. Nahar, S.N. Database NORAD-Atomic-Data for atomic processes in plasma. *Atoms* **2020**, *8*, 68. [CrossRef]
26. Nahar, S.N.; Eissner, W.; Chen, G.X.; Pradhan, A.K. Atomic data from the Iron Project–LIII. Relativistic allowed and forbidden transition probabilities for Fe XVII. *Astron. Astrophys.* **2003**, *408*, 789–801. [CrossRef]
27. Eissner, W.; Jones, M.; Nussbaumer, H. Techniques for the calculation of atomic structures and radiative data including relativistic corrections. *Comput. Phys. Commun.* **1974**, *8*, 270–306. [CrossRef]
28. Eissner, W. Atomic structure calculations in Breit-Pauli approximation. In *The Effects of Relativity on Atoms, Molecules, and the Solid State*; Wilson, S., Grant, I.P., Gyorffy, B.L., Eds.; Plenum Press: New York, NY, USA, 1991; pp. 55–64.
29. Nahar, S.N. Atomic data from the Iron Project. *Astron. Astrophys.* **2004**, *413*, 779.
30. Nahar S.N. Modern Trends in Physics Research. In Proceedings of the 4th International Conference on MTPR-10, Sharm El Sheikh, Egypt, 12–16 December 2010; El Nadi, L., Ed.; World Scientific: Singapore, 2013; pp. 275–285.

Disclaimer/Publisher's Note: The statements, opinions and data contained in all publications are solely those of the individual author(s) and contributor(s) and not of MDPI and/or the editor(s). MDPI and/or the editor(s) disclaim responsibility for any injury to people or property resulting from any ideas, methods, instructions or products referred to in the content.

Review

Photoionization and Opacity

Anil Pradhan

Chemical Physics Program, Biophysics Graduate Program, Department of Astronomy, The Ohio State University, Columbus, OH 43210, USA; pradhan.1@osu.edu; Tel.: +1-614-292-5850

Abstract: Opacity determines radiation transport through material media. In a plasma source, the primary contributors to atomic opacity are bound–bound line transitions and bound-free photoionization into the continuum. We review the theoretical methodology for state-of-the-art photoionization calculations based on the R-matrix method as employed in the Opacity Project, the Iron Project, and solution of the heretofore unsolved problem of plasma broadening of autoionizing resonances due to electron impact, Stark (electric microfields), Doppler (thermal), and core-excitations. R-matrix opacity calculations entail huge amount of atomic data and calculations of unprecedented complexity. It is shown that in high-energy-density (HED) plasmas, photoionization cross sections become 3-D energy–temperature–density-dependent owing to considerable attenuation of autoionizing resonance profiles. Hence, differential oscillator strengths and monochromatic opacities are redistributed in energy. Consequently, Rosseland and Planck mean opacities are affected significantly.

Keywords: photoionization; opacity; autoionization; resonances; plasma broadening

1. Introduction

Physically, the opacity of an object depends on all possible intrinsic light–atom interactions that may absorb, scatter, or re-emit photons emanating from the source and received by the observer. In addition, the opacity depends on external conditions in the source and the medium. In recent years, there have been a number of theoretical and experimental studies of opacities (viz. [1–3]). Whereas photoionization and opacity are linked in all plasma sources, we focus specifically on high-energy-density (HED) environments such as stellar interiors and laboratory fusion devices, that are characterized by temperatures and densities together, typically $T > 10^6 K$ and densities $N > 10^{15}$ cm^{-3}. Computed atomic cross sections and transition proabilities are markedly perturbed by plasma effects.

Monochromatic opacity consist of four terms, namely, bound–bound (bb)), bound-free (bf), free–free (ff), and scattering (sc):

$$\kappa_{ijk}(\nu) = \sum_k A_k \sum_j F_j \sum_{i,i'} [\kappa_{bb}(i,i';\nu) + \kappa_{bf}(i,\epsilon'_i;\nu) + \kappa_{ff}(\epsilon_i,\epsilon'_{i'};\nu) + \kappa_{sc}(\nu)] . \quad (1)$$

In Equation (1), A_k is element abundance k, its ionization fraction F_j, i and initial bound and final bound/continuum states i, i', of a given atom; the ϵ represents electron energy in the continuum. To determine emergent radiation, a harmonic mean κ_R, is defined, Rosseland Mean Opacity (RMO), with monochromatic opacity $\kappa_{ijk}(\nu)$

$$\frac{1}{\kappa_R} = \frac{\int_0^\infty g(u)\kappa_\nu^{-1} du}{\int_0^\infty g(u) du} \quad \text{with} \quad g(u) = u^4 e^{-u}(1-e^{-u})^{-2}. \quad (2)$$

Here, $g(u)$ is the derivative of the Planck function including stimulated emission, $\kappa_{bb}(i,i') = (\pi e^2/m_e c) N_i f_{ii'} \phi_\nu$, and $\kappa_{bf} = N_i \sigma_\nu$. The κ_ν then depends on bb oscillator strengths, bf, photoionization cross sections σ_ν, on the equation-of-state (EOS) that gives level populations N_i. We describe large-scale computations using the coupled channel or close coupling (hereafter CC) approximation implemented via the R-matrix (RM) method for opacity in Equation (1) primarily for: (i) the bb transition probabilities and (ii) the bf photoionization cross sections.

In this review, we focus on the bf-opacity, and in particular on resonant phenomena manifest in myriad series of autoionizing resonances that dominate photoionization cross sections throughout the energy ranges of interest in practical applications.

2. Photoionization

Photoionization (PI) of an ion X^{+z} with ion charge z into the (e + ion) continuum is

$$X^{+z} + h\nu \rightarrow X^{+z+1} + e. \qquad (3)$$

PI also entails the indirect process of resonances via formation of autoionizing (AI) doubly excited states, and subsequent decay into the continuum, as

$$h\nu + X^{+Z} \rightleftharpoons (X^{+Z})^{**} \rightleftharpoons X^{+Z+1} + e \qquad (4)$$

Infinite series of AI resonances are distributed throughout the photoionization cross section and generally dominate at lower energies encompassing and converging on to ionization thresholds corresponding to excited levels of the residual ion in the (e + ion) continua. A large number of photoionization cross-section values for all bound levels are needed to compute plasma opacities. Total photoionization cross section (σ_{PI}) of each bound level of the (e + ion) system are required, from the ground state as well as from all excited states. Practically, however, we consider $n(SLJ) < 10$, and approximate relatively small number of energies below thresholds. Total σ_{PI} corresponds to summed contribution of all ionization channels leaving the residual ion in the ground and various excited states.

AI resonances in photoionization cross sections are dissolved by plasma density and temperature, resulting in an enhanced continuum background, as discussed later. However, the strong and isolated resonances can be seen in absorption spectra. Moreover, a sub-class of AI resonances corresponding to strong dipole transitions within the core ion, known as Photoexcitation-of-core (PEC) or Seaton resonances, correspond to the inverse process of dielectronic recombination [4,5].

Transition matrix for photoionization $S = < \Psi_F ||\mathbf{D}||\Psi_B >$ is obtained from bound and continuum wave functions which give the line strength using the expression above. Photoionization cross section is obtained as

$$\sigma_{PI} = \frac{4\pi}{3c}\frac{1}{g_i}\omega S, \qquad (5)$$

where ω is the incident photon energy in Rydberg units.

2.1. The Opacity Project and R-Matrix Method

Astrophysical opacity calculations using the RM method were intitated under the Opacity Project (circa 1983) [4–7]. The RM opacity codes were developed to compute large-scale and accurate bound–bound (bb) transition oscillator strengths, and bound-free (bf) photoionization cross sections, Considerable effort was devoted to precise delineation of the *intrinsic* AI resonance profiles in terms of shapes, heights, energy ranges, and magnitudes determined by numerous coupled channels of the (e + ion) system.

In the CC-RM method, the total (e + ion) system is expressed in terms of the eigenfunctions of the target or core states and a free-electron

$$\Psi(E) = \mathcal{A}\sum_i \chi_i \theta_i + \sum_j c_j \Phi_j. \qquad (6)$$

The χ_i are target ion wavefunctions in a specific $S_i L_i$ state, θ_i is the free-electron wavefunction, and Φ_j are bound channel correlation functions with coefficient c_j (viz. [4,5]). The coupled channel labeled as $S_i L_i k_i^2 \ell_i (SL\pi)$; k_i^2 is the incident kinetic energy. In contrast, the distorted wave approximation used in current opacity models neglects the summation over channels in Equation (6), and therefore coupling effects are not considered as in the RM method in an *ab initio* manner, due to possibly hundreds to thousands of coupled channels for complex ions. In principle, this approximation implies neglect of quantum superposition in the distorted wave method, and interference that manifests in autoionizing resonance profiles.

The bb, bf transition matrix elements for the (e + ion) wave functions $\Psi_B(SL\pi; E)$ and $\Psi_F(SL\pi; E')$, respectively, bound state B and B' line strengths (a.u.) are given by

$$S(B; B') = |\langle \Psi_B(E_B) || \mathbf{D} || \Psi_{B'}(E_{B'}) \rangle|^2. \tag{7}$$

For opacity computations, we consider the $\mathbf{D}$ dipole operator, since non-dipole transitions do not generally act as significant contributors. With the final continuum state represented by $\Psi_F(E')$ and the initial state by $\Psi_B(E)$, the photoionization cross section is

$$\sigma_\omega(B; E') = \frac{4}{3} \frac{\alpha \omega}{g_i} |\langle \Psi_B(E_B) || \mathbf{D} || \Psi_F(E') \rangle|^2. \tag{8}$$

The ω is photon frequency and E' is the photoelectron energy of the outgoing electron. The Breit–Pauli R-matrix (BPRM) incorporates relativistic effects using the the Breit–Pauli (BP) Hamiltonian for the (e + ion) system in BPRM codes in intermediate coupling with a pair-coupling scheme $S_i L_i (J_i) l_i (K_i) s_i (J\pi)$ [8], whereby states $S_i L_i$ split into fine-structure levels $S_i L_i J_i$. Consequently, the number of channels becomes several times larger than the corresponding *LS* coupling case. The IP work is generally based on BPRM codes, as for example the large amount of radiative and collisional data in the database NORAD [9].

2.2. R-Matrix Calculations for Opacities

The *R*-Matrix codes employed in opacities calculations are considerably different and extensions of the original *R*-Matrix codes [4–6]. The OP codes were later extended under the Iron Project [10] to incorporate relativistic effects and fine structure in the Breit–Pauli approximation [8]. The RM opacity codes were further adapted with new extensions at Ohio State University for complete RM opacity calculations [3,11]. Figure 1 shows the flowchart of the RM codes at the Ohio Supercomputer Center (OSC). The atomic structure codes SUPERSTRUCTURE [12] and CIV3 [13], are first utilized to obtain an accurate configuration-interaction representation of the core-ion states. Next, the two *R*-Matrix codes STG1 and STG2 are employed to generate multipole integrals and algebraic coefficients for the (e + ion) Hamiltonian corresponding to coupled integro-differential equations in the CC approximation. In the BPRM codes, the code RECUPD recouples the *LSJ* pair coupling representation incluing fine structure explicitly. The total (e + ion) Hamiltonian matrix is diagonalized in STGH. The *R*-Matrix basis functions and dipole matrix elements thus obtained are input to code STGB for bound state wavefunctions B, code STGF for continuum wavefunctions, *bb* transitions code STGBB, and code STGBF to compute photoionization cross sections. Code STGF(J) may also be used to obtain electron impact excitation collision strengths.

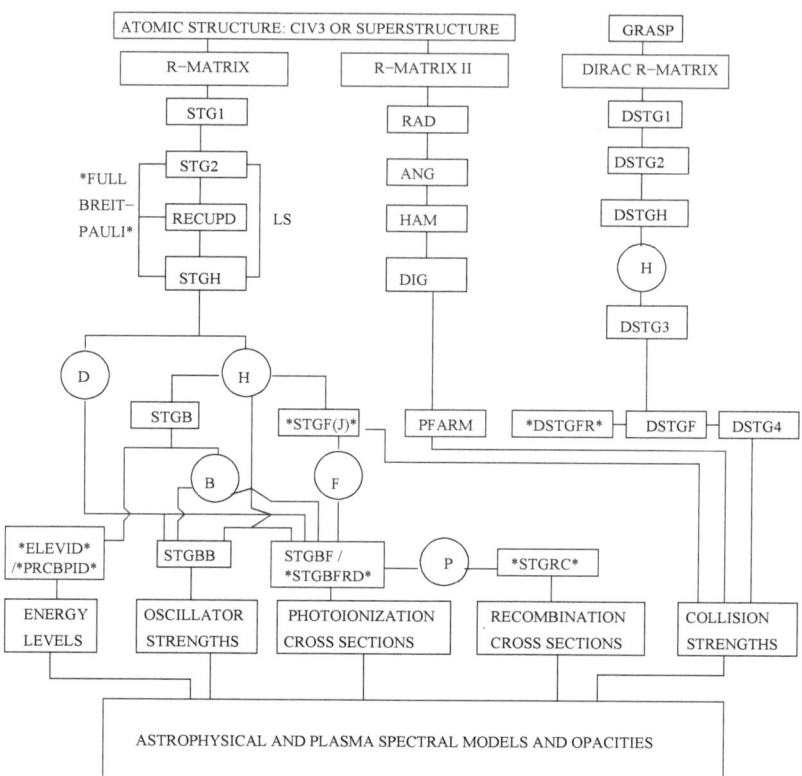

Figure 1. The R-matrix codes for opacities calculations (codes with * were developed or extended at OSU). Atomic data produced are further processed via a suite of equation-of-state, plasma broadening, and opacity codes to obtain monochromatic and mean opacities at each temperature and density [14].

The immense complexity of RM calculations, compared to DW method and atomic structure calculations, requires substantial computational effort and resources. In particular, inner-shell transitions are often dominant contributors to opacity. However, those could not be completed in OP work, except for outer-shell radiative transtions using the RM or BPRM methods due to computational constraints and then available high-performance computing platforms. Therefore, the simpler DW method was used for most of the OP opacity calculations, such as in DW-type methods in other opacity models that also neglect channel couplings and hence initio consideration of autoionizing resonances in the bound-free continua. A prominent exemplar is the extensive role of *photoexcitation-of-core* (PEC) resonances, or Seaton resonances [4,5], associated with strong dipole transitions (viz. [3,11] for Fe XVII).

Despite unprecedented effort and advances, the OP-RM work faced several then intractable difficulties that limited the scope of atomic calculations. Primarily, the limitations were due to computational constraints which, in turn, did not enable accounting for important physical effects and a complete RM calculation of atomic opacities. The main features and deficiencies of OP are as follows: (I) The calculations were in LS coupling neglecting relativistic fine structure. (II) The close coupling wavefunction expansion for the target or the core ion in the (e + ion) system included only a few ground configuration LS terms. (III) Inner-shell excitations could not be included owing to the restricted target-ion expansion that precluded photoexcitation of levels from inner shells into myriad resonances in the continua of the residual (e + ion) system. (IV) Autoionizing resonances in bound-free photoionization cross sections were delineated within the few excited target terms. (V) Total angular and spin (e + ion) symmetries with large orbital angular-spin quantum numbers were not computed. All of these factors are crucial for a complete, converged and accurate opacity calculation. As mentioned, the OP work initially began with the R-matrix codes, albeit with very small wavefunction expansions (e + ion) system, usually limited to the ground configuration of the core ion. Thus, OP opacities incorporated a small subset of RM data. Rather, most of the opacities contributions were obtained using atomic structure codes and the Distorted Wave (hereafter DW) approximation, similar to other opacity models [5,6,9,10,15].

The first complete RM calculation leading up to the calculation of opacities was carried out for the ion Fe XVII that is of considerable importance in determining the opacity at the base of the solar convection zone (BCZ) ([11], hereafter NP16). The solar radius of the BCZ has been accurately determined through Helioseismology to be 0.713 ± 0.001 $R_\odot$.

Other new physical issues also emerged in RM calculations for opacities. There are three major problems that need to be solved: (A) Convergence of large-coupled channel wavefunction expansions necessary to include sufficient atomic structures manifest in opacity spectra. (B) Completeness of high $n\ell$ contributions up to $n \equiv \infty$, and (C) attenuation of resonance profiles due to *intrinsic* autoionization broadening (included in RM calculations in an ab initio manner) and *extrinsic* plasma effects due to temperature and density, as generally considered for bound–bound line opacity.

RM photoionization calculations have been carried for several Fe ions [16]. In particular, large-scale computations of cross sections and transition probabilities have been performed for Fe ions that determine iron opacity at the solar BCZ: Fe XVII, Fe XVIII, Fe XIX, Fe XX and Fe XXI (to be published; S.N. Nahar, private communication).

2.3. R-Matrix and Distorted Wave Methods

Current opacity models employ the DW approximation or variants thereof, based on an atomic structure calculation coupled to the continuum. Oscillator strengths and photoionization cross sections are computed for all possible bound–bound and bound-free transitions among levels specified by electronic configurations included in the atomic calculation. However, since the DW approximation includes only the coupling between initial and final states, the complexity of interference between the bound and continuum wavefunction expansions involving other levels is neglected, and so are the detailed profiles of autoionizing resonances embedded in the continua. DW models employ the independent resonance approximation that treats the bound–bound transition probability independently from coupling to the continuum. Apart from relative simplicity of atomic computations, the advantage of using DW models is that well-established plasma line broadening treatments may be used.

On the other hand, RM opacities calculations are computationally laborious and time-consuming. However, as demonstrated in the erstwhile OP-RM work, albeit severely limited in scope, coupling effects are important. Opacity in the bound-free continuum is dominated by autoionizing resonances, as shown in recently completed works (viz. [3,11,17]). The most important consequence of neglecting detailed resonance profiles in DW models and missing opacity is that *intrinsic* autoionizing broadening and *extrinsic* plasma broadening thereof are not fully accounted for. It has now been shown that AI resonances are broadened much wider in the continuum than lines, and thereby enhance opacity significantly [3,11].

Recent work ([18], D21) extended Fe XVII RM calculations by including more configurations than NP16a. Although this confirmed our earlier results for photoionization cross sections, D21 do not consider plasma broadening of autoionizing resonances and therefore do not obtain a complete description of bound-free opacity from RM calculations (discussed below).

The unbroadened cross sections in D21 appear similar to ours; however, they did not compare those in detail with previously published data in [11] for Fe XVII, and publicly available from the electronic database NORAD [9]. Furthermore, D21 report 10% lower Rosseland mean opacities than OP2005, which is at variance with other DW models that are higher by up to a factor of about 1.5 [3,11], possibly due to an incomplete number of bound Fe XVII levels.

3. Inner- and Outer-Shell Excitations

Being simpler and based on pre-specified electronic configurations as in atomic structure calculations, inner-shell excitation DW data may be readily computed treating resonances as bound levels in the continuum. Although OP opacities were computed using DW data, OP atomic codes were originally developed to implement the RM methodology that could not be carried through owing to computational constraints. Most importantly, it could not be employed for opacities due to inner-shell excitations that are dominant contributors because most electrons in complex ions are in closed shells and bear excitation energies that lie above the first ionization threshold, giving rise to series of autoionizing resoances, and in particular PEC resonances due to strong dipole inner-shell trasitions in the core ion [11,17]. On the other hand, the much simpler DW treatment in opacity models is readily implemented but is inaccurate in the treatment of important resonance phenomena. Extensive comparison of RM and DW calculations for Fe XVII considered herein, and implications for plasma opacities, are given in [11,19].

4. Plasma Broadening of Resonances

Whereas line broadening has long been studied and its treatments are generally and routinely incorporated in opacity models (viz. [4]), plasma broadening of autoionizing resonance profiles is not heretofore considered. Attenuation of shape, height, energies, and magnitude of autoionizing resonances in photoionization cross sections must be delineated in detail, as in the RM method, as a function of density and temperature in order to determine the distribution of total differential oscillator strength and structure of the bound-free continua.

AI resonances are fundamentally different from bound–bound lines as related to quasi-bound levels with *intrinsic* quantum mechanical autoionization widths. Broadening has significant contribution to mean opacities, enhancing the Rosseland mean opacity by factors ranging from 1.5 to 3, as shown in other works and discussed below [17]. However, line-broadening processes and formulae may be to develop a theoretical treatment and computational algorithm outlined herein (details to be presented elsewhere). The convolved bound-free photoionization cross section of level i may be written as:

$$\sigma_i(\omega) = \int \tilde{\sigma}(\omega')\phi(\omega',\omega)d\omega', \qquad (9)$$

where σ and $\tilde{\sigma}$ are the cross sections with plasma-broadened and unbroadened AI resonance structures, ω is the photon energy (Rydberg atomic units are used throughout), and $\phi(\omega',\omega)$ is the normalized Lorentzian profile factor in terms of the *total* width Γ due to all AI broadening processes included:

$$\phi(\omega',\omega) = \frac{\Gamma(\omega)/\pi}{x^2 + \Gamma^2}, \qquad (10)$$

where $x \equiv \omega - \omega'$. The crucial difference with line broadening is that AI resonances in the (e + ion) system correspond to and are due to quantum mechanical interference between discretized continua defined by excited core ion levels in a multitude of channels. The RM method (viz. [4–6]), accounts for AI resonances in an (e + ion) system with generally asymmetric profiles (unlike line profiles that are usually symmetric).

Given N core-ion levels corresponding to resonance structures,

$$\sigma(\omega) = \sum_i^N \left[\int \tilde{\sigma}(\omega') \left[\frac{\Gamma_i(\omega)/\pi}{x^2 + \Gamma_i(\omega)} \right] d\omega' \right]. \quad (11)$$

With $x \equiv \omega' - \omega$, the summation is over all excited thresholds E_i included in the N-level RM wavefunction expansion, and corresponding to total damping width Γ_i due to all broadening processes. The profile $\phi(\omega', \omega)$ is centered at each continuum energy ω, convolved over the variable ω' and relative to each excited core ion threshold i. In the present formulation we associate the energy to the effective quantum number relative to each threshold $\omega' \to \nu_i$ to write the total width as:

$$\begin{aligned}\Gamma_i(\omega, \nu, T, N_e) &= \Gamma_c(i, \nu, \nu_c) + \Gamma_s(\nu_i, \nu_s^*) \\ &\quad + \Gamma_d(A, \omega) + \Gamma_f(f - f; \nu_i, \nu_i'),\end{aligned} \quad (12)$$

pertaining to collisional Γ_c, Stark Γ_s, Doppler Γ_d, and free–free transition Γ_f widths respectively, with additional parameters as defined below. We assume a Lorentzian profile factor that subsumes both collisional broadening due to electron impact, and Stark broadening due to ion microfields, that dominate in HED plasmas. This approximation should be valid since collisional profile wings extend much wider as x^{-2}, compared to the shorter range $exp(-x^2)$ for thermal Doppler, and $x^{-5/2}$ for Stark broadening (viz. [4,17]). In Equation (11), the limits $\mp \infty$ are then replaced by $\mp \Gamma_i/\sqrt{\delta}$; δ is chosen to ensure the Lorentzian profile energy range for accurate normalization. Convolution by evaluation of Equations (1)–(3) is carried out for each energy ω throughout the tabulated mesh of energies used to delineate all AI resonance structures, for each cross section, and each core ion threshold. We employ the following expressions for computations:

$$\Gamma_c(i, \nu) = 5 \left(\frac{\pi}{kT} \right)^{1/2} a_0^3 N_e G(T, z, \nu_i)(\nu_i^4/z^2), \quad (13)$$

where T, N_e, z, and A are the temperature, electron density, ion charge and atomic weight, respectively, and ν_i corresponds to a given core ion threshold $i : \omega \equiv E = E_i - \nu_i^2/z^2$ is a continuous variable. The Gaunt factor [17] $G(T, z, \nu_i) = \sqrt{3}/\pi[1/2 + ln(\nu_i kT/z)]$ Another factor $(n_x/n_g)^4$ is introduced for Γ_c to allow for doubly excited AI levels with excited core levels n_x relative to the ground configuration n_g (e.g., for Fe XVIII $n_x = 3, 4$ relative to the ground configuration $n_g = 2$). A treatment of the Stark effect for complex systems entails two approaches, one where both electron and ion perturbations are combined, or separately (viz. [4,17]) employed herein. Excited Rydberg levels are nearly hydrogenic, the Stark effect is linear and ion perturbations are the main broadening effect, though collisional broadening competes increasingly with density as ν_i^4 (Equation (13)). The total Stark width of a given n-complex is $\approx (3F/z)n^2$, where F represents the plasma electric microfields. Assuming the dominant ion perturbers to be protons and density equal to electrons, we take $F = [(4/3)\pi a_0^3 N_e)]^{2/3}$, consistent with the Mihalas–Hummer–Däppen equation-of-state formulation [4].

$$\Gamma_s(\nu_i, \nu_s^*) = [(4/3)\pi a_0^3 N_e]^{2/3} \nu_i^2. \quad (14)$$

In employing Equation (12), a Stark ionization parameter $\nu_s^* = 1.2 \times 10^3 N_e^{-2/15} z^{3/5}$ is introduced such that AI resonances may be considered fully dissolved into the continuum for $\nu_i > \nu_s^*$ (analogous to the Inglis–Teller series limit for plasma ionization of bound levels). Calculations are carried out with and without ν_s^*, as shown in [17]. The Doppler width is:

$$\Gamma_d(A, T, \omega) = 4.2858 \times 10^{-7} \sqrt{(T/A)}, \quad (15)$$

where ω is *not* the usual line center but taken to be each AI resonance energy. The last term Γ_f in Equation (5) accounts for free–free transitions among autoionizing levels with ν_i, ν'_i such that

$$X_i + e(E_i, \nu_i) \longrightarrow X'_i + e'(E'_i, \nu'_i). \qquad (16)$$

The large number of free–free transition probabilities for $+ve$ energy AI levels $E_i, E'_i > 0$ may be computed using RM or atomic structure codes (viz. [20]).

We utilize new results from an extensive Breit–Pauli R-Matrix (BPRM) calculation with 218 fine structure levels dominated by $n \leq 4$ levels of the core ion Fe XVIII (to be reported elsewhere). A total of 587 Fe XVII bound levels ($E < 0$) are considered, dominated by configurations $1s^2 2s^2 2p^6 (^1S_0), 1s^2 2s^p 2p^q n\ell, [SLJ]$ ($p, q = 0 - -2, n \leq 10, \ell \leq 9, J \leq 12$). The core Fe XVII levels included in the RM calculation for the (e + Fe XVIII) →Fe XVII system are: $1s^2 2s^2 2p^5 (^2P^o_{1/2,3/2}), 1s^2 2s^2 2p^q, n\ell, [S_i L_i J_i]$ ($p = 4, 5, n \leq 4, \ell \leq 3$). The Rydberg series of AI resonances correspond to $(S_i L_i J_i)$ $n\ell$, $n \leq 10, \ell \leq 9$, with effective quantum number defined as a continuous variable $\nu_i = z/\sqrt{(E_i - E)}$ ($E > 0$), throughout the energy range up to the highest 218th Fe XVIII core level; the $n = 2, 3, 4$ core levels range from E = 0–90.7 Ry ([11]). The Fe XVII BPRM calculations were carried out resolving the bound-free cross sections at ∼40,000 energies for 454 bound levels with AI resonance structures. Given 217 excited core levels of Fe XVIII, convolution is carried out at each energy or approximately 10^9 times for each (T, N_e) pair.

Figure 2 displays detailed results for unbroadened photoionization cross section (black) and plasma broadened (red and blue, without and with Stark ionization cut-off) The excited bound level of Fe XVII is $2s^2 2p^2\ ^3D_2$ at temperature–density $T = 2 \times 10^6$ K and $N_e = 10^{23}$ cm^{-3}. The cross section is shown on the Log$_{10}$ scale in the top panel, and on a linear scale in the bottom panel isolating the energy region of highest and strongest AI resonances. The main features evident in the figure are as follows. (i) AI resonances show significant plasma broadening and smearing of a multitude of overlapping Rydberg series at The narrower high-$n\ell$ resonances dissolve into the continua but stronger low-$n\ell$ resonance retain their asymmetric shapes with attenuated heights and widths. (ii) At the $N_e = 10^{23}$ cm^{-3}, close to that at the solar BCZ, resonance structures not only broaden but their strengths shift and are redistributed over a wide range determined by total width $\Gamma(\omega, \nu_i, T, N_e)$ at each energy $\hbar\omega$ (Equation (12)). (iii) Stark ionization cut-off (blue curve) results in step-wise structures that represent the average due to complete dissolution into continua. (iv) Integrated AI resonance strengths are conserved, and are generally within 5–10% of each other for all three curves in Figure 2. It is found that the ratio of RMOs with and without plasma broadening may be up to a factor of 1.6 or higher ([17]); recent work for other ions shows the ratio may be up to factor of 3.

The scale and magnitude of new opacity calculations is evident from the fact that photoionization cross sections of 454 bound levels of Fe XVII are explicitly calculated using the RM opacity codes, 1154 levels of Fe XVIII, and 899 levels Fe XIX. Plasma broadening is then carried out for for each temperature and density of interest throughout the solar and stellar interiors or HED plasma sources.

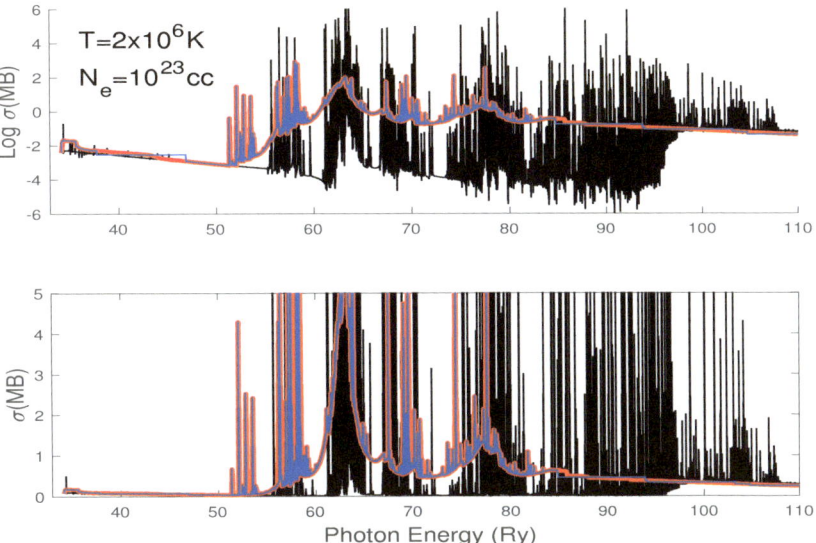

Figure 2. Energy–temperature–density-dependent photoionization cross section of highly excited bound level $2s^2 2p^5 3p\ ^2D_2$ of Fe XVII ⟶ e + Fe XVIII, due to plasma broadening of autoionizing resonances: unbroadened—black curve, broadened—red and blue (see text). Top panel: Log σ (MB) in the full energy range up to the highest ionization threshold of core ion Fe XVIII, bottom panel: Linear-scale σ_{PI} in the energy range of the largest AI structures.

5. Energy Dependence

Photoionization cross sections vary widely in different approximations used to calculate opacities. Simple methods such as the *quantum defect method* and the central-field approximation, yield a feature-less background cross section. High-n levels in a Rydberg series of levels behave hydrogenically at sufficiently high energies, and the photoionization cross section may be approximated using Kramer's formula (discussed in [5])

$$\sigma_{PI} = \frac{8\pi}{3^{1.5} c} \frac{1}{n^5 \omega^3}. \qquad (17)$$

Equation (17) is used in OP work to extrapolate photoionization cross sections in the high-energy region. However, it is not accurate, as seen in Figure 3. At high energies, inner shells and sub-shells are ionized, and their contribution must also be included in total photoionization cross sections. At inner (sub-)shell ionization thresholds, there is a sharp upward jump or edge and enhancement of the photoionization cross section. Figure 3 shows results from a relativistic distorted wave (RDW) calculation and Kramer's fomula Equation (17). The RDW results do not include resonances, and differ from the OP results with resonance structures in the relatively small energy region near the ioniization threshold.

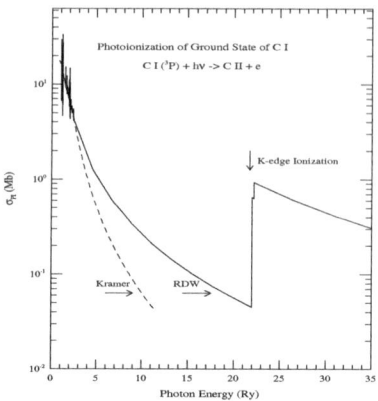

Figure 3. Photoionization cross section σ_{PI} of the ground state of CI, $1s^2 2s^2 2p^2\ ^3P$, computed using the relativistic distorted wave (RDW) code by H.L. Zhang (discussed in [5]) compared with the Kramer's hydrogenic formula Equation (17). The large jump is due to photoionization of the inner 1s—shell or the K—edge. The resonance structures at very low energies are obtained from the coupled channel RM calculations in the Opacity Project.

6. From Convergence to Completeness

The NP16 work [11] also addressed an important point that a reasonably complete expansion of target configurations and levels in BPRM photoionization calculations is necessary to ensure converged bound-free opacities. The criteria for accuracy and completeness are: (i) *convergence* of the wavefunction expansion (Equation (6)), and (ii) *completeness* of PI cross sections, monochromatic and mean opacities with respect to possibly large number of multiply excited configurations.

While NP16 demonstrated convergence with respect to $n = 2, 3, 4$ levels of the Fe XVIII target ion included in the RM calculations, more highly excited configurations that might affect high-energy behavior were not included. Subsequent work using and comparing with the DW method was therefore carried out to ascertain the effect of high-$n\ell$ configurations on opacities [20].

Specifying excited configurations is straightforward in an atomic structure-DW calculation, but it is more complex and indirect in RM calculations. For example, in order to investigate the role of more excited configurations the NP16 BPRM calculations that yield 454 bound levels for Fe XVII, were complemented with >50,000 highn, ℓ "topup" levels to compute opacities and RMOs. Photoionization cross sections of the 454 strictly bound levels computed (negative eigenenergies) take into account embedded autoionizing resonances that are treated as distinct levels in DW calculations; therefore, in total there are commensurate number of levels to ensure completeness.

However, the large number of highly excited configurations made only a small contribution to opacities, relative to the main BPRM cross sections, and only to the background cross sections. without resonances. Therefore, the simpler DW method may be used for topup contributions without loss of accuracy as to supplement RM calculations. Recent work has shown that the topup contribution to RM opacities does not exceed 5% to RMOs [14].

7. Sum Rule and Oscillator Strength Distribution

The total bb and integrated bf oscillator strength, when summed over all possible bb and bf transitions, must satisfy the definition of the oscillator strength as fractional excitation probability, i.e., $\sum_j f_{ij} = N$, where N is the number of active electrons. However, while the f-sum rule ensures completeness, it does not ensure accuracy of atomic calculations *per se*. That depends on the precise energy distribution of differential oscillator df/dE, strength or photoionization cross section σ_{PI}. To wit: the hydrogenic approximation, if used for complex atoms would satisfy the f-sum rule but would clearly be inaccurate. As discussed herein, the RM method is concerned primarily with df/dE in the bf-continuum based on full delineation of autoionizing resonance profiles.

As an end result, the RMO depends on energy distribution of monochromatic opacity, convolved over the Planck function at a given temperature. Compared with OP results, the distribution of RM Fe XVII monochromatic opacity is quite different, and much more smoothed out without sharp variations that stem mainly from the treatment of resonances as bb lines, even with limited autoionization broadening included perturbatively in DW opacity models. Experimentally, a flatter opacity distribution is also observed, in contrast to theoretical opacity models that exhibit larger dips in opacity at "opacity windows" [3,11,21,22].

8. Conclusions

This review describes photoionization work related to opacities. The state-of-the-art R-matrix calculations are discussed in comparison with the distorted wave data currently employed in opacity models. Atomic and plasma effects such as channel coupling, broadening of autoionizing resonances, high-energy behavior, and oscillator strength sum-rule are described.

Existing OP and IP radiative data for photoionization and transition probabilities for astrophysically abundant elements have been archived in databases TOPbase and TIPbase. OP opacities and radiative accelerations are available online from OPserver [15]. R-matrix data for nearly 100 atoms and ions from up-to-date and more accurate calculations are available from the database NORAD at OSU [9].

Funding: This research received no external funding.

Data Availability Statement: All data reported in this work is available from the database NORAD [9] created by S.N. Nahar and maintained at the Ohio State University.

Acknowledgments: I would like to thank Sultana Nahar for their contributions. The work reported in this review was partially supported by grants from the U.S. NASA, NSF, DOE, and the computational work was carried out at the Ohio Supercomputer Center.

Conflicts of Interest: The author declares no conflict of interest.

References

1. Mendoza, C. Computation of Atomic Astrophysical Opacities. *Atoms* **2018**, *6*, 28. [CrossRef]
2. Pain, J.-C.; Croset, P. Ideas and Tools for Error Detection in Opacity Databases. *Atoms* **2023**, *11*, 27. [CrossRef]
3. Pradhan, A.K. Recalculation of Astrophysical Opacities: Overview, Methodology, and Atomic Calculations. In Proceedings of the Workshop on Astrophysical Opacities, Kalamazoo, MI, USA, 1–4 August 2017; Volume 515, pp. 79–88.
4. *The Opacity Project*; The Opacity Project Team, Institute of Physics Publishing: Bristol, UK, 1995; Volume 1.
5. Pradhan, A.K.; Nahar, S.N. *Atomic Astrophysics and Spectroscopy*; Cambridge University Press: Cambridge, UK, 2011.
6. *R-Matrix Theory of Atomic Collisions*; Springer Series on Atomic, Optical and Plasma Physics; Springer: Berlin/Heidelberg, Germany, 2011.
7. Seaton, M.J.; Yu, Y.; Mihalas, D.; Pradhan, A.K. Opacities for Stellar Envelopes. *Mon. Not. R. Astr. Soc.* **1994**, *266*, 805. [CrossRef]
8. Berrington, K.A.; Eissner, W.; Norrington, P.H. RMATRIX1: Belfast atomic R-matrix codes. *Comput. Phys. Commun.* **1995**, *92*, 290–420. [CrossRef]
9. Nahar, S.N. NORAD: Nahar-OSU-Radiative-Atomic-Data, The Ohio State University. Available online: http://norad.astronomy.osu.edu (accessed on 4 February 2023).
10. Hummer, D.G.; Berrington, K.A.; Eissner, W.; Pradhan, A.K.; Saraph, H.E.; Tully, J.A. Atomic data from the IRON Project. 1: Goals and methods. *Astron. Astrophys.* **1993**, *279*, 298–309.
11. Nahar, S.N.; Pradhan, A.K. Large Enhancement in high-energy photoionization of Fe XVII and missing continuum plasma opacity. *Phys. Rev. Lett.* **2016**, *116*, 249502–294307. [CrossRef] [PubMed]
12. Eissner, W.; Jones, M.; Nussbaumer, H. Techniques for the calculation of atomic structures and radiative data including relativistic corrections. *Comput. Phys. Commun.* **1974**, *8*, 270–306. [CrossRef]
13. Hibbert, A. CIV3—A general program to calculate configuration interaction wave functions and electric-dipole oscillator strengths. *Comput. Phys. Commun.* **1975**, *9*, 141. [CrossRef]

14. Pradhan, A.K.; Nahar, S.N.; Eissner, W.; Zhao, L. Chemical Physics Program, Biophysics Graduate Program. Department of Astronomy, The Ohio State University: Columbus, OH, USA, 2023; *manuscript in preparation*.
15. Mendoza, C.; Seaton, M.J.; Buerger, P.; Bellorin, P.; Melendez, M.; Gonzalez, J.; Rodriguez, L.S.; Palacios, E.; Pradhan, A.K.; Zeippen, C.J. Interactive online computations of opacities and radiative accelerations. *Mon. Not. R. Astron. Soc.* **2007**, *378*, 1031. Available online: http://cdsweb.u-strasbg.fr/topbase/topbase.html (accessed on 5 February 2023). [CrossRef]
16. Nahar, S.N. The IRON Project: Photoionization of Fe Ions. *ASP Conf. Ser.* **2018**, *515*, 93–103.
17. Pradhan, A.K. Plasma Broadening of Autoionzing Resonances. 2023. Available online: https://arxiv.org/pdf/2301.07734.pdf (accessed on 6 February 2023).
18. Delahaye, F.D.; Ballance, C.P. Smyth RT and Badnell NR, Quantitative comparison of opacities calculated using the R-matrix methods: Fe XVII. *Mon. Not. R. Astr. Soc.* **2021**, *508*, 421–432. [CrossRef]
19. Nahar, S.N.; Pradhan, A.K.; Chen, G.-X.; Eissner, W. Highly Excited Core Resonances in Photoionization of Fe XVII: Implications for Plasma Opacities. *Phys. Rev. A* **2011**, *83*, 053417. [CrossRef]
20. Zhao, L.; Eissner, W.; Nahar, S.N.; Pradhan, A.K. Converged R-matrix calculations of the Photoionization of Fe XVII in Astrophysical Plasmas: From Convergence to Completeness. *ASP Conf. Ser.* **2018**, *515*, 89–92.
21. Bailey, J.E.; Nagayama, T.; Loisel, G.P.; Rochau, G.A.; Blancard, C.; Colgan, J.; Cosse, P.; Faussurier, G.; Fontes, C.J.; Gilleron, F.; et al. A higher-than-predicted measurement of iron opacity at solar interior temperatures. *Nature* **2015**, *517*, 56–59. [CrossRef] [PubMed]
22. Nagayama, T.; Bailey, J.; Loisel, G.; Dunham, G.; Rochau, G.; Blancard, C.; Colgan, J.; Cossé, P.; Faussurier, G.; Fontes, C.; et al. Systematic Study of L-Shell Opacity at Stellar Interior Temperatures. *Phys. Rev. Lett.* **2019**, *122*, 235001. [CrossRef] [PubMed]

Disclaimer/Publisher's Note: The statements, opinions and data contained in all publications are solely those of the individual author(s) and contributor(s) and not of MDPI and/or the editor(s). MDPI and/or the editor(s) disclaim responsibility for any injury to people or property resulting from any ideas, methods, instructions or products referred to in the content.

Review

Photoionization and Electron–Ion Recombination in Astrophysical Plasmas

D. John Hillier

Department of Physics and Astronomy & Pittsburgh Particle Physics, Astrophysics and Cosmology Center (PITT PACC), University of Pittsburgh, 3941 O'Hara Street, Pittsburgh, PA 15260, USA; hillier@pitt.edu; Tel.: +1-412-624-9000

Abstract: Photoionization and its inverse, electron–ion recombination, are key processes that influence many astrophysical plasmas (and gasses), and the diagnostics that we use to analyze the plasmas. In this review we provide a brief overview of the importance of photoionization and recombination in astrophysics. We highlight how the data needed for spectral analyses, and the required accuracy, varies considerably in different astrophysical environments. We then discuss photoionization processes, highlighting resonances in their cross-sections. Next we discuss radiative recombination, and low and high temperature dielectronic recombination. The possible suppression of low temperature dielectronic recombination (LTDR) and high temperature dielectronic recombination (HTDR) due to the radiation field and high densities is discussed. Finally we discuss a few astrophysical examples to highlight photoionization and recombination processes.

Keywords: photoionization; recombination; massive stars; Wolf–Rayet stars; local thermodynamic equilibrium (LTE); nLTE

Citation: Hillier, D.J. Photoionization and Electron–Ion Recombination in Astrophysical Plasmas. *Atoms* **2023**, *11*, 54. https://doi.org/10.3390/atoms11030054

Academic Editors: Sultana N. Nahar and Guillermo Hinojosa

Received: 24 January 2023
Revised: 3 March 2023
Accepted: 3 March 2023
Published: 9 March 2023

Copyright: © 2023 by the authors. Licensee MDPI, Basel, Switzerland. This article is an open access article distributed under the terms and conditions of the Creative Commons Attribution (CC BY) license (https://creativecommons.org/licenses/by/4.0/).

1. Introduction

One of the most important ways we learn about the Universe is through spectroscopy. From spectroscopy we can typically deduce important stellar parameters such as a star's effective temperature,[1] surface gravity (=GM_*/R_*^2) and abundances. These in turn provide insights into stellar evolution, galactic evolution, and the evolution of the Universe. To perform analyses of stellar data requires atomic data, although the amount and type of atomic data needed varies greatly with the application. In the most extreme cases, in which local thermodynamic equilibrium does not hold (discussed below), we require, for example, photoionization cross-sections, oscillator strengths for bound–bound transitions, line-broadening data, collisional cross-sections, autoionization rates, chemical reaction rates, and charge exchange cross-sections. For some simple species, such as hydrogen, we have excellent atomic data while for other important species, such as Fe group elements, crucial atomic data is lacking. Unfortunately, for many ionization stages even basic information, such as accurate energy levels, is also missing.[2] However, invaluable work by the NIST and Imperial college atomic spectroscopy groups is helping to rectify this situation for some important astrophysical ions (e.g., [2,3]).

In this review we discuss the importance of photoionization cross-sections for astronomical applications, with an emphasis on massive stars. Such a discussion will necessarily consider recombination processes—the inverse of photoionization processes. Before doing so it is necessary to define some important physical and astronomical terms.

A star (or any other astrophysical object emitting radiation) is not in thermal equilibrium. However, in some cases, and at some locations, it is well justified to assume that the plasma in the star is in "local" thermal equilibrium. Below the atmosphere (the thin layer that emits the radiation we observe) the material in most stars can be considered to be in "local" thermal equilibrium (LTE). In such cases, the state of the gas (e.g., the thermodynamical properties, the ionization state, and the populations of atomic levels) are set by the density and electron temperature via thermodynamic arguments. The ionization state of the gas and the level populations are determined, for example, by the Saha and Boltzmann equations (e.g., [4]). Moreover, there is only one temperature—the electron temperature, the ion temperature, the excitation temperature, and the radiation temperature are all identical. The temperature of the gas varies with location (it must, since radiation is propagating outwards) but the scale on which it varies does not affect the thermodynamic state of the gas. Unfortunately, much of the radiation we observe comes from gas that is NOT in LTE—typically referred to as nLTE or non-LTE.

At the stellar surface the assumption of LTE becomes less valid. This is not surprising—at the stellar surface radiation is escaping from the star and there is no incident radiation (at least for single stars), and hence the radiation density must drop from its blackbody value by a factor of (at least) ~ 2 (since there is no incident radiation—see (see [4], p. 120). Further, because radiation can now travel significant distances, regions of different temperatures are directly coupled, potentially making the radiation field at a given location strongly non-Planckian.

Fortunately, in some cases the densities are high enough that collisional processes can still strongly couple the level populations and the ionization state of the gas to the local electron temperature, allowing us to use the Saha and Boltzmann equations to compute level populations. In such cases the electron temperature (which will be the same as the ion temperature) determines the state of the gas and it is this temperature that we normally state. In general, however, the electron and radiation temperatures will be different.[3]

The Boltzmann equation, which relates the populations of two levels within the same ionization state, is

$$n_u^* = \frac{g_u n_l^*}{g_l} exp(-E_{lu}/kT) \qquad (1)$$

where n_l^* and n_u^* are the LTE population densities of the lower and upper states, respectively, g_l and g_u are the level degeneracies, and E_{lu} is the difference energy between the two levels (e.g., [4]).

The Saha equation, which relates the ground state populations of two consecutive ionization equations, is

$$n_{1,i}^* = n_{1,i+1}^* N_e \frac{g_{1,i} C_I}{g_{1,i+1} T^{3/2}} \exp\left(-\psi_i/kT\right) \qquad (2)$$

where T is the electron temperature in Kelvin, $C_I = 2.07 \times 10^{-16}$ (cgs units), N_e is the electron density, ψ_i is the ionization energy, and the subscript i $(i+1)$ is used to denote the ionization stage.

In many stars, and especially those with lower density gas (nebulae, stellar winds, supernovae) the departures from LTE are significant and MUST be allowed for. Gaseous nebulae, which typically have densities less than 10^6 atoms cm^{-3}, provide an excellent example in which the departures from LTE are extreme. The radiation field that ionizes the nebula typically emanates from a hot star ($T_{\text{eff}} \gtrsim 25{,}000$ K) and is strongly diluted since the nebula is typically very distant from the star (Figure 1). The equilibrium temperature of the gas is typically around 10,000 K (which is primarily determined by the chemical composition of the nebula) and is insensitive to the effective temperature of the star. The nebula, longward (i.e., at larger wavelengths) of the H I Lyman jump at 912 Å, is transparent to most radiation. Because of the non-Planckian radiation field and the low densities, collisions with electrons cannot drive the gas into LTE.

Figure 1. A composite image of the Helix nebula obtained using the HST and the Cerro Tololo Inter-American Observatory in Chile. Hα+[N II] ($\lambda\lambda$6548.1, 6583.4 Å), emission is shown in red, an average of Hα+[N II] and [O III] ($\lambda\lambda$4958.9, 5006.8 Å) is shown in green, and forbidden [O III] emission is shown in blue [5]. The nebula lies at a distance of about 200 pc, and the size of the semi-major axis is approximately 5.5 arc minutes or 66,000 AU [5]. The central star has an effective temperature of 104,000 K, $L \approx 80\, L_\odot$, and $R \approx 0.028\, R_\odot$ [6]. Using the semi-major axis as a representative size scale, the dilution factor ($0.25(r/R_*)^2$) is of order 10^{-18}. Detailed insights into the structure and morphology of the Helix nebula can be found in many works (e.g., [5,7]). Image credit: NASA, ESA, C.R. O'Dell (Vanderbilt University), and M. Meixner, P. McCullough, and G. Bacon (Space Telescope Science Institute).

When LTE no longer holds (i.e., nLTE) we are forced to solve for the ionization state of the gas and the level populations from first principles. That is, we need to consider all the processes, and inverse processes, that populate a given level. These processes include photoionization and recombination, bound–bound emission and absorption, collisional excitation and de-excitation, collisional ionization and collisional recombination, dielectronic recombination and autoionization, charge exchange reactions and, in "cooler gas" ($T \lesssim 6000$ K), chemical reactions, and dust chemistry.[4] Determining the state of the gas is a complicated problem. Many of the processes above depend on the radiation field which in turn depends on the level populations. Further, the radiation field couples the gas to regions of different temperatures. This is a highly non-linear problem and can only be solved by iterative techniques. In many cases we can consider the population numbers to be static (and the equations are referred to as the equations of statistical equilibrium) but in other cases (e.g., supernovae) we may need to allow for time dependence (and solve the kinetic equations). Another major issue is correcting for plasma effects that limit the number of levels in atoms and ions (i.e., crudely, an atom/ion cannot be larger than the inter-atom spacing). One approach is probabilistic and was developed in a series of papers by Hummer, Mihalas, and Dappen [12–14], and is used, for example, in the nLTE radiative transfer codes TLUSTY [15] and CMFGEN [16]. In this approach, levels are assigned an occupation probability that varies smoothly with the level energy, density and temperature, and that leads to a finite partition function.

Thus, a vast amount of atomic data is needed. Sadly, despite heroic efforts by Bob Kurucz [17,18], members of the Opacity Project [19] and Iron Project [20], Sultana Nahar [21], and many others, much of the needed data is still missing. Extensive photoionization data is available, for example, through TOPbase [22], TIPbase [23], and NORAD [21]. As discussed later in this article, details can matter, and it is not always obvious, a priori, which data are essential for accurate analyses. As this paper is concerned with photoionization/recombination, this paper does not generally elaborate on other important processes. Information on these additional processes can be found in, e.g., [24–27].

Below we discuss photoionization and recombination as relevant to astrophysical applications. Much of the following discussion will be based on my own experiences in developing CMFGEN, a nLTE radiative transfer code originally designed to model hot massive stars ($M > 20\,M_\odot$, $T_{\text{eff}} \geq 20,000$ K) and their stellar winds [16,28]. The winds in these stars are driven by radiation pressure acting on bound–bound transitions belonging, for example, to C, N, O, Ar, and Fe (e.g., [29–31]). Since its initial development the code has undergone considerable revisions and improvements. It has been successfully used to model O stars (e.g., [32–36]), Wolf–Rayet (WR) stars [37,38], luminous blue variables (e.g., [39–41]), B stars (e.g., [42]), the central stars of planetary nebulae (e.g., [43]), and A stars. Over the last decade CMFGEN was adapted to treat time-dependent radiation transfer, and to solve the time-dependent kinetic equations [44], and it has been used to model spectra resulting from a variety of SN explosions (e.g., [45–48]).

The review is organized as follows: In Section 2 we briefly discuss the importance of photoionization processes for stellar interiors and introduce the Rosseland mean opacity. We then consider photoionization processes in Section 3 with an emphasis on inner shell ionizations in Section 3.1. Recombination processes are discussed in Section 4 while suppression of dielectronic recombination by collisions and the radiation field is discussed in Section 5.1 and Section 5.2 respectively. Specific examples of where photoionization/recombination processes are important are then discussed—direct recombination (Section 6.1), the Sun (Section 6.2), O stars, WR stars, luminous blue variables (LBVs), (Section 6.3), Of and WN stars (Section 6.3.1), carbon lines in WC stars (Section 6.3.2), C II in a [WC] star (Section 6.4), and supernovae (Section 6.5).

2. Stellar Interiors

In stellar interiors energy is transported by radiation and by convection, and in degenerate stars by conduction. As LTE holds, photoionization processes have no direct influence on the level populations (since the populations are determined by the Saha and Boltzmann equations), but they do help to determine the temperature of the gas and they do help to set the continuous radiation field that we observe. Due to the small mean-free-path of photons, radiation transport is diffusive and in this regime a single quantity, the Rosseland mean opacity, is required to describe the transport of radiative energy. At depth in the star the radiation diffuses and the specific intensity (I_ν) at frequency ν is given by

$$I_\nu = B_\nu - \frac{\mu}{\chi_\nu} \frac{dB_\nu}{dr} \qquad (3)$$

(e.g., [4]) where B_ν is the Planck (i.e., blackbody) function given by

$$B_\nu = \frac{2h\nu^3}{c^2} \frac{1}{\exp(h\nu/kT) - 1}, \qquad (4)$$

χ_ν is the opacity, and μ is the angle between the radius and the direction of radiation propagation. The radiative flux is simply given by

$$F_\nu = \frac{-4\pi}{3\chi_\nu} \frac{dB_\nu}{dr}. \qquad (5)$$

The negative sign in the expression for the flux arises because T, and hence B_ν, decrease with increasing r. Integrating over all frequencies yields a total radiative flux, F, given by

$$F = -16\sigma \frac{T^3}{3\chi_R} \frac{dT}{dr} \qquad (6)$$

where σ is the Stefan–Boltzmann constant and χ_R is the Rosseland mean opacity as defined by

$$\frac{1}{\chi_R} = \frac{\int_0^\infty \frac{1}{\chi_\nu} dB_\nu/dT\, d\nu}{\int_0^\infty dB_\nu/dT\, d\nu} . \tag{7}$$

In stellar interiors the Rosseland mean opacity is a function of density, temperature, and composition, and is primarily determined by the most abundant species, with H, He, C, N, O, Ne, and Fe being the most crucial. As it is a harmonic mean it can be strongly influenced by regions of low opacity. Thus, it is crucial to take into account all opacity sources, particularly contributions in regions of otherwise low opacity. For a given species, the opacity is determined by photo-ionization processes, bound–bound transitions, and free–free processes. The required cross-sections are non-trivial to compute, especially for Fe group elements (with a partial filled 3d shell). Fortunately, because it is a broad integral, the Rosseland mean opacity is insensitive to small random errors (e.g., bound–bound transitions slightly offset from their correct wavelength) in the atomic data.

The computation of the Rosseland mean opacity is further complicated by the need to account for plasma effects—atoms/ions in a star are not isolated but experience a time-varying electric field due to their neighbors. This broadens bound–bound transitions which enhances the Rosseland mean opacity since the influence of the line is spread over a broader band into regions which may have lower opacity. Further, the size of the atoms/ions (and hence the number of levels) will be truncated since atoms/ions can only occupy a finite volume. The latter can be thought of as a lowering of the ionization potential but more rigorous approaches, for example, use probabilistic arguments [12,49]. An extensive discussion of some of the issues related to plasma effects on the equation of state, and additional references, are given by [50]. Extensive efforts have been made to provide LTE opacity libraries for stellar astrophysics that take into account different physical effects with varying degrees of fidelity. These include the Opacity Project [51], opacities computed using the OPAL code (e.g., [52,53]), the OPAS code [54], and a suite of codes developed at The Los Alamos National Laboratory [55].

3. Photoionization

The photoionization rate from a level l (in an arbitrary ion of arbitrary charge) can be written as

$$\left(\frac{dn_l}{dt}\right)_{PR} = -n_l \int_{\nu_o}^\infty \left(\frac{4\pi}{h\nu}\right) \sigma_\nu J_\nu d\nu \tag{8}$$

where n_l is the population density of level l, σ_ν is the frequency dependent photoionization cross-section (units are cm^2), ν_o the threshold frequency for ionization, h is Planck's constant, and J_ν is the mean intensity (erg cm^{-2} s^{-1} Hz^{-1}). J_ν is defined by

$$J_\nu = \frac{1}{4\pi} \oint I_\nu \, d\Omega \tag{9}$$

where $d\Omega$ is an increment in solid angle. If the radiation field is Planckian and isotropic,

$$J_\nu = I_\nu = B_\nu . \tag{10}$$

The photoionization cross-sections are generally obtained from numerical calculations —it is not feasible to measure all the cross-sections in the laboratory. Rather, laboratory measurements are used to test the accuracy of theoretical calculations. The accuracy of the cross-section varies greatly being dependent on both assumptions used in the modeling and on the complexity of the model atom. For example, it is much easier to compute atomic data for atoms/ions with a partially filled p shell than it is for the lanthanides and actinides which have a partially filled 4f or 5f shell, respectively (e.g., [56]). The lanthanides are believed to be created in neutron–neutron star mergers and are an important opacity source in the outburst spectra that result from such mergers [57–62].

Typically in model atmosphere codes the rates (integrals) are evaluated using numerical quadrature. Thus

$$\left(\frac{dn_l}{dt}\right)_{PR} = -n_l \sum_i \left(\frac{4\pi}{h\nu_i}\right) w_i \sigma_i J_i \qquad (11)$$

where w_i is the quadrature weight at frequency ν_i.

In a simple species such as hydrogen, the photoionization process is simply

$$H + h\nu \to H^+ + e^-.$$

In more electron-rich species the process is more complicated since there are multiple photoionization routes. For example, there are two direct photoionization routes from C III(2s 2p):

$$C\ III(2s\ 2p^1\ P^o) + h\nu \to C\ IV(2s^2 S) + e^- \quad \&$$
$$C\ III(2s\ 2p^{1\ P^o}) + h\nu \to C\ IV(2p^2 P^o) + e^-.$$

The first process occurs provided the photon energy exceeds the ionization energy of the C III(2s 2p $^1P^o$) state[5]. The second process occurs when the photon energy exceeds the sum of the ionization energy and the difference in energy between the 2s and 2p states in C IV. Of course, photons of sufficient energy may also ionize C^{2+} by ejecting an inner (1s) electron—a process of great importance when X-rays are present.

There may also be multiple indirect photoionization routes such as:

$$C\ III(2p^2\ {}^1D) + h\nu \to C\ III(2p4d^1F^o) \to C\ IV(2s^2 S) + e^-.$$

The above produces a relatively "narrow" resonance in the photoionization cross-section—it is narrow since the photon has to have the right energy to excite one of the 2p electrons into the 4d state (Figure 2). The energy of this state lies above the C IV ground state. The last step in this process is referred to as autoionization.

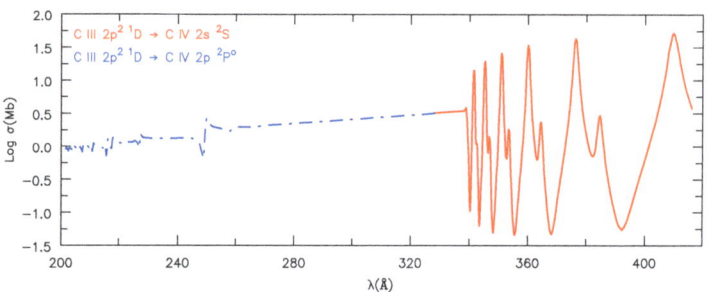

Figure 2. Illustration of the photoionization cross-section (in megabarn, with $1\,\text{Mb} = 10^{-18}\,\text{cm}^2$) of C III $2p^2\ {}^1D$. Ionization to the C IV ground state occurs via autoionizing levels such as C III 2p 4d $^1F^o$. Shortward of ≈ 325 Å, photons have sufficient energy to ionize directly to C IV 2p $^2P^o$. The data were convolved with a Gaussian profile with a full width at half maximum of $\sim 600\,\text{km\,s}^{-1}$ (i.e., $\sigma = 250.0\,\text{km\,s}^{-1}$). The photoionization cross-sections for C III were computed by P. J. Storey (private communication). In other photoionization cross-sections the resonances are often much narrower than those shown here.

In LTE the final state arising from the photoionization process is irrelevant—only the total opacity matters. In general, in nLTE, the final state matters, since each process contributes to the population of a different state whose population needs to be determined from first principles. In practice this is generally not a crucial concern for most spectral modeling since the rates for processes connecting states within an ion are generally much larger than the photoionization and recombination rates. However, there are cases where the final-state-dependent cross-sections are important.

As a first example, we again consider the C III / C IV system in WR stars[6]. In the photosphere the strong C IV $\lambda\lambda 1548, 1551$ doublet (due to $1s^2\, 2s$–$1s^2\, 2p$) is optically thick, and hence the $1s^2\, 2p$ state is strongly coupled to the $1s^2\, 2s$ ground state through collisional de-excitation and excitation (i.e., the 2p state is in LTE (computed using the local electron temperature) with respect to the 2s state). Consequently we can treat all ionizations/recombinations as occurring to/from the C IV ground state. However, as the density declines photon escape in the resonance line will lead to a decoupling of $1s^2\, 2p$ from the $1s^2\, 2s$ state, in which case we should treat recombinations from the $1s^2\, 2p$ state separately from those occurring from the $1s^2\, 2s$ state. Fortunately, because the $1s^2\, 2p$ state lies ≈ 8 eV above the ground state, recombinations from the $1s^2\, 2p$ state are generally not very important for the temperatures and densities appropriate to WR stars in the regime important for spectrum formation. This may not be the case in other regimes, and for other ions with states closer in energy to the ground state.

One crucial area where state-dependent photoionization cross-sections are important is in X-ray fluorescence where the ejection of an inner shell electron leads to the ion being in a highly excited state, and the emission of characteristic X-rays or the subsequent ejection of one or more additional electrons (Auger ionization) (e.g., [68,69]). The subsequent decay of these more highly charged ion gives rise to lines which can be detected and their strength is dependent on the details of the autoionization processes that occurred after the inner shell electron was ejected (e.g., [70–72]).

3.1. Inner Shell Ionization

Typically ionization from the inner shell of an ion (e.g., from the $1s^2$ shell in O I–O V) is not very important for modeling stellar spectra, since very little flux will be emitted at the relevant energies when the ionization stages are abundant. An exception occurs when there is a significant source of X-rays, as can occur when a star has a corona or when there is a compact object with an accretion disk. For massive stars, X-rays can arise in shocks generated by a wind–wind collision in a binary system or in shocks generated from instabilities in the driving of the wind by radiation pressure (e.g., [73–75]). For O stars, the observed X-ray fluxes generated by these two processes are typically in the range of 10^{-5} to $10^{-8}\, L_*$ (e.g., [76,77]).

With the discovery of X-ray emission from O stars it was realized that X-rays could explain the presence of both O VI $\lambda\lambda 1032, 1038$ and N V $\lambda\lambda 1238, 1243$ P Cygni profiles[7] in the UV spectra of O stars. An example P Cygni profile is shown in Figure 3. Since O stars typically have effective temperatures of $<\sim 40,000$ K, the photospheric radiation field cannot produce sufficient O VI to explain the observed O VI profile. However, X-rays, through Auger ionization, can produce sufficient O VI [78]. In the case of O VI, the crucial reaction is:

$$\text{O IV}(1s^2\, 2s^2\, 2p) + \text{X-ray} \rightarrow \text{O V}(1s\, 2s^2\, 2p) + e^- \rightarrow \text{O VI}(1s^2\, 2s) + 2e^- \ .$$

Typically two electrons are ejected (i.e, one by interaction with the photon and one by the Auger process) in Auger ionization for CNO elements but for heavier elements more than two electrons can be ejected (e.g., [68]). In CMFGEN we assume all inner-shell ionizations only eject two electrons and the intermediate states are omitted.[8] Many studies have shown that inner shell ionization of X-rays can successfully explain the presence of O VI and N V in O stars (e.g., [79,80]). Auger ionization complicates the kinetic equations since more than two ionization stages are directly coupled.

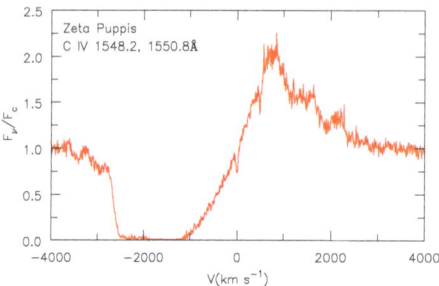

Figure 3. The C IV $\lambda\lambda$1548.2, 1550.8 P Cygni profile in the O4 I(n)fp star Zeta Puppis. The x-axis was computed using $v = c(\lambda/\lambda_o - 1.0)$ where $\lambda_o = 1548.187$. The spectrum has been normalized by the continuum spectrum (F_c)—i.e., a smooth curve drawn through spectral regions showing no evidence for bound–bound absorption or emission. Strong blue shifted absorption is seen, indicating an outflowing stellar wind with a terminal velocity (V_∞) in excess of 2600 km s^{-1}. The redshifted emission primarily arises from continuum photons that were emitted in other directions, absorbed by C IV, and subsequently scattered to the observer. The two narrow absorptions near 0 and 500 km s^{-1} are due to absorption by C IV in the interstellar medium.

4. Recombination

The recombination rate is given by

$$\left(\frac{dn_l}{dt}\right)_{RR} = n_K \left(\frac{n_l}{n_K}\right)^* \int_{v_o}^{\infty} \left(\frac{4\pi}{hv}\right) \sigma_v \left(\frac{2hv^3}{c^2} + J_v\right) \exp(-hv/kT) dv \tag{12}$$

(e.g., [4]) where the subscript K refers to the recombining ion and the LTE population is computed using the actual electron density. The quantity $(n_l/n_K)^*$ (for a given level) is only a function of the electron density and temperature. When the gas is in LTE, and when $J_v = B_v$, the photoionization and recombination rates (absolute values) are identical.

In my work I treat recombination as the reverse process of photoionization and hence in CMFGEN recombination rates are computed using the photoionization cross-sections. As noted earlier, rates are evaluated using numerical quadrature, and identical weights are used for both the forward and reverse process. At high densities it is desirable to treat both processes identically since small differences can cause erroneous populations to be determined when solving the kinetic equations. At depth, where LTE conditions apply, it is important that they identically cancel. Generally the weights are evaluated using the trapezoidal rule—more accurate quadrature schemes are generally not feasible because of the complex frequency dependence (and depth dependence) of the radiation field, and because the same quadrature scheme must be used to compute the rates for both photoionization and recombination. Care must be taken near bound-free edges, since the integrand in the recombination rate can vary rapidly with frequency—especially true for highly ionized states at low temperatures since the recombination rate at frequency v scales as $\exp[-h(v-v_o)/kT]$.

For low densities, such as those found in H II regions, planetary nebulae, and many collisionally ionized plasmas, recombination rates are often evaluated separately, and treated as a distinct process. At "low" densities most transitions are optically thin, and recombination into high states simply cascade into the ground state and metastable levels. In a H II region, for example, the ionization of H is maintained through photoionizations from the ground state and photoionizations from excited states can be ignored. However transitions to the ground state can be optically thick. Consequently two limiting cases are considered when computing H line strengths—Case A, in which all transitions are assumed to be optically thin, and Case B, in which only the Lyman transitions are optically thick (e.g., [25]). Under the optically thick assumption the rate of decays in a transition is assumed to be exactly balanced by the rate of radiative excitations in the transition.

4.1. Direct Radiative Recombination

This process is taken to refer to simple recombination processes such as

$$H^+ + e^- \rightarrow H(nl) + h\nu \text{ and}$$

$$C\ \text{IV}(2s\ ^2S) + e^- \rightarrow C\ \text{III}(2s\ nl) + h\nu \ .$$

In the above n and l refer to the principal and angular momentum quantum numbers of the electron.

The cross-sections for the processes are "smooth" and easily integrated. However, state-of-the-art photoionization cross-sections, such as those available through TOPbase and NORAD, often include indirect photoionization routes or resonances. When discussing the reverse process it is often convenient to split the resonances into two classes—low temperature dielectronic recombination (LTDR) resonances and high temperature dielectronic recombination (HTDR) resonances.

4.2. Low Temperature Dielectronic Recombination (LTDR)

LTDR refers to recombination through double-excited states that lie close to, but above, the ionization energy of the ion ground state [81,82]. It is treated separately from HTDR since it is very species/ionization state specific—the energy levels involved in LTDR need to be known accurately since the distance of these states from the ionization limit is a crucial factor in determining the recombination rate. Dielectronic recombination is the inverse process of autoionization.

We can understand the LTDR process as follows by using an example. Consider the C III $2p\,4f\ ^1D$ state (Figure 4) which can autoionize (the inverse process to dielectronic recombination) to give C IV and a free electron. For this state the autoionization rate coefficient is large ($>10^{13}$ s^{-1}).[9] However, the level can also undergo a "stabilizing transition", leading to a recombination, with the most important stabilizing transition being

$$2p\,4d\ ^1F^o \rightarrow 2p^2\ ^1D + h\nu \ .$$

The Einstein A coefficient for this process (A_{ul}) is $\sim 6.2 \times 10^9$ s^{-1} (P. J. Storey, private communication), much lower than the autoionization probability. Consequently the $2p\,4f\ ^1F^o$ state will be in LTE with respect to C IV and hence the LTDR recombination rate for this single transition is $n_u^* A_{ul}$ where n_u^* (in cgs units) is given by

$$n_u^* = \frac{2.07 \times 10^{-16}}{T^{3/2}} \frac{g_u}{g_{CIV}} N_e N_{CIV} \exp\left(-\psi_l/kT\right) \tag{13}$$

(e.g., [4]). In the above formula g_u is the statistical weight for the $2p\,4d\ ^1F^o$ state, g_{CIV} is the statistical weight of the C IV ground state ($2s\ ^2S$), N_e is the electron density, N_{CIV} is the ground state population of C IV, and ψ_l is the energy of the $2p\,4d\ ^1F^o$ state above the ground state of C IV. At 10^4 K that single transition leads to a LTDR recombination coefficient (defined as the (LTDR rate)/N_e/N_{CIV}) of 3.1×10^{-12} cm^3 s^{-1} (P. J. Storey, private communication) which is essentially identical to the direct recombination rate of 3.2×10^{-12} cm^3 s^{-1} [83].

Thus, we see the following:

1. The LTDR rate is very sensitive to ψ when ψ_l/kT is of order unity or larger.
2. When $\psi_l/kT << 1$, the LTDR recombination rate scales as $T^{-3/2}$ and thus increases more quickly with decreasing temperature than the radiative recombination rate, which typically scales as $T^{-\alpha}$ with $\alpha \sim 0.7$. (see, e.g., [83]).
3. The LTDR process will be most important for those states with a large Einstein A coefficient and for those states lying closest to, but above, the ion ground state.
4. The process is very dependent on the details of the atomic structure. In the above case, the energy of the $2p\,4d\ ^1F^o$ state is crucial for determining the LTDR rate. As the LTDR autoionizing states lie well above the C III ground state, and can have large energy

widths, the energies of the states are not necessarily known. Theoretical calculations can provide estimates, but will have difficulties for states that lie "very close" to the ionization limit since a small error in the energy level can make a big difference in the recombination rate, particularly at low temperatures.

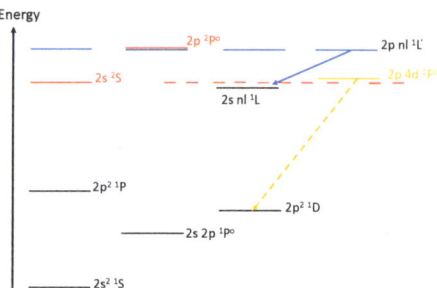

Figure 4. Simplified pseudo-Grotrian diagram for C III and C IV to help illustrate LTDR and HTDR. Five bound C III levels are shown in black and two C IV levels are shown in red. Example HTDR autoionizing levels (e.g., 2p 50p, 2p 50d, etc.) that converge on the C IV 2p ^{2}P^o state are shown in blue. One of the most important LTDR autoionizing levels is shown in orange—it has been moved up slightly in energy to separate it more clearly from the C IV ground state—the horizontal dashed red line is used to indicate the energy of the C IV ground state. The blue line shows a HTDR transition, which has a wavelength approximately equal to that of the C IV resonance transitions (1548, 1551 Å), while the dashed orange line shows a LTDR transition (at ~412 Å). Triplet levels also experience both LTDR and HTDR.

The LTDR rate can exceed the direct recombination rate, and in many cases plays a crucial role in determining nLTE level populations, and observed line strengths (e.g., [81,82,84]).

The LTDR process is complicated by states that are forbidden to autoionize in LS coupling, such as the 2p 4d ^{3}D^o state in C III or the quartet states in C II. In such cases the populations of these levels must be determined by solving the rate equations. These levels will be collisionally and radiatively coupled to states that can autoionize and, because of departures from LS coupling, they can also have non-zero autoionization rates which are larger than the radiative decay routes from the state. Thus, these levels can be an important additional recombination channel.

In CMFGEN we handle the quartet states in C II as part of our atomic models while the doublet autoionizing states are assumed to be in LTE with respect to the ground state of C III and are not directly treated. Recombination through the quartet states is treated via the line transitions connecting them to lower levels, while transitions for the autoionizing states are treated via the photoionization cross-sections. Generally we assume the states within a term are populated according to their statistical weights, although this will not be valid for some levels since the autoionizing rates can depend strongly on their total angular momentum. For example, the autoionizing probabilities for the C II 2s 2p(^{3}P^o) 4s ^{4}P^o j = 1/2, 3/2, and 5/2 states are 5.3×10^9, 1.3×10^{10}, and $<3\,\text{s}^{-1}$. These were obtained from the full width at half maximum tabulated by [84]. One issue, potentially important at high densities, is that we do not have accurate collisional cross-sections for the states not permitted to autoionize in LS coupling.

For C III we typically assume, following [82], that all low lying states can autoionize. LTDR is easily taken into account via the photoionization cross-sections; however, the assumption is only necessarily valid for those states in which the autoionization rates (greatly) exceed other processes populating/depopulating the autoionizing state.

A potential problem in nLTE calculations is that, due to difficulties of current atomic codes to compute accurate energies, the resonances in the photoionization cross-sections are offset from their true positions. Such offsets are probably unimportant when computing the Rosseland mean opacity, but can be important for spectral studies. First, an inaccurate energy will influence the location of observable resonances in stellar spectra. Second, it can have an effect at "low" temperatures due to the scaling of the LTDR rate with temperature ($\exp(-\psi/kT)/T^{3/2}$). Third, complicated nLTE effects could arise. For example, a wrong resonance wavelength can potentially cause issues if a strong resonance coincides (or now does not coincide) with another bound–bound transition (since a strong resonance can affect the radiation field in the transition and vice versa).

The direct inclusion of resonances in photoionization cross-sections also has other potential issues. First, the resonances vary much more rapidly than the background cross-section and hence a very fine frequency grid needs to be used—this is particularly true for narrow resonances. For computational expedience, we typically sample the continuum cross-sections in CMFGEN every 500 km s^{-1} (but finer near level edges). To avoid aliasing[10] we smooth the cross-sections. In early versions of the atomic data the cross-sections were smoothed to a resolution of 3000 km s^{-1} but in new data sets we no longer store the smoothed cross-sections. Instead, newer cross-sections can be smoothed to the desired resolution, set by a control parameter, when they are read in.

Second, a narrow resonance can mean that the autoionization lifetime of the upper level may be comparable to, or even larger than, radiative transitions from the same level. As a consequence the upper level may not be in LTE with respect to the ion and hence the photoionization cross-section should not be used to compute the recombination rate. When identified, such a resonance should be clipped out and the upper levels treated as a bound state.

Third, photoionization cross-sections are usually computed in LS coupling. This means, for example, that the multiplet structure of the resonances is not treated—a problem more crucial when the resonances are "narrow".

4.3. High Temperature Dielectronic Recombination (HTDR)

HTDR involves high Rydberg states [85,86] and it is very difficult to treat accurately in stellar atmosphere codes. The easiest way to visualize HTDR is to discuss a specific example.

Consider for example C III whose ground state is $2s^2$ ^{1}S where we have omitted the complete inner shell for simplicity. The states contributing to HTDR are the Rydberg states of the form 2p nl that converge on the C IV state 2p ^{2}P^o (Figure 4). Such states can autoionize to give C IV 2s ^{2}S or the 2p electron can radiatively decay giving rise to C III 2s nl. At "low" densities the nl electron will decay to a lower level, producing a "real" recombination. Since the autoionization rates decay slowly with n, and since the 2p $\rightarrow$ 2s transition probability is approximately constant, high n values (e.g., up to n = 100) determine the net recombination rates. Such levels are not typically included in nLTE calculations. The process is further complicated because the autoionizing rates strongly depend on the angular momentum—low "l" states have much higher autoionization probabilities than do higher angular-momentum states [85,87].

At low densities the HTDR recombination will scale roughly as $\exp(-E/kT)/T^{3/2}$ where E is the energy of the 2s-2p transition in C IV. Since the energy of the 2s-2p transition is well known, and since the energy of the high Rydberg states is easily approximated, the accuracy of the energy levels is not a crucial factor determining HTDR rates. For some species, several Rydberg series may contribute to the HTDR rate, yielding a more complicated temperature dependence than the simple expression provided above.

In Figure 5, the different recombination coefficients for C III are plotted. The radiative recombination rate and the HTDR rate are from [83], while the LTDR rate is from [81]. The crucial importance of dielectronic recombination for C III is clearly demonstrated by this figure.

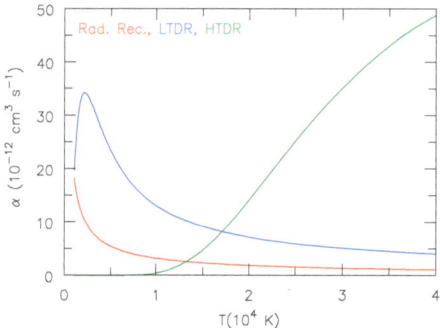

Figure 5. Comparison of the recombination coefficients for normal radiative recombination, for LTDR and for HTDR for recombination from C^{3+} to C^{2+}. Only at low temperatures ($\lesssim$1000 K) does radiative recombination dominate. Above this temperature LTDR dominates until a temperature of $\sim$17,000 K, at which time HTDR becomes the dominant recombination mechanism.

5. Suppression of Dielectronic Recombination

5.1. Collisional Processes

The classic HTDR formula only applies at low densities. As the density rises, collisional ionization by electrons can significantly suppress HTDR. This has been explicitly considered for HTDR of C^{3+} to C^{2+} (i.e., C IV to C III) by [88,89]. The authors of [89] found suppression factors of $\sim$0.7 at 10^4, 0.4 at 10^6, 0.2 at 10^8, and 0.1 at 10^{10} cm^{-3} (these data were read from Figure 1 in [90]). The reduction in the HTDR rate arises because the stabilizing transition (2p-2s in the case of C III) from the autoionizing states leaves the electron in a high nl state. As the electrons cascade down to lower nl states, they can be collisionally ionized by electrons.

5.2. The Importance of the Radiation Field

The radiation field is typically regarded as unimportant in the dielectronic process, but in stellar atmospheres and winds, the process could effectively suppress recombination, as illustrated below. In principle, there are two suppression routes. The radiation field can directly suppress the stabilizing transition or the radiation field can directly ionize an electron out of the high nl state which the stabilizing transitions have left the ion in. The latter is similar to collisional suppression, except it is the radiation field, rather than collisions with electrons, which is reionizing the atom/ion.

For simplicity we treat the resonance as a line transition between two bound states, with the upper level being the autoionizing level. The net recombination rate will be given by

$$NR = n_u A_{ul}(1 - \bar{J}/S_{lu}) \qquad (14)$$

where $\bar{J}$ is the mean intensity in the line and is given by

$$\bar{J} = \int_0^\infty \phi(\nu) J_\nu \, d\nu \, , \qquad (15)$$

$\phi(\nu)$ is the line absorption/emission profile (which, in general, is determined by the finite lifetimes of the levels involved in the transitions, thermal motions of the atoms, and the interaction of the radiating atom/ion with its neighbors (see [4], Chapter 9)) and S_{lu} is the line source function given by

$$S_{lu} = \frac{2h\nu}{c^2}\left(\frac{n_u/g_u}{n_l/g_l - n_u/g_u}\right) \, . \qquad (16)$$

In LTE $n_u = (g_u/g_l)n_l \exp(-h\nu/kT)$ and hence S_{lu} simplifies to the Planck function. As is readily apparent the net rate does not directly depend on the optical depth—such a dependence only occurs indirectly through the dependence of $\bar{J}$ on the optical depth.

In nebula conditions $\bar{J}/S_{lu}$ is typically $<< 1$ (since the nebula is very distant from the star the radiation field is greatly diluted) and the contribution to the recombination rate by this single transition is simply $n_u A_{ul}$. When the rates are summed over all resonances you recover the LTDR/HDTR recombination rate. However, such a rate is typically an upper limit since the radiation field can reduce this rate.

At depth in a stellar atmosphere $\bar{J} \equiv S_{lu} \equiv B_\nu$, and thus the net LTDR and HDTR rates are identically zero—that is, every downward transition in the stabilizing transition is balanced by an upward transition. However, above the atmosphere the temperature of the radiation field and the electrons are not the same. Typically $\bar{J}$ will fall below S_{lu}; however, in a wind $\bar{J}$ can be greater than S_{lu} in some transitions. In Figure 6 we show the mean intensity (in the comoving frame) and the blackbody mean intensity at a temperature of 1.6×10^4 K and a density of $\sim 10^{11}$ electrons cm^{-3}—roughly 50% of the emission in the line referred to as C III $\lambda 2297$ originates above that density. From that figure we see that the radiation field at the wavelength of the C IV resonance transition, and at/near the stabilizing transition, is close to a blackbody at the local electron temperature. Thus, the radiation can act to suppress HTDR.

In Figure 7, we show the recombination and photoionization rates for n = 26 through 30 singlet states of C III (treated as a single level) for a test calculation in which we included HTDR transitions for levels up to n = 30, and with no suppression of the recombination rate with the angular orbital quantum number. The resulting model spectrum is almost identical to the spectrum computed without HTDR—a consequence of the low temperatures in the C III line-formation region and the suppression of HTDR via the radiation field in the stabilizing transition. This result is model dependent—in practice the importance of HTDR needs to be examined on a case-by-case basis.

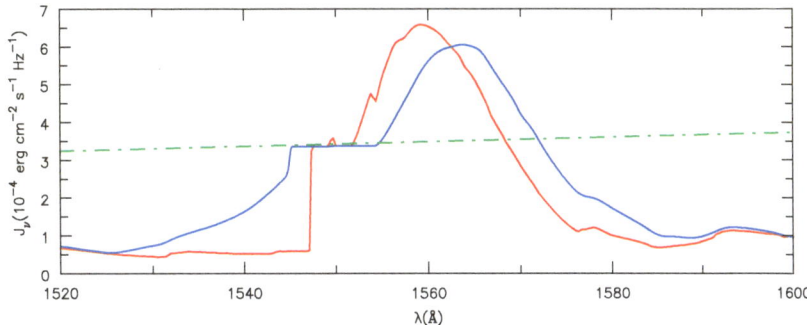

Figure 6. Comparison of the mean intensity (in the comoving frame, red curve) and the blackbody mean intensity (green dashed curve) at a temperature of 1.6×10^4 K and a density of $\sim 10^{11}$ electrons cm^{-3}—roughly 50% of the emission in C III $\lambda 2297$ originates above that density. At this density the C IV 2s and 2p states are collisionally coupled and as a result the radiation field near the wavelengths of the C IV transitions at 1548, 1551 Å is also close to that of a blackbody. The blue curve is similar to the red curve except that we used a Voigt profile for the line absorption/emission profile (see Equation (15)), rather than the simple Doppler profile (see [4]) which is generally used when computing the atmospheric structure and level populations. The use of a Voigt profile is crucial for explaining the observed profile of the C IV resonance doublet at 1548, 1551 Å in WC stars [91].

The influence of the radiation field is not an issue if the resonance is included as part of the photoionization cross-section as the influence of the radiation field is automatically taken into account. It is also not an issue for "autoionizing" states treated as bound levels, since the radiation field is again taken into account. However, it is a potential issue if the LTDR or HTDR rate is included as a separate process, and the inverse process is not included.

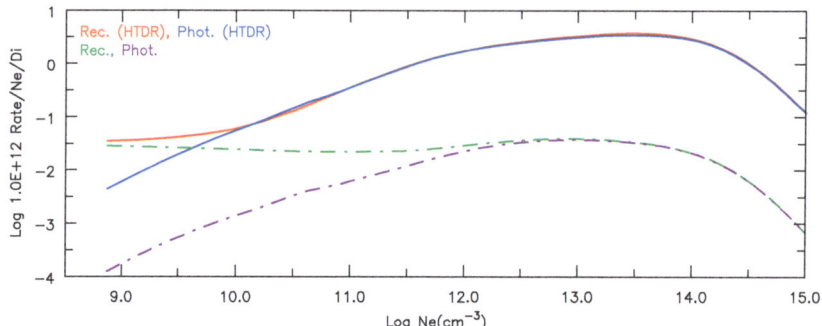

Figure 7. Illustration of the photoionization and recombination rates to/from a combined level (n = 26, . . . , 30; singlet levels only) in C III. The solid curves show the rates with HTDR included, while the dashed curves show the rates without HTDR included. At most densities the recombination and photoionization rates agree closely in the model with HTDR. In such regions the influence of HTDR on the ionization structure is suppressed. Towards the outer boundary the HTDR recombination rate converges to that of a model without HTDR due to the decrease in electron temperature.

6. Astrophysical Examples

Below we discuss some examples of where photoionization data is crucial. It is unrealistic to discuss all cases, since photoionization data is crucial in any photoionized plasma and is crucial for nLTE analyses. For some plasmas, in which collisional processes dominate, photoionization is less important but the inverse process (recombination) is still critical.

6.1. Recombination Processes

The importance of photoionization/recombination processes depends critically on the application. Here we discuss photoionized plasmas and gradually work our way up in density.

In ionized nebulae and H II regions H and He lines are produced by recombination, while the ionization is typically maintained via ionizations from the ground state. The strongest metal lines (i.e., those not due to H and He), such as O III and N II, typically arise via collisional excitation from the ground state. Metal recombination lines are much weaker, simply because the metal abundances are typically a factor of 10^3 (or more) lower than that of H. Some other lines are produced by line fluorescence, where the radiation field in a bound–bound transition in one species gives rise to line emission in another species (the Bowen mechanism). For example, some O III lines in planetary nebulae (gaseous nebulae surrounding stars with effective temperatures in excess of 30,000 K) are produced by the chance overlap of an O III line with He II Lyα [25,92]. With high signal-to-noise spectra metal recombination lines are seen (e.g., [93,94]) and for most lines their strength is effectively set by optically thin recombination (radiative and dielectronic) theory.

As the density increases, processes become increasingly complex. More transitions become optically thick, affecting the cascade process and hence line strengths. Collisional coupling between the levels also becomes more important. If the radiation field is not too diluted, photoionization from excited states also becomes increasingly important. For example, in O and WR stars, the ionization of He^+ to He^{++} occurs predominantly from the n = 2 state whose population is maintained via the intense radiation field in Lyα [95]. Similarly, the ionization of C IV is maintained from the n = 3 levels.

As the density further increases, lines become thicker and photoionization/recombination can become more important since continuous processes at many wavelengths remain optically thin. In hot stars the departure coefficients ($=n/n^*$) for H and He II levels typically rise above the photosphere. Bound–bound transitions are optically thick, preventing cascades. The radiation is diluted, hence recombinations into a level typically exceed photoionizations from that level. Eventually, however, photon escape in lines becomes important and the departure coefficients decrease. However, one must be careful with generalizations—at some wavelengths (particularly in the Wien regime of the blackbody curve) the rapid fall of the electron temperature with height above the photosphere means that the energy density in the radiation field may initially exceed the radiation energy density predicted by the blackbody formula for the local electron temperature.

Bound–bound processes are crucial for determining line strengths. However, it is ultimately photoionization and recombination that determines the ionization state. In some cases, charge exchange processes are crucial [96,97]. Particularly important are charge exchange process of neutral H with, for example, Fe^{2+} and O^+. The reaction

$$O^+ + H \rightleftharpoons O + H^+$$

is resonant, has a total rate coefficient of order 10^{-9} cm^3 s^{-1} [98], and is crucial for determining the O^+/O ratio in regions where the neutral hydrogen fraction exceeds roughly 10^{-3}.

6.2. The Sun

For the Sun, our nearest star, we can determine its structure in two ways. First, we can use the observed Solar parameters (M, L, R_*, abundances) to construct a theoretical model of the Sun. Second, we can use helioseismology observations to constrain the internal structure[11]. Unfortunately, the structure determined from theoretical models and that determined from the helioseismology observations are inconsistent. They can be reconciled if the adopted opacities (for the relevant temperatures) are too low. This could arise if the adopted O abundance is too low or alternatively it could arise from inaccuracies in the opacities (i.e., inaccuracies in the photoionization cross-sections, oscillator strengths, etc). The resolution of the problem is still unclear [50,99,100].

6.3. O, WR, and LBV Stars

O stars are the most luminous hydrogen-core burning stars known. They have masses in the range 30 to $\sim$100 $M_\odot$ and luminosities typically greater than 10^5 $L_\odot$. Due to nuclear processing H is being converted to He in the core. At the same time most of the C and O initially present have been converted to N. Mass loss, and mixing, then operate to reveal this nuclear process at the stellar surface. During later evolution stages He is converted to C and O, and mass loss can also reveal this material at the stellar surface.

All massive stars are losing mass in a stellar wind. In O stars and their descendants (e.g., LBVs, WR stars) the winds are driven by radiation pressure. Due to their high luminosities the stars are close to the Eddington limit[12]. Consequently, it is relatively easy for radiation pressure acting through bound–bound transitions to drive material off the surface of the star via a stellar wind. Due to instabilities in the line driving, it is believed that the winds are highly clumped (e.g., [73–75]). Additional evidence for clumping comes from variability studies (e.g., [101,102]), from the anomalously low strength of some UV resonance transitions relative to the level of Hα emission (e.g., [33–35]), and the weakness of electron scattering wings associated with strong emission lines in P Cygni stars and WR stars[13] [103,104].

The wind density in massive stars varies considerably. For main sequence O stars the winds are relatively weak and only affect a few spectral features. Their photosphere is geometrically thin (i.e., $<<$radius of the star) and, in principle, can be modeled using plane-parallel model atmospheres (i.e., the curvature of the star's atmosphere can be ignored), although the wind may still have an influence at some wavelengths. As the stars evolve, the wind density tends to arise and become increasingly important, and the use of a plane-parallel atmosphere is no longer valid. Indeed, in WR stars, the wind is so dense that spectrum formation occurs in the stellar wind and nLTE spherical models that treat the wind are essential.

6.3.1. N III and N IV lines in Of and WN stars

Of stars are evolved O stars that show emission in N III and He II λ4686 [105,106]. First computations of model atmospheres suggested that the N III lines are driven into emission by LTDR [107]. However, more recent work that includes line blanketing (by lines of iron group elements) and winds reduces the importance of dielectronic recombination, and continuum fluorescence acting through UV resonance transitions plays a crucial role [108].[14]

WN stars, which are a type of WR star, are evolved O stars which show abundances which have been influenced by the CNO nuclear burning cycle—H is depleted (in many it is absent), He is enhanced, and much of the C and O has been converted to N. In a WN star such as HD 50896, several N IV lines are seen. The formation of these lines is complex, but typically their strength is determined by a combination of dielectronic recombination and continuum fluorescence [110]; while the models used by [110] did not include iron group elements, more recent models with iron-group elements confirm the importance of LTDR for WN stars. In Figure 8, we illustrate the influence of LTDR on several N IV emission lines for a model appropriate to an early-type WN star (such as HD 50896).

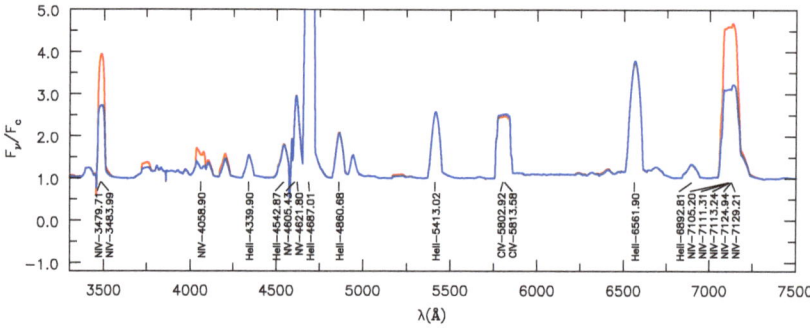

Figure 8. Illustration of the influence of LTDR on the strength of several N IV emission lines. The model in red is the full calculation, while for the blue spectrum we omitted LTDR transitions in the calculation of the nLTE populations. A N IV transition at 1718 Å is also influenced by LTDR. For these calculations we used smooth N IV photoionization cross-sections and the LTDR data of [81,82]. Calculations using the opacity photoionization cross-sections of [111], that were obtained from [22], yield a spectrum very similar to that shown in red. The model has a luminosity of $3 \times 10^5 \, L_\odot$, a radius (at a Rosseland optical depth of 2/3) of 2.9 $R_\odot$, an effective temperature of $\sim$78,000 K, a mass-loss rate of $\dot{M} = 1.5 \times 10^{-5} \, M_\odot \, \mathrm{yr}^{-1}$, and a volume filling factor (which characterizes the degree of clumping in the wind) of 0.1.

6.3.2. Carbon in WC stars

WC stars are the evolved descendants of WN stars. Due to extensive mass loss, and nuclear processing in the interior of the star, their atmospheres are devoid of H, and are primarily composed of He, C and O (with similar mass fractions). Due to their dense stellar winds, and high C and O abundance, emission lines of He, C, and O dominate the spectrum. In the optical region, most of these arise from recombination, although optical depth effects greatly complicate line formation [91].

Below we discuss the spectrum of the WC4-type star, BAT99-9, which has recently been discussed by [37]. In most ways its spectrum, and parameters, are typical of other WC4 stars in the Large Magellanic Cloud (LMC). However, it does differ in one important aspect—it still exhibits one N V and two N IV emission lines. Nitrogen is expected to disappear rapidly as a star transitions from WN to WC because N in the interior of the star is converted to ^{22}Ne.

The electron temperature structure and wind velocity of a model for the LMC WC4 star, BAT99-9, is shown in Figure 9. The non-Planckian nature of the radiation field at two depths in the wind is illustrated in Figure 10.

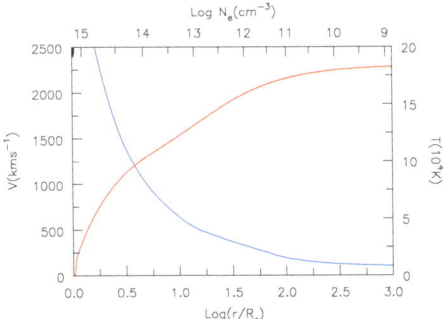

Figure 9. Illustration of the electron temperature (blue) and velocity structure (red) in the model for the WC4-type star, BAT99-9. The model has $\log L/L_\odot = 5.48$, an effective temperature of 84,000 K, and a mass-loss rate of $1.4 \times 10^{-5} \, M_\odot \, \mathrm{yr}^{-1}$.

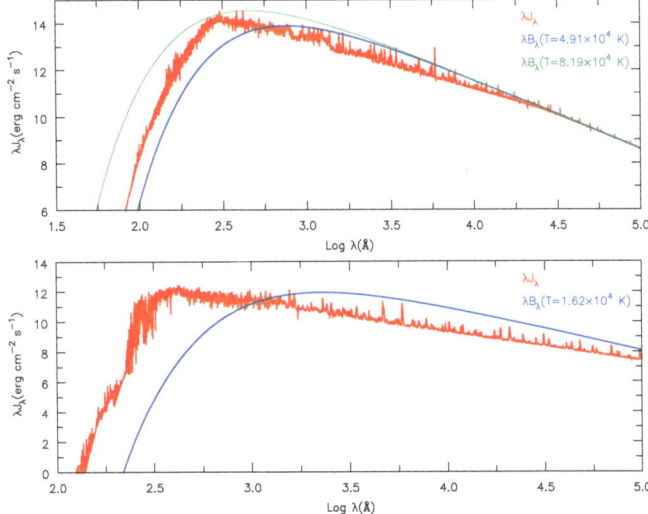

Figure 10. An illustration of the radiation field in a WC4 star at a depth where $V(r) = 0.67 \times V_\infty$ ($N_e = 1.7 \times 10^{13} \, \mathrm{cm}^{-3}$) (top panel) and at a depth $V(r) = 0.934 \times V_\infty$ ($N_e = 1.2 \times 10^{11} \, \mathrm{cm}^{-3}$) (lower panel). Shown also is the blackbody spectrum at the local electron temperatures ($T_e = 4.90 \times 10^4$ K and $T_e = 1.61 \times 10^4$ K). In the top panel (where $V(r) = 0.67 \times V_\infty$) we also illustrate the blackbody spectrum at the effective temperature ($T_e = 8.19 \times 10^4$ K) that has been normalized to match the model spectrum at 5.0 µm. As readily apparent, the local radiation field is very different from that defined by the blackbody at the local electron temperature or that defined by the effective temperature. Consequently, and because of the low electron densities, the ionization state of the gas, and the level populations, are far from their LTE values. Due to the strong departures from LTE accurate atomic data is crucial for determining the state of the gas and hence for predicting the stellar spectrum.

A characteristic of WC stars is the stratified ionization structure—as we move farther out in the wind the ionization decreases. The complex ionization structure for C and O is illustrated in Figure 11. Because of stratified ionization structure many different species need to be included to model the spectrum. In BAT99-9 we see emission from four stages of

O (O III through O VI). To understand driving at the base of the wind additional ionization stages are needed—in some models we include Fe IV through Fe XVII.

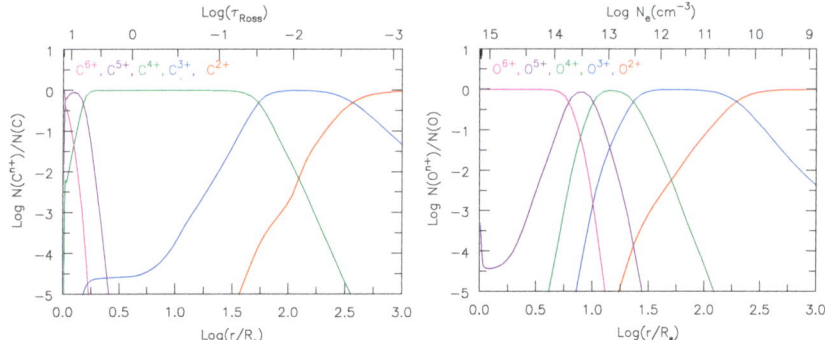

Figure 11. Illustration of the stratified ionization structure in the WC4 star, BAT99-9. On the top axis we show the Rosseland optical depth (**left plot**) and the electron density in units of electrons per cm^3 (**right plot**). The top scales are not linear. As we move out in the wind, the ionization decreases. This variation is verified by the emission profile line widths—O VI lines are narrower than O III lines.

The stratified ionization structure is a consequence of several factors. First, the winds of WR stars are not transparent. For example, the He II Lyman continuum (shortward of 228 Å) is optically thick. Further, the transparency is a strong function of wavelength. Second, as we move out in the wind the radiation field becomes diluted. Third, the intense radiation field in some spectra bands can pump low lying levels. Because of this, and because of the high densities which reduce cascades, ionizations from excited states can play an important role in determining the ionization state of the gas.

In Figure 12 we illustrate the photoionization rate, normalized by the total recombination rate to all (included) levels. The normalization was primarily chosen to emphasize the important process in the line formation region. In the inner regions of these dense winds photoionizations and recombinations to each level will be in detailed balance. As we move out in the wind the photoionization from most levels will decrease due to dilution of the radiation, although for some levels the photoionization and recombination rates may maintain equality if the continua are optically thick. For C III we see that three levels, in order of importance, control the ionization—$2s\,2p\,^3P^o$, $2s^2\,^1S$, and $2s\,2p\,^1P^o$. On the other hand, for C IV it is the n = 3 levels (3s, 3p, and 3d) that help to determine the C IV/C V ionization ration. The ionization eventually shifts because the radiation is becoming diluted (as $1/r^2$) and the populations of the n= 3 levels are also declining. One reason for the difference in behavior of C III and C IV is there is often a rapid decline in the strength of the radiation field shortward of $\sim$228 Å.

The presence of multiple ionization stages in the wind results in emission from multiple ionization stages. For example, in the case of BAT99-9 we see emission from two ionization stages of carbon (C III & C IV)[15] and four ionization stages of oxygen (O III through O VI) with the characteristic line width (after allowance for blending and for the formation mechanism) decreasing as the ionization increases. The origin of one O and two C lines is shown in Figure 13—it shows that a given emission line originates over a range of radii and that lower ionization features form farther out in the wind.

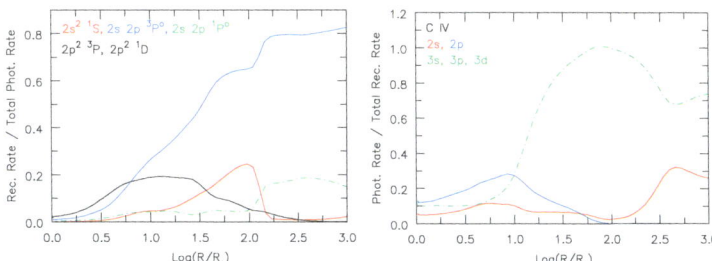

Figure 12. Illustration of the model photoionization rates for the WC4 star, BAT99-9. For illustration purposes, the rates for the $3l$ states have been combined in the right plot.

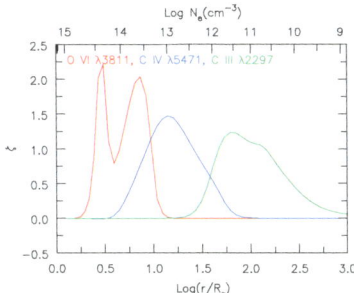

Figure 13. Illustration of where O VI $\lambda 3811$, C IV $\lambda 5471$, and C III $\lambda 2297$ originate in the wind. $\int \zeta \, d\ln r$ is proportional to the emission. High excitation lines originate in the inner denser wind, while lower excitation lines originate in the outer wind.

6.4. C II in [WC] Stars

The LMC star J060819.93-715737.4 (hereafter J0608) has an exquisite C II spectrum—over 150 lines can be identified in the optical [112,113]. It is classified as a [WC11] star with the [] denoting that it is associated with a low mass star ($<$a few$M_\odot$) rather than the product of the evolution of a massive star.[16] The star is probably devoid of H (the observed spectrum exhibits H emission but these probably arise in circumstellar material and not in the stellar wind) and the C abundance is substantially enhanced (atmospheric mass fractions of He and C are approximately 0.4 and 0.6, respectively); while the rich C II spectrum is predominately produced by recombination, it cannot be explained by classical optically thin cascades—optical depth effects play a crucial role in determining the relative C II line strengths.

The spectrum of J0608 is similar to the [WC11] star CPD–56° 8032 whose spectrum has been extensively discussed and analyzed [114–117]. Those studies show the importance of LTDR in producing the spectrum and identify several optical lines that arise from autoionizing levels. The spectrum of J0608 has slightly lower ionization than CPD–56° 8032 and has a lower terminal wind speed, and as a consequence provides a more ideal object by which to explore the C II spectrum.

A small section of the rich C II spectrum is shown in Figure 14. The authors of [113] argue that some of the lines are formed via fluorescence processes, but our own modeling suggests that the spectrum can be explained by allowing for the optical depth effects and by allowing for a transition from ionized to neutral carbon in the outer wind. The latter truncates the emission of the strongest C III lines.

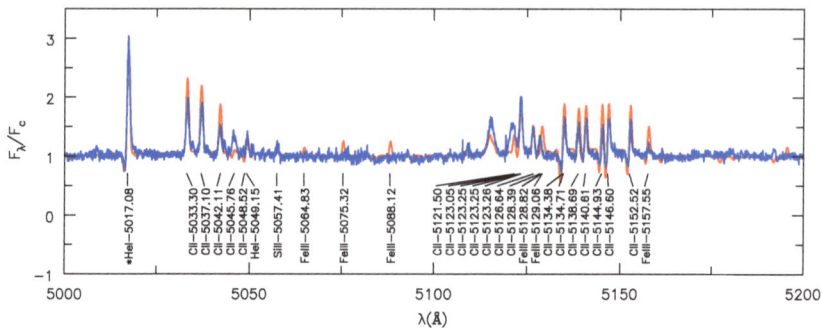

Figure 14. A small section of the spectrum of the WC 11 star J0608 (blue) which was obtained by Nidia Morrell (private communication). A model fit (in red) is shown and model lines are identified. The broad feature near 5115 Å (the first unidentified feature) is due to a resonance in the C II free–free (i.e., Bremsstrahlung) cross-section. It arises from the $2s\,2p(^3P^o)3d\,^2P^o - 2s\,2p(^3P^o)4f\,^2D$ transition [117–119]. Both these states are autoionizing.

In Figure 14, we also show a free–free resonance ($\lambda \sim 5115$ Å) previously identified in CPD–56° 8032 [117]. The observations were obtained with a resolution of $\approx 7\,\text{km}\,\text{s}^{-1}$ and hence the line is resolved. In CMFGEN the resonance is treated as a free–free resonance since both levels involved in the transition are autoionizing (with $A \sim 10^{11}\,\text{s}^{-1}$; [117]). In the case of this free–free resonance it was trivial to omit it from a "continuum" calculation. The latter is needed so we can rectify the spectrum (i.e., normalize the continuum to unity). However, this is not the case for bound-free resonances that appear in the photoionization cross-sections. Such resonances can appear in the computed continuum, distorting an otherwise smooth spectrum. These resonances riddle the UV continuum spectrum. However, in practice they are difficult to discern because of the rich forest of bound–bound transitions which mask the continuum spectrum.

6.5. Supernovae

Supernovae are fascinating objects. They represent the end points of evolution for many stars and are an important source of metals (astronomical jargon for all elements more massive than He) in the Universe. Broadly speaking there are two classes of supernovae —those arising from the core collapse of a massive star (e.g., [120]) and those arising from the thermonuclear detonation of a white dwarf (WD) star (a compact object of stellar origin with a mass less than $\sim 1.4\,M_\odot$ that is supported by electron-degeneracy pressure). The latter class is designated as a Type Ia SN and, while we know that it involves a WD, we do not know in what type of binary system the explosion occurs.

An extensive discussion of the possible progenitors of Type Ia SNe is given by [121]. Type Ia SN could arise when the WD accretes hydrogen-rich material from a "normal" star (e.g., a red supergiant or a main sequence star). As a WD accretes mass its radius shrinks (assuming it does not eject the accreted mass via a surface explosion in an event called a nova, which is believed to occur in many systems). As it approaches the Chandrasekhar mass of $1.414\,M_\odot$ (the upper mass limit or a WD star) it will undergo a thermonuclear explosion. Another possibility is that the WD star accretes He rich material from a WD companion. This material undergoes a surface thermonuclear explosion which triggers an inward propagating shock that triggers the detonation of the accreting WD. A third possibility involves collisions and mergers of two WD stars. In the first scenario the exploding WD has a mass of $1.4\,M_\odot$, while in the other two cases the mass of the exploding WD is (typically) less than $1.4\,M_\odot$. The different scenarios predict different chemical compositions for the ejecta and thus determining the chemical composition of the ejecta offers a potential means of determining the nature of exploding WD.

At late times (say 200 days) the spectra of Type Ia supernovae ejecta are dominated by emission lines of Fe, although lines due to Ni, Ca, and S are also present. One issue with current models of the ejecta is that they fail to yield an iron spectrum in agreement with observation (e.g., [45,122]). Basically, the Fe II lines are too weak relative to Fe III (Figure 15) and this limits our ability to interpret ejecta observations. Is the issue related to a problem in the ejecta explosion models, is it due to the ejecta being clumped (which enhances recombination and hence lowers the ionization), is it due to issues with the iron atomic data, is it due to problems treating the thermalization of high-energy electrons [122], or are we missing additional physics? Unfortunately in these systems the iron atomic data is of crucial importance, since Fe II/ Fe III is of order unity and the lines of both species probably form in the same region. Thus, a factor of 2 error in the Fe II recombination rate will make a factor of 2 error in the Fe II/ Fe III ionization fraction and will change the relative line strengths (which are produced via collisional excitations) by a factor of 2. Fortunately, for Fe II and Fe III, HTDR is unimportant at the relevant temperatures, as can be gleaned from the rates provided by [123].

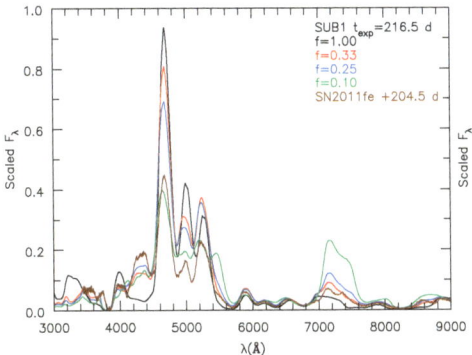

Figure 15. Illustration of a small section of the spectrum of a Ia SN showing the influence of clumping which enhances recombination and hence reduces the ionization state of the gas. In the unclumped model, with $f = 1$, Fe III lines near 5000 Å are too strong. Increasing the clumping (i.e., decreasing f) lowers the ionization, thus improving the agreement. However, the emission feature near 7400 Å (a blend of Fe II, Ni II, and Ca II lines) is worse. The shape of the feature near 5000 Å is also sensitive to the adopted Fe II photoionization cross-sections and updated calculations could also improve the fit.

7. Conclusions

Accurate photoionization cross-sections are essential for many areas of astrophysics. The required quality varies greatly with the application. The biggest needs are at "intermediate" densities where nLTE is relevant, and where complex density and radiation processes directly affect level populations. This is the case for many astrophysical phenomena associated with, for example, stellar winds, accretion disks, and supernovae ejecta.

Funding: Partial support for the work was provided by NASA theory grant 80NSSC20K0524 and STScI Grant No HST-AR-16131.001-A. STScI is operated by the Association of Universities for Research in Astronomy, Inc., under NASA contract NAS 5-26555.

Data Availability Statement: CMFGEN, and the atomic data used by CMFGEN, are available at www.pitt.edu/~hillier (accessed on 1 February 2023). This site also contains some older O star models. Data from supernova calculations can be downloaded from Zenodo or requested from the appropriate author. A grid of spectra are available from the Pollux data base [124]. A large grid of CMFGEN spectra models has been constructed and is being made available [125,126]. Hillier will also provide CMFGEN models upon request.

Acknowledgments: The author would like to thank P. J. Storey for extensive discussions on atomic data, and for supplying atomic data on C II and C III that was directly used in this review. He would also like to thank the numerous workers who have undertaken extensive atomic data calculations and made their work freely available on the internet. A special thanks to Nidia Morrell who obtained the high resolution spectrum of J0608. The invaluable comments made by the referees are also greatly appreciated. This paper has made use of NASA's Astrophysics Data System Bibliographic Services.

Conflicts of Interest: The authors declare no conflict of interest.

Abbreviations

The following abbreviations are used in this manuscript:

ADS	Astrophysics Data System
ESA	European Space Agency
HST	Hubble Space Telescope
LTE	Local thermodynamic equilibrium
nLTE	Non-local thermodynamic equilibrium
LTDR	Low temperature dielectronic recombination
HTDR	High temperature dielectronic recombination
LBV	Luminous blue variable
LS coupling	Total orbital angular momentum (L) is coupled with the total spin (S)
NASA	National Aeronautics and Space Administration
SN	Supernova
WD	White dwarf
WR star	Wolf–Rayet star
WN	Wolf–Rayet star belonging to the nitrogen sequence
WC	Wolf–Rayet star belonging to the carbon sequence

Notes

[1] The effective temperature of the star is defined by the relation $L = 4\pi R_*^2 \sigma_{SB} T_{\text{eff}}^4$ where L is the stellar luminosity (energy emitted per second), σ_{SB} the Stefan–Boltzmann constant, and R_* is the radius of the star.

[2] The meaning of "accurate" is highly context dependent. Atomic data calculations can, in some cases, give energy levels accurate to 1% and for some purposes this is sufficient. However, for spectral modeling such energy levels cannot be used to compute transition wavelengths—a 1% shift (which will be potentially larger if the levels are close in energy) will move a line far from its correct location, influencing spectral synthesis calculations. Moreover, in non-LTE a wrong wavelength will influence how a line interacts with neighboring transitions. In O supergiants two weak Fe IV lines, that overlap with He I $\lambda 304$, influence the strength of He I singlet transitions in the optical and accurate wavelengths (and oscillator strengths) of these Fe IV lines are crucial for understanding the He I singlet transitions [1].

[3] Strictly speaking, a radiation temperature is only well defined if the radiation field is Planckian. However, astronomers often use color temperatures, defined by fitting a scaled blackbody ratio to the flux at two wavelengths, to characterize the nature of the radiation field in some pass band. In nLTE, astronomers may also use the excitation temperature to characterize the excitation or ionization state of a gas. In general these will not be the same as the local electron temperature, and will vary with level and ionization stage (though possibly in a systematic way).

[4] The Sun's atmosphere is cool and dense enough for molecules to form and 50 molecular species have been identified [8]. In the solar spectrum, spectral features due, for example, to CO, SiO, H_2, OH, CH, C_2, and CN, have been identified [8,9]. Dust formation in red giants and supergiants is common but not well understood [10], and in some cases may be associated with non-equilibrium chemistry [11].

[5] Throughout the article we neglect full shells when providing the electron configuration. We use the principal quantum number (n), orbital angular momentum number ($l = 0, \ldots, n-1$ = s, p, d, f, g $\ldots$), and spin ($\pm 1/2$) to describe the state of an electron. Thus, 2p indicates an electron with n = 2 and l = 1. LS-coupling (in which the orbital angular momenta are coupled and the electron spins are coupled) is used to provide the term designation. A term designation has the format $^{2S+1}L^x$ where S is the sum of the (valence) electron spins, L is the total orbital angular momentum, and "o" is used to indicate that the arithmetic sum of the electron orbital angular momenta is odd (o) or even (in which case e is omitted by convention). An excellent primer on atomic spectroscopy is provided by [63].

[6] WR stars are a class of massive stars that evolved from O stars (stars with initial masses $\gtrsim 15\,M_\odot$). They are experiencing mass loss via a stellar wind (induced by radiation pressure acting through bound–bound transitions) with a mass-loss rate typically in excess of $10^{-5}\,M_\odot\,\text{yr}^{-1}$ and a terminal wind speed of ~ 1000 to $3500\,\text{km}\,\text{s}^{-1}$ [64,65]. In many WR stars the wind is sufficiently

7   dense that the entire spectrum we observe originates in the wind—the hydrostatic core of the star is not seen. There are two main WR classes: WN stars exhibit N and He (and sometimes H) emission lines, and exhibit enhanced N and He at the stellar surface due to the CNO cycle (the main H-fusion chain in massive stars). WC stars exhibit emission lines of He, C, and O, with a C abundance comparable to that of He (e.g., [66,67]). They have lost all of their hydrogen envelope, with the enhanced C abundance arising from the triple alpha process ($3\,^4\text{He} \rightarrow\,^{12}\text{C}$).

7   A P Cygni profile is formed when continuum radiation is absorbed and scattered by outflowing material. Outflowing gas along the line of sight absorbs continuum radiation and scatters it out of the line of sight, producing blue-shifted absorption. Radiation absorbed in other directions can be scattered into the line of sight and, for a spherically symmetric expanding gas, the combination with the blue-shifted absorption will give rise to red-shifted emission.

8   The ejecta of Type Ia SNe are composed primarily of intermediate mass elements (Ca, Si) and iron group (Fe, Ni, Co) elements. In such ejecta we may need to treat Auger ionization more rigorously since it could potentially affect the ionization state of the gas and the thermalization of non-thermal electrons. The non-thermal electrons are initially produced via Compton scattering of gamma-ray photons produced from decay of radioactive ^{56}Ni and ^{56}Co. In this case inner shell ionization will most likely occur via non-thermal electrons. However the subsequent Auger ionization and fluorescence are independent of how the K-shell hole was created.

9   From AUTOSTRUCTURE calculations made by a collaboration of researchers at Auburn University, Rollins College, the University of Strathclyde, and other universities. Tables produced by N. R. Badnell and are available at Atomic Data from AUTOSTRUCTURE.

10   A signal processing term that refers to the distortion of data due to sampling which is too coarse. In the present case a narrow but strong resonance could be missed in the photoionization cross-section when the frequency sampling is too coarse. Alternatively, its influence could be artificially enhanced if it is not fully resolved.

11   The Sun is simultaneously oscillating in thousands of different vibration modes. The frequency and strength of these modes depends on the internal structure of the Sun (e.g., the depth of the convection, the sound speed).

12   At the Eddington limit the force arising from the scattering of radiation by free electrons matches the gravitational force.

13   The strength of most emission lines in WR stars is proportional to the density squared. Thus, a clumped wind can yield the same line strengths for a lower mass-loss rate (i.e., for a lower average density). On the other hand electron scattering line wings arise from Thomson scattering of line photons by free electrons and hence scale with density. Thus, the strength of electron scattering wings relative to their neighboring emission line can act as a global diagnostic of clumping. In WR stars the wings are offset to the red from their originating transition because of the large outflow velocities.

14   In this process a strong transition (typically in the UV) absorbs continuum photons, a process whose efficiency is enhanced by the velocity field which allows the UV transitions to intercept more continuum radiation. In many cases the absorbed photons will typically be re-emitted in the same transition. However, in some cases the upper levels have an alternate decay route—decay via this transition can then lead to emission in this bound–bound transition. This is also known as the Swings mechanism [109].

15   C II emission is also predicted but this is masked by blending with other lines.

16   The 11 appended to WC denotes the ionization class of the star—in this case, a spectrum dominated by C II with little evidence for C III.

References

1. Najarro, F.; Hillier, D.J.; Puls, J.; Lanz, T.; Martins, F. On the sensitivity of He I singlet lines to the Fe IV model atom in O stars. *Astron. Astrophys.* **2006**, *456*, 659–664. [CrossRef]
2. Nave, G.; Johansson, S. The Spectrum of Fe II. *Astrophys. J.* **2013**, *204*, 1. [CrossRef]
3. Clear, C.P.; Pickering, J.C.; Nave, G.; Uylings, P.; Raassen, T. Wavelengths and Energy Levels of Singly Ionized Nickel (Ni II) Measured Using Fourier Transform Spectroscopy. *Astrophys. J.* **2022**, *261*, 35. [CrossRef]
4. Mihalas, D. *Stellar Atmospheres*, 2nd ed.; W. H. Freeman and Company: San Francisco, CA, USA, 1978.
5. O'Dell, C.R.; McCullough, P.R.; Meixner, M. Unraveling the Helix Nebula: Its Structure and Knots. *Astrophys. J.* **2004**, *128*, 2339–2356. [CrossRef]
6. Benedict, G.F.; McArthur, B.E.; Napiwotzki, R.; Harrison, T.E.; Harris, H.C.; Nelan, E.; Bond, H.E.; Patterson, R.J.; Ciardullo, R. Astrometry with the Hubble Space Telescope: Trigonometric Parallaxes of Planetary Nebula Nuclei NGC 6853, NGC 7293, Abell 31, and DeHt 5. *Astrophys. J.* **2009**, *138*, 1969–1984. [CrossRef]
7. Meaburn, J.; López, J.A.; Richer, M.G. Optical line profiles of the Helix planetary nebula (NGC 7293) to large radii. *Mon. Not. R. Astron. Soc.* **2008**, *384*, 497–503. [CrossRef]
8. Jørgensen, U.G. Molecules in Stellar and Star-Like Atmospheres. In *Stellar Atmosphere Modeling*; Astronomical Society of the Pacific Conference Series; Hubeny, I., Mihalas, D., Werner, K., Eds.; Astronomical Society of the Pacific: San Francisco, CA, USA, 2003; Volume 288, p. 303.
9. Grevesse, N.; Sauval, A.J. Molecules in the Sun and Molecular Data. In *IAU Colloq. 146: Molecules in the Stellar Environment*; Jorgensen, U.G., Ed.; Springer: Berlin/Heidelberg, Germany, 1994; Volume 428, p. 196. [CrossRef]

10. Cherchneff, I.; Sarangi, A. New Insights on What, Where, and How Dust Forms in Evolved Stars. In *The B[e] Phenomenon: Forty Years of Studies*; Astronomical Society of the Pacific Conference Series; Miroshnichenko, A., Zharikov, S., Korčáková, D., Wolf, M., Eds.; Astronomical Society of the Pacific: San Francisco, CA, USA, 2017; Volume 508, p. 57.
11. Gobrecht, D.; Cherchneff, I.; Sarangi, A.; Plane, J.M.C.; Bromley, S.T. Dust formation in the oxygen-rich AGB star IK Tauri. *Astron. Astrophys.* **2016**, *585*, A6. [CrossRef]
12. Hummer, D.G.; Mihalas, D. The equation of state for stellar envelopes. I—An occupation probability formalism for the truncation of internal partition functions. *Astrophys. J.* **1988**, *331*, 794–814. [CrossRef]
13. Mihalas, D.; Dappen, W.; Hummer, D.G. The equation of state for stellar envelopes. II—Algorithm and selected results. *Astrophys. J.* **1988**, *331*, 815–825. [CrossRef]
14. Daeppen, W.; Mihalas, D.; Hummer, D.G.; Mihalas, B.W. The equation of state for stellar envelopes. III—Thermodynamic quantities. *Astrophys. J.* **1988**, *332*, 261–270. [CrossRef]
15. Hubeny, I.; Lanz, T. Non-LTE line-blanketed model atmospheres of hot stars. 1: Hybrid complete linearization/accelerated lambda iteration method. *Astrophys. J.* **1995**, *439*, 875–904. [CrossRef]
16. Hillier, D.J.; Miller, D.L. The Treatment of Non-LTE Line Blanketing in Spherically Expanding Outflows. *Astrophys. J.* **1998**, *496*, 407–427. [CrossRef]
17. Kurucz, R.; Bell, B. Atomic Line Data. In *Atomic Line Data (R.L. Kurucz B. Bell) Kurucz CD-ROM No. 23.*; Smithsonian Astrophysical Observatory: Cambridge, MA, USA, 1995; Volume 23.
18. Kurucz, R.L. Including All the Lines. *Am. Inst. Phys. Conf.* **2009**, *1171*, 43–51. [CrossRef]
19. Seaton, M.J. Atomic data for opacity calculations. I—General description. *J. Phys. B At. Mol. Phys.* **1987**, *20*, 6363–6378. [CrossRef]
20. Hummer, D.G.; Berrington, K.A.; Eissner, W.; Pradhan, A.K.; Saraph, H.E.; Tully, J.A. Atomic data from the IRON Project. 1: Goals and methods. *Astron. Astrophys.* **1993**, *279*, 298–309.
21. Nahar, S. Database NORAD-Atomic-Data for Atomic Processes in Plasma. *Atoms* **2020**, *8*, 68. [CrossRef]
22. Cunto, W.; Mendoza, C.; Ochsenbein, F.; Zeippen, C.J. Topbase at the CDS. *Astron. Astrophys.* **1993**, *275*, L5.
23. Mendoza, C. TOPbase/TIPbase. In *Atomic and Molecular Data and Their Applications, ICAMDATA*; American Institute of Physics Conference Series; Berrington, K.A., Bell, K.L., Eds.; American Institute of Physics: Melville, NY, USA, 2000; Volume 543, pp. 313–315. [CrossRef]
24. Rybicki, G.B.; Lightman, A.P. *Radiative Processes in Astrophysics*; John Wiley & Sons: Hoboken, NJ, USA, 1979.
25. Osterbrock, D.E.; Ferland, G.J. *Astrophysics of Gaseous Nebulae and Active Galactic Nuclei*; University Science Books: Mill Valley, CA, USA, 2006.
26. Pradhan, A.K.; Nahar, S.N. *Atomic Astrophysics and Spectroscopy*; Cambridge University Press: New York, NY, USA, 2015.
27. Ralchenko, Y. *Modern Methods in Collisional-Radiative Modeling of Plasmas*; Springer International Publishing: Cham, Switzerland, 2016.
28. Hillier, D.J. An iterative method for the solution of the statistical and radiative equilibrium equations in expanding atmospheres. *Astron. Astrophys.* **1990**, *231*, 116–124.
29. Castor, J.I.; Abbott, D.C.; Klein, R.I. Radiation-driven winds in Of stars. *Astrophys. J.* **1975**, *195*, 157–174. [CrossRef]
30. Pauldrach, A.; Puls, J.; Kudritzki, R.P. Radiation-driven winds of hot luminous stars—Improvements of the theory and first results. *Astron. Astrophys.* **1986**, *164*, 86–100.
31. Sundqvist, J.O.; Björklund, R.; Puls, J.; Najarro, F. New predictions for radiation-driven, steady-state mass-loss and wind-momentum from hot, massive stars. I. Method and first results. *Astron. Astrophys.* **2019**, *632*, A126. [CrossRef]
32. Martins, F.; Schaerer, D.; Hillier, D.J. On the effective temperature scale of O stars. *Astron. Astrophys.* **2002**, *382*, 999–1004. [CrossRef]
33. Crowther, P.A.; Hillier, D.J.; Evans, C.J.; Fullerton, A.W.; De Marco, O.; Willis, A.J. Revised Stellar Temperatures for Magellanic Cloud O Supergiants from Far Ultraviolet Spectroscopic Explorer and Very Large Telescope UV-Visual Echelle Spectrograph Spectroscopy. *Astrophys. J.* **2002**, *579*, 774–799. [CrossRef]
34. Bouret, J.C.; Lanz, T.; Hillier, D.J.; Heap, S.R.; Hubeny, I.; Lennon, D.J.; Smith, L.J.; Evans, C.J. Quantitative Spectroscopy of O Stars at Low Metallicity: O Dwarfs in NGC 346. *Astrophys. J.* **2003**, *595*, 1182–1205. [CrossRef]
35. Hillier, D.J.; Lanz, T.; Heap, S.R.; Hubeny, I.; Smith, L.J.; Evans, C.J.; Lennon, D.J.; Bouret, J.C. A Tale of Two Stars: The Extreme O7 Iaf+ Supergiant AV 83 and the OC7.5 III((f)) star AV 69. *Astrophys. J.* **2003**, *588*, 1039–1063. [CrossRef]
36. Bouret, J.C.; Hillier, D.J.; Lanz, T.; Fullerton, A.W. Properties of Galactic early-type O-supergiants. A combined FUV-UV and optical analysis. *Astron. Astrophys.* **2012**, *544*, A67. [CrossRef]
37. Hillier, D.J.; Aadland, E.; Massey, P.; Morrell, N. BAT99-9—a WC4 Wolf-Rayet star with nitrogen emission: Evidence for binary evolution? *Mon. Not. R. Astron. Soc.* **2021**, *503*, 2726–2732. [CrossRef]
38. Aadland, E.; Massey, P.; Hillier, D.J.; Morrell, N.I.; Neugent, K.F.; Eldridge, J.J. WO-type Wolf-Rayet Stars: The Last Hurrah of Massive Star Evolution. *Astrophys. J.* **2022**, *931*, 157. [CrossRef]
39. Najarro, F. Spectroscopy of P Cygni. In *P Cygni 2000: 400 Years of Progress*; Astronomical Society of the Pacific Conference Series; de Groot, M., Sterken, C., Eds.; Astronomical Society of the Pacific: San Francisco, CA, USA, 2001; Volume 233, p. 133.
40. Groh, J.H.; Hillier, D.J.; Damineli, A.; Whitelock, P.A.; Marang, F.; Rossi, C. On the Nature of the Prototype Luminous Blue Variable Ag Carinae. I. Fundamental Parameters During Visual Minimum Phases and Changes in the Bolometric Luminosity During the S-Dor Cycle. *Astrophys. J.* **2009**, *698*, 1698–1720. [CrossRef]

41. Groh, J.H.; Hillier, D.J.; Damineli, A. On the Nature of the Prototype Luminous Blue Variable AG Carinae. II. Witnessing a Massive Star Evolving Close to the Eddington and Bistability Limits. *Astrophys. J.* **2011**, *736*, 46. [CrossRef]
42. Puebla, R.E.; Hillier, D.J.; Zsargó, J.; Cohen, D.H.; Leutenegger, M.A. X-ray, UV and optical analysis of supergiants: ϵ Ori. *Mon. Not. R. Astron. Soc.* **2016**, *456*, 2907–2936. [CrossRef]
43. Herald, J.E.; Bianchi, L. Far-Ultraviolet Spectroscopic Analyses of Four Central Stars of Planetary Nebulae. *Astrophys. J.* **2004**, *609*, 378–391. [CrossRef]
44. Hillier, D.J.; Dessart, L. Time-dependent radiative transfer calculations for supernovae. *Mon. Not. R. Astron. Soc.* **2012**, *424*, 252–271. [CrossRef]
45. Wilk, K.D.; Hillier, D.J.; Dessart, L. Understanding nebular spectra of Type Ia supernovae. *Mon. Not. R. Astron. Soc.* **2020**, *494*, 2221–2235. [CrossRef]
46. Dessart, L.; Hillier, D.J. Radiative-transfer modeling of nebular-phase type II supernovae. Dependencies on progenitor and explosion properties. *Astron. Astrophys.* **2020**, *642*, A33. [CrossRef]
47. Dessart, L.; Hillier, D.J.; Sukhbold, T.; Woosley, S.E.; Janka, H.T. The explosion of 9-29 $M_\odot$ stars as Type II supernovae: Results from radiative-transfer modeling at one year after explosion. *Astron. Astrophys.* **2021**, *652*, A64. [CrossRef]
48. Dessart, L.; Hillier, D.J.; Sukhbold, T.; Woosley, S.E.; Janka, H.T. Nebular phase properties of supernova Ibc from He-star explosions. *Astron. Astrophys.* **2021**, *656*, A61. [CrossRef]
49. Hubeny, I.; Hummer, D.G.; Lanz, T. NLTE model stellar atmospheres with line blanketing near the series limits. *Astron. Astrophys.* **1994**, *282*, 151–167.
50. Christensen-Dalsgaard, J. Solar structure and evolution. *Living Rev. Sol. Phys.* **2021**, *18*, 2. [CrossRef]
51. Seaton, M.J.; Yan, Y.; Mihalas, D.; Pradhan, A.K. Opacities for stellar envelopes. *Mon. Not. R. Astron. Soc.* **1994**, *266*, 805. [CrossRef]
52. Iglesias, C.A.; Rogers, F.J. Updated Opal Opacities. *Astrophys. J.* **1996**, *464*, 943. [CrossRef]
53. Colgan, J.; Kilcrease, D.P.; Magee, N.H.; Sherrill, M.E.; Abdallah, J., J.; Hakel, P.; Fontes, C.J.; Guzik, J.A.; Mussack, K.A. A New Generation of Los Alamos Opacity Tables. *Astrophys. J.* **2016**, *817*, 116. [CrossRef]
54. Blancard, C.; Cossé, P.; Faussurier, G. Solar Mixture Opacity Calculations Using Detailed Configuration and Level Accounting Treatments. *Astrophys. J.* **2012**, *745*, 10. [CrossRef]
55. Magee, N.H.; Abdallah, J.; Clark, R.E.H.; Cohen, J.S.; Collins, L.A.; Csanak, G.; Fontes, C.J.; Gauger, A.; Keady, J.J.; Kilcrease, D.P.; et al. Atomic Structure Calculations and New Los Alamos Astrophysical Opacities. In *Astrophysical Applications of Powerful New Databases*; Astronomical Society of the Pacific Conference Series; Adelman, S.J., Wiese, W.L., Eds.; Astronomical Society of the Pacific: San Francisco, USA, 1995 ; Volume 78, p. 51.
56. Badnell, N.R.; Ballance, C.P.; Griffin, D.C.; O'Mullane, M. Dielectronic recombination of W^{20+} ($4d^{10}4f^8$): Addressing the half-open f shell. *Phys. Rev. A* **2012**, *85*, 052716. [CrossRef]
57. Metzger, B.D.; Martínez-Pinedo, G.; Darbha, S.; Quataert, E.; Arcones, A.; Kasen, D.; Thomas, R.; Nugent, P.; Panov, I.V.; Zinner, N.T. Electromagnetic counterparts of compact object mergers powered by the radioactive decay of r-process nuclei. *Mon. Not. R. Astron. Soc.* **2010**, *406*, 2650–2662. [CrossRef]
58. Kasen, D.; Badnell, N.R.; Barnes, J. Opacities and Spectra of the r-process Ejecta from Neutron Star Mergers. *Astrophys. J.* **2013**, *774*, 25. [CrossRef]
59. Barnes, J.; Kasen, D. Effect of a High Opacity on the Light Curves of Radioactively Powered Transients from Compact Object Mergers. *Astrophys. J.* **2013**, *775*, 18. [CrossRef]
60. Fontes, C.J.; Fryer, C.L.; Hungerford, A.L.; Wollaeger, R.T.; Korobkin, O. A line-binned treatment of opacities for the spectra and light curves from neutron star mergers. *Mon. Not. R. Astron. Soc.* **2020**, *493*, 4143–4171. [CrossRef]
61. Tanaka, M.; Kato, D.; Gaigalas, G.; Kawaguchi, K. Systematic opacity calculations for kilonovae. *Mon. Not. R. Astron. Soc.* **2020**, *496*, 1369–1392. [CrossRef]
62. Fontes, C.J.; Fryer, C.L.; Wollaeger, R.T.; Mumpower, M.R.; Sprouse, T.M. Actinide opacities for modelling the spectra and light curves of kilonovae. *Mon. Not. R. Astron. Soc.* **2023**, *519*, 2862–2878. [CrossRef]
63. Martin, W.C.; Wise, W.L. Atomic Spectroscopy: An Introduction. 2016. Available online: https://www.nist.gov/system/files/documents/2016/10/03/atspec.pdf (accessed on 1 February 2023).
64. Hillier, D. Wolf-Rayet Stars. In *Encyclopedia of Astronomy and Astrophysics*; Murdin, P., Ed.; Institute of Physics Publishing: Bristol, UK, 2000; p. 1895. [CrossRef]
65. Crowther, P.A. Physical Properties of Wolf-Rayet Stars. *Annu. Rev. Astron. Astrophys.* **2007**, *45*, 177–219. [CrossRef]
66. Sander, A.; Hamann, W.R.; Todt, H. The Galactic WC stars. Stellar parameters from spectral analyses indicate a new evolutionary sequence. *Astron. Astrophys.* **2012**, *540*, A144. [CrossRef]
67. Aadland, E.; Massey, P.; Hillier, D.J.; Morrell, N. The Physical Parameters of Four WC-type Wolf-Rayet Stars in the Large Magellanic Cloud: Evidence of Evolution. *Astrophys. J.* **2022**, *924*, 44. [CrossRef]
68. Weisheit, J.C. X-Ray Ionization Cross-Sections and Ionization Equilibrium Equations Modified by Auger Transitions. *Astrophys. J.* **1974**, *190*, 735–740. [CrossRef]
69. Kaastra, J.S.; Mewe, R. X-ray emission from thin plasmas. I—Multiple Auger ionisation and fluorescence processes for Be to Zn. *Astron. Astrophys.* **1993**, *97*, 443–482.
70. McGuire, E.J. K-Shell Auger Transition Rates and Fluorescence Yields for Elements Be-Ar. *Phys. Rev.* **1969**, *185*, 1–6. [CrossRef]

71. McGuire, E.J. K-Shell Auger Transition Rates and Fluorescence Yields for Elements Ar-Xe. *Phys. Rev. A* **1970**, *2*, 273–278. [CrossRef]
72. Bambynek, W.; Crasemann, B.; Fink, R.W.; Freund, H.U.; Mark, H.; Swift, C.D.; Price, R.E.; Rao, P.V. X-Ray Fluorescence Yields, Auger, and Coster-Kronig Transition Probabilities. *Rev. Mod. Phys.* **1972**, *44*, 716–813. [CrossRef]
73. Owocki, S.P.; Castor, J.I.; Rybicki, G.B. Time-dependent models of radiatively driven stellar winds. I—Nonlinear evolution of instabilities for a pure absorption model. *Astrophys. J.* **1988**, *335*, 914–930. [CrossRef]
74. Feldmeier, A. Time-dependent structure and energy transfer in hot star winds. *Astron. Astrophys.* **1995**, *299*, 523.
75. Sundqvist, J.O.; Owocki, S.P.; Puls, J. 2D wind clumping in hot, massive stars from hydrodynamical line-driven instability simulations using a pseudo-planar approach. *Astron. Astrophys.* **2018**, *611*, A17. [CrossRef]
76. Chlebowski, T. X-ray emission from O-type stars—Parameters which affect it. *Astrophys. J.* **1989**, *342*, 1091–1107. [CrossRef]
77. Berghoefer, T.W.; Schmitt, J.H.M.M.; Cassinelli, J.P. The ROSAT all-sky survey catalogue of optically bright OB-type stars. *Astron. Astrophys.* **1996**, *118*, 481–494. [CrossRef]
78. Cassinelli, J.P.; Olson, G.L. The effects of coronal regions on the X-ray flux and ionization conditions in the winds of OB supergiants and Of stars. *Astrophys. J.* **1979**, *229*, 304–317. [CrossRef]
79. Pauldrach, A.W.A.; Hoffmann, T.L.; Lennon, M. Radiation-driven winds of hot luminous stars. XIII. A description of NLTE line blocking and blanketing towards realistic models for expanding atmospheres. *Astron. Astrophys.* **2001**, *375*, 161–195. [CrossRef]
80. Zsargó, J.; Hillier, D.J.; Bouret, J.C.; Lanz, T.; Leutenegger, M.A.; Cohen, D.H. On the Importance of the Interclump Medium for Superionization: O VI Formation in the Wind of ζ Puppis. *Astrophys. J.* **2008**, *685*, L149–L152. [CrossRef]
81. Nussbaumer, H.; Storey, P.J. Dielectronic recombination at low temperatures. *Astron. Astrophys.* **1983**, *126*, 75–79.
82. Nussbaumer, H.; Storey, P.J. Dielectronic recombination at low temperatures. II Recombination coefficients for lines of C, N, O. *Astron. Astrophys.* **1984**, *56*, 293–312.
83. Aldrovandi, S.M.V.; Pequignot, D. Radiative and Dielectronic Recombination Coefficients for Complex Ions. *Astron. Astrophys.* **1973**, *25*, 137.
84. Sochi, T.; Storey, P.J. Dielectronic recombination lines of C^+. *At. Data Nucl. Data Tables* **2013**, *99*, 633–650. [CrossRef]
85. Burgess, A. Dielectronic recombination in the corona. *Ann. D'Astrophysique* **1965**, *28*, 774.
86. Burgess, A. A General Formula for the Estimation of Dielectronic Recombination Co-Efficients in Low-Density Plasmas. *Astrophys. J.* **1965**, *141*, 1588–1590. [CrossRef]
87. Burgess, A. Delectronic Recombination and the Temperature of the Solar Corona. *Astrophys. J.* **1964**, *139*, 776–780. [CrossRef]
88. Davidson, K. Dielectronic recombination and abundances near quasars. *Astrophys. J.* **1975**, *195*, 285–291. [CrossRef]
89. Badnell, N.R.; Pindzola, M.S.; Dickson, W.J.; Summers, H.P.; Griffin, D.C.; Lang, J. Electric Field Effects on Dielectronic Recombination in a Collisional-Radiative Model. *Astrophys. J.* **1993**, *407*, L91. [CrossRef]
90. Nikolić, D.; Gorczyca, T.W.; Korista, K.T.; Ferland, G.J.; Badnell, N.R. Suppression of Dielectronic Recombination due to Finite Density Effects. *Astrophys. J.* **2013**, *768*, 82. [CrossRef]
91. Hillier, D.J. WC stars—Hot stars with cold winds. *Astrophys. J.* **1989**, *347*, 392–408. [CrossRef]
92. Bowen, I.S. The Excitation of the Permitted O III Nebular Lines. *Publ. Astron. Soc. Pac.* **1934**, *46*, 146–148. [CrossRef]
93. Fang, X.; Liu, X.W. Very deep spectroscopy of the bright Saturn nebula NGC 7009—II. Analysis of the rich optical recombination spectrum. *Mon. Not. R. Astron. Soc.* **2013**, *429*, 2791–2851. [CrossRef]
94. Peimbert, M.; Peimbert, A.; Delgado-Inglada, G. Nebular Spectroscopy: A Guide on Hii Regions and Planetary Nebulae. *Publ. Astron. Soc. Pac.* **2017**, *129*, 082001. [CrossRef]
95. Hillier, D.J. An empirical model for the Wolf-Rayet star HD 50896. *Astrophys. J.* **1987**, *63*, 965–981. [CrossRef]
96. Field, G.B.; Steigman, G. Charge Transfer and Ionization Equilibrium in the Interstellar Medium. *Astrophys. J.* **1971**, *166*, 59. [CrossRef]
97. Williams, R.E. The ionization structure of planetary nebulae–X. The contribution of charge exchange and optically thick condensations to [O I] radiation. *Mon. Not. R. Astron. Soc.* **1973**, *164*, 111. [CrossRef]
98. Stancil, P.C.; Schultz, D.R.; Kimura, M.; Gu, J.P.; Hirsch, G.; Buenker, R.J. Charge transfer in collisions of O^+ with H and H^+ with O. *Astron. Astrophys.* **1999**, *140*, 225–234. [CrossRef]
99. Nagayama, T.; Bailey, J.E.; Loisel, G.P.; Dunham, G.S.; Rochau, G.A.; Blancard, C.; Colgan, J.; Cossé, P.; Faussurier, G.; Fontes, C.J.; et al. Systematic Study of L -Shell Opacity at Stellar Interior Temperatures. *Phys. Rev. Lett.* **2019**, *122*, 235001. [CrossRef]
100. Bailey, J.E.; Nagayama, T.; Loisel, G.P.; Rochau, G.A.; Blancard, C.; Colgan, J.; Cosse, P.; Faussurier, G.; Fontes, C.J.; Gilleron, F.; et al. A higher-than-predicted measurement of iron opacity at solar interior temperatures. *Nature* **2015**, *517*, 56–59. [CrossRef]
101. Eversberg, T.; Lepine, S.; Moffat, A.F.J. Outmoving Clumps in the Wind of the Hot O Supergiant zeta Puppis. *Astrophys. J.* **1998**, *494*, 799–805. [CrossRef]
102. Lépine, S.; Moffat, A.F.J. Direct Spectroscopic Observations of Clumping in O-Star Winds. *Astrophys. J.* **2008**, *136*, 548–553. [CrossRef]
103. Hillier, D.J. The effects of electron scattering and wind clumping for early emission line stars. *Astron. Astrophys.* **1991**, *247*, 455–468.
104. Hillier, D.J.; Miller, D.L. Constraints on the Evolution of Massive Stars through Spectral Analysis. I. The WC5 Star HD 165763. *Astrophys. J.* **1999**, *519*, 354–371. [CrossRef]

105. Conti, P.S.; Alschuler, W.R. Spectroscopic Studies of O-Type Stars. I. Classification and Absolute Magnitudes. *Astrophys. J.* **1971**, *170*, 325. [CrossRef]
106. Walborn, N.R. Some Spectroscopic Characteristics of the OB Stars: An Investigation of the Space Distribution of Certain OB Stars and the Reference Frame of the Classification. *Astrophys. J.* **1971**, *23*, 257. [CrossRef]
107. Mihalas, D.; Hummer, D.G.; Conti, P.S. On the N III $\lambda\lambda$4640, 4097 lines is Of stars. *Astrophys. J.* **1972**, *175*, L99–L104. [CrossRef]
108. Rivero González, J.G.; Puls, J.; Najarro, F. Nitrogen line spectroscopy of O-stars. I. Nitrogen III emission line formation revisited. *Astron. Astrophys.* **2011**, *536*, A58. [CrossRef]
109. Swings, P. Anomalies in the earliest spectral types. *Ann. D'Astrophysique* **1948**, *11*, 228–246.
110. Hillier, D.J. The formation of nitrogen and carbon emission lines in HD 50896 (WN5). *Astrophys. J.* **1988**, *327*, 822–839. [CrossRef]
111. Tully, J.A.; Seaton, M.J.; Berrington, K.A. Atomic data for opacity calculations. XIV—The beryllium sequence. *J. Phys. B At. Mol. Phys.* **1990**, *23*, 3811–3837. [CrossRef]
112. Margon, B.; Manea, C.; Williams, R.; Bond, H.E.; Prochaska, J.X.; Szymański, M.K.; Morrell, N. Discovery of a Rare Late-type, Low-mass Wolf-Rayet Star in the LMC. *Astrophys. J.* **2020**, *888*, 54. [CrossRef]
113. Williams, R.; Manea, C.; Margon, B.; Morrell, N. Line Identification and Excitation of Autoionizing States in a Late-type, Low-mass Wolf-Rayet Star. *Astrophys. J.* **2021**, *906*, 31. [CrossRef]
114. Leuenhagen, U.; Hamann, W.R.; Jeffery, C.S. Spectral analyses of late-type WC central stars of planetary nebulae. *Astron. Astrophys.* **1996**, *312*, 167–185.
115. De Marco, O.; Barlow, M.J.; Storey, P.J. The WC10 central stars CPD-56 deg8032 and He 2-113—I. Distances and nebular parameters. *Mon. Not. R. Astron. Soc.* **1997**, *292*, 86–104. [CrossRef]
116. De Marco, O.; Crowther, P.A. The WC10 central stars CPD-56 deg8032 and He2-113—II. Model analysis and comparison with nebular properties. *Mon. Not. R. Astron. Soc.* **1998**, *296*, 419–429. [CrossRef]
117. De Marco, O.; Storey, P.J.; Barlow, M.J. The WC10 central stars CPD-56 deg8032 and He2-113—III. Wind electron temperatures and abundances. *Mon. Not. R. Astron. Soc.* **1998**, *297*, 999–1014. [CrossRef]
118. Barlow, M.J.; Storey, P.J. The Wind Temperature and C/He and O/He Ratios of the WC10 Central Star CPD560 8032. In *Planetary Nebulae*; IAU Symposium; Weinberger, R., Acker, A., Eds.; Kluwer Academic: Dordrecht, Germany, 1993; Volume 155, p. 92.
119. Storey, P.J.; Sochi, T. Electron temperatures and free-electron energy distributions of nebulae from C II dielectronic recombination lines. *Mon. Not. R. Astron. Soc.* **2013**, *430*, 599–610. [CrossRef]
120. Foglizzo, T. Explosion Physics of Core-Collapse Supernovae. In *Handbook of Supernovae*; Alsabti, A.W., Murdin, P., Eds.; Springer International Publishing: Berlin/Heidelberg, Germany, 2017; p. 1053. [CrossRef]
121. Hoeflich, P. Explosion Physics of Thermonuclear Supernovae and Their Signatures. In *Handbook of Supernovae*; Alsabti, A.W., Murdin, P., Eds.; Springer International Publishing: Berlin/Heidelberg, Germany, 2017; p. 1151. [CrossRef]
122. Shingles, L.J.; Flörs, A.; Sim, S.A.; Collins, C.E.; Röpke, F.K.; Seitenzahl, I.R.; Shen, K.J. Modelling the ionization state of Type Ia supernovae in the nebular phase. *Mon. Not. R. Astron. Soc.* **2022**, *512*, 6150–6163. [CrossRef]
123. Shull, J.M.; van Steenberg, M. The ionization equilibrium of astrophysically abundant elements. *Astrophys. J.* **1982**, *48*, 95–107. [CrossRef]
124. Palacios, A.; Gebran, M.; Josselin, E.; Martins, F.; Plez, B.; Belmas, M.; Lèbre, A. POLLUX: A database of synthetic stellar spectra. *Astron. Astrophys.* **2010**, *516*, A13. [CrossRef]
125. Zsargó, J.; Rosa Fierro-Santillán, C.; Klapp, J.; Arrieta, A.; Arias, L.; Mendoza Valencia, J.; Sigalotti, L.D.G. Creating and using large grids of pre-calculated model atmospheres for rapid analysis of stellar spectra. *arXiv* **2021**. [CrossRef]
126. Zsargó, J.; Fierro-Santillán, C.R.; Klapp, J.; Arrieta, A.; Arias, L.; Valencia, J.M.; Sigalotti, L.D.G.; Hareter, M.; Puebla, R.E. Creating and using large grids of precalculated model atmospheres for a rapid analysis of stellar spectra. *Astron. Astrophys.* **2020**, *643*, A88. [CrossRef]

Disclaimer/Publisher's Note: The statements, opinions and data contained in all publications are solely those of the individual author(s) and contributor(s) and not of MDPI and/or the editor(s). MDPI and/or the editor(s) disclaim responsibility for any injury to people or property resulting from any ideas, methods, instructions or products referred to in the content.

Review

Atomic Processes, Including Photoabsorption, Subject to Outside Charge-Neutral Plasma

Tu-Nan Chang [1,*], Te-Kuei Fang [2], Chensheng Wu [3] and Xiang Gao [4,*]

1. Department of Physics, University of Southern California, Los Angeles, CA 90089-0484, USA
2. Department of Physics, Fu Jen Catholic University, Taipei 242, Taiwan; 051420@mail.fju.edu.tw
3. Institution of Applied Physics and Computational Mathematics, Beijing 100088, China; wuchensheng89@foxmail.com
4. Beijing Computational Science Center, Beijing 100193, China
* Correspondence: tnchang@usc.edu (T.-N.C.); xgao@csrc.ac.cn (X.G.)

Abstract: We present in this review our recent theoretical studies on atomic processes subject to the plasma environment including the α and β emissions and the ground state photoabsorption of the one- and two-electron atoms and ions. By carefully examining the spatial and temporal criteria of the Debye–Hückel (DH) approximation based on the classical Maxwell–Boltzmann statistics, we were able to represent the plasma effect with a Debye–Hückel screening potential V_{DH} in terms of the Debye length D, which is linked to the ratio between the plasma density N and its temperature kT. Our theoretical data generated with V_{DH} from the detailed non-relativistic and relativistic multiconfiguration atomic structure calculations compare well with the limited measured results from the most recent experiments. Starting from the quasi-hydrogenic picture, we were able to show qualitatively that the energy shifts of the emission lines could be expressed in terms of a general expression as a function of a modified parameter, i.e., the reduced Debye length λ. The close agreement between theory and experiment from our study may help to facilitate the plasma diagnostics to determine the electron density and the temperature of the outside plasma.

Keywords: atomic processes in plasma; Debye–Hückel; α and β emissions; multiconfiguration method

Citation: Chang, T.-N.; Fang, T.-K.; Wu, C.; Gao, X. Atomic Processes, Including Photoabsorption, Subject to Outside Charge-Neutral Plasma. *Atoms* **2022**, *10*, 16. https://doi.org/10.3390/atoms10010016

Academic Editors: Sultana N. Nahar and Guillermo Hinojosa

Received: 15 December 2021
Accepted: 26 January 2022
Published: 29 January 2022

Publisher's Note: MDPI stays neutral with regard to jurisdictional claims in published maps and institutional affiliations.

Copyright: © 2022 by the authors. Licensee MDPI, Basel, Switzerland. This article is an open access article distributed under the terms and conditions of the Creative Commons Attribution (CC BY) license (https://creativecommons.org/licenses/by/4.0/).

1. Introduction

Reliable data for many of the atomic processes subject to the outside charge-neutral plasma are important for the numerical modeling of the evolution of many processes for the energy-related controlled fusion program and also some of the astrophysical systems [1–4]. With the help of the high-speed large-scale computational facilities, currently, the theoretical atomic structure calculations by including all the electron–electron interactions between atomic electrons are capable of generating highly reliable atomic data in close agreement with the experimental observations in the plasma-free environment. However, a detailed theoretical understanding of the atomic process in the plasma environment would need to include the practically unattainable efforts to cover the long-range interactions between the atomic electrons and all the positively charged ions and the negatively charged electrons in the plasma. Over the years, by including the interactions between the atomic electrons and the outside plasma, attempts have been made with somewhat detailed theoretical methods [5–13] to generate data that may understand the limited experimental measurements. One such example is the application of the ion sphere (IS) approach [8–10]. Whereas the calculated redshifts of the α emission of the He-like Al ion based on an analytical IS model [8,9] are in agreement with the recent picosecond time-resolved measurement [14], its estimated redshifts at a fixed temperature are substantially greater than the measured data from a high-resolution satellite line-free measurement of the β line of the He-like Cl ions (see Figure 4 of [15]). In addition, shortly after the He-like Cl ion measurement, the estimated energy shift from an average atom ion sphere (AIS) model calculation was reported to be in good agreement [10]. However, the subsequent application of this AIS model to the α emission of the He-like Al ion has led to redshifts substantially smaller than the earlier experimentally observed data (see Figure 2 of [16]). The disagreement between the IS theories and experiments likely results from the less-than-adequate representation of the interaction between the outside plasma and the atomic electrons.

The main objective of this review is to summarize a series of recent studies based on the Debye–Hückel (DH) model, proposed before the quantum mechanics was fully developed [17], on the atomic processes in the plasma environment. The DH approximation, based on the classical Maxwell–Boltzmann statistics for an electron–ion collision-less plasma, is best known to work for the gas-discharged plasma at relatively low density [18]. To apply the DH approximation to the atomic processes subject to plasma with higher density, one needs to consider two important key criteria. First, temporally, the time scale of the atomic process (e.g., lifetime) should be substantially different from its correlation time t_p, or the inverse of the plasma frequency ($f_p = 8.977 \times 10^3 \, N^{1/2}$Hz) of the outside plasma with density N. For a plasma with its density of the order of 1×10^{22} cm^{-3}, t_p is of the order of 10^{-15} s. This is substantially shorter than the lifetime T_{2p} of the Lyman-α emission line of the hydrogen atom from its $2p$ state, which is of the order of nanoseconds, or more precisely, 1.6 ns. It is also known that T_{2p} scales as the inverse of Z^4, and it would decrease to a similar order of magnitude of t_p as Z for the H-like ion increases up to about 18. The values of T_{2p} will again be substantially shorter than t_p with Z greater than 50. The other important time factor is the time for an electron moving around the nucleus. Again, take the Lyman-α emission line of the hydrogen atom as an example: the time for the $2p$ electron moving around the nucleus is about 10^{-15} s, which is similar for t_p. Since this revolving speed scales as Z^2 and it would be an order of magnitude higher than t_p with $Z > 5$, as a result, the DH approximation works for the Lyman-α emission line of the H-like ions only with Z between six and eighteen or greater than fifty. Second, spatially, the atomic orbitals involved in the transition should only be affected marginally by the outside plasma to retain most of their atomic characteristic. For example, for the transitions involving the electron in its ground state with the radius of its electron orbital sufficiently short in range, the contribution to the overlapping transition matrix between the lower and upper atomic orbitals of the atomic processes is mostly from a critical interacting region with the influence of the outside plasma included and the amplitude of the lower atomic orbital mostly non-zero. In other words, the DH approximation should only be applied to those transitions involving the ground state or the low-n states. Detailed discussions about these two key criteria were also given elsewhere [19–23]. In essence, the complicated many-body interaction between the atomic electrons and the outside charge-neutral plasma is effectively represented by a simple potential $V_{DH}(r)$ depending on two key parameters. The first one is the Debye length D, which can be expressed in the Bohr radius a_0 in terms of the ratio of the temperature kT and the electron density N (in units of eV and 10^{22} cm^{-3}, respectively) of the outside plasma as $D = 1.4048 \, (kT/N)^{1/2}$ [17,18]. For the plasma-free environment, the density N equals zero and D goes to infinity. Effectively, the Debye length D, which appears in the form of $e^{-r/D}$ or a Debye screening, modifies the attractive nuclear interaction $-Z/r$ in the potential V_{DH} due to the outside plasma. The second one is an ad hoc parameter, i.e., the radius A of the Debye sphere, which separates the plasma-induced Debye potential V_{DH} outside the Debye sphere and the slightly modified close-in region where the atomic characteristic dominates. More details on these two parameters, as well as the original approach of the DH approximation leading to the effective potential V_{DH} are presented in Section 2. A brief summary of the multiconfiguration non-relativistic and relativistic calculational procedures is also given in Section 2.

Due to its simplicity, the DH model has been applied extensively to study the atomic processes subject to the outside plasma [24–35]. As we show in Section 2, the atomic electrons are subject to an effective DH potential derived from the Gauss theorem by assuming an infinitely heavy nuclear charge Z located at $r = 0$ with screening from the fast free-moving plasma electrons of high mobility. In contrast, with the relatively low mobility of the plasma ions, one could not assume a substantial presence of the plasma ions between the fast-moving atomic electrons. In other words, there is little theoretical justification to apply the Debye screening to the two-body Coulomb interaction between atomic electrons. By including the Debye screening, which effectively reduces the Coulomb repulsive interaction between atomic electrons, a high-precision theoretical calculation has indeed led to the spurious conclusion that the only known bound state of H^- between two loosely bound electrons would stay bound even in the presence of a strong outside plasma [34,35].

We review in detail in Section 3 our recent applications of the DH approximation to the α-emission lines of the H-like and He-like ions subject to the outside plasma [20–23]. Our calculated redshifts in transition energy are in agreement both with the experimentally observed data [14,15,36], as well as the results from some of the more elaborate simulations based on the quantum mechanical approaches [5–13]. Interestingly, our studies led to a simple scaling feature for the redshifts of the transition energy and the oscillator strength as functions of a related parameter, the *reduced Debye length* $\lambda = Z_{eff} D$, defined as the product of the Debye length D and the effective nuclear charge $Z_{eff} = Z - 1$ of the atomic ion [22,23]. Specifically, the ratio between the shifts in the transition energy $\Delta \omega$ and the plasma-free transition energy ω_0, i.e., $R_\omega = \Delta \omega / \omega_0$, could be expressed by a simple polynomial in terms of this new parameter λ for all ions with applicable nuclear charge Z. Indeed, our calculations with the non-relativistic and the relativistic multiconfiguration calculations have confirmed such a general scaling feature (e.g., see Figure 5 of [21] and Figure 1 of [23]). By introducing this new parameter λ, we were able to focus our application of this slightly modified DH approximation for the general features of the atomic transition data that could be extended to all applicable ions from a single theoretical calculation.

We focus our review on the atomic photoabsorption from the ground state of the one- and two-electron atoms in Section 4. First, we should point out that we modified the term photoionization from our originally published works to photoabsorption due to the realization that the speed of the outgoing ionized electron after the absorption of the incoming photon is generally less than the speed of the outside plasma electrons, and experimentally, the resulting outgoing photoelectron could not be measured, such as the one in a plasma-free photoionization experiment. The application of the DH approximation would also be limited to the process involving mostly the ground state of the target atoms to meet the spatial criterion when the transition rate is only affected by the overlap between the initial and final state wavefunctions at small r when the plasma effect is well represented by the V_{DH}, as we discussed earlier.

Finally, we summarize briefly in Section 5 the implications of our studies and the further interplays between the theoretical estimate of the atomic data based on the application of the DH approximation and the more advanced experimental works. In particular, to fully take advantage of the general feature in terms of the reduced Debye length for the transition energy shifts as a plasma diagnostic possibility, further high-precision experiments are necessary for a better determination of the range of the radius A of the Debye sphere.

2. Debye–Hückel Approximation and the Calculational Procedure

The theoretical methods were presented in detail in our recent works [21,23]. In this section, we repeat some of the key equations to facilitate a self-contained discussion.

Based on the original Debye–Hückel model [17], the potential $V_o(r)$ for an electron–ion collision-less plasma at a distance r far from the atomic nucleus Z outside a Debye sphere of radius A can be derived from Poisson's equation based on the Gauss law:

$$\nabla^2 V_o(r) = -\frac{\rho(r)}{\epsilon}, \quad r \geq A, \tag{1}$$

where ϵ is the dielectric constant of the electron–ion gas and ρ is its total charge density at r. Starting from the Boltzmann distribution and assuming a charge density ρ_o and a zero potential at $r = \infty$, the charge densities of the negative charge $-q$ and the equal positive charge q at r could be expressed as $\rho_-(r) = \rho_o e^{qV_o(r)/kT}$ and $\rho_+(r) = \rho_o e^{-qV_o(r)/kT}$, respectively, where k is the Boltzmann constant and T the absolute temperature. The total charge density at r is then given by:

$$\rho(r) = \rho_o(e^{-\frac{qV_o(r)}{kT}} - e^{\frac{qV_o(r)}{kT}}) = -2\rho_o \sinh\left(\frac{qV_o(r)}{kT}\right). \qquad (2)$$

Additionally, if qV_o is relatively small compared to kT, Equation (2) for the potential in the outer region of the Debye sphere could then be approximated by the *linear* Poisson–Boltzmann equation, i.e.,

$$\nabla^2 V_o(r) = \left(\frac{1}{D^2}\right) V_o(r), \quad r \geq A, \qquad (3)$$

where $D > A$ is the Debye length in Bohr radius a_0 defined in terms of the ratio of the density N ($\sim q\rho_o$) in units of 10^{22} cm^{-3} and the temperature kT in units of eV of the outside plasma given earlier, i.e.,

$$D = 1.4048\left(\frac{kT}{N}\right)^{1/2} a_0. \qquad (4)$$

The potential inside the Debye sphere is derived from the Gauss law, i.e.,

$$V_i(r < A) = -\frac{Ze^2}{r} + \text{Constant}. \qquad (5)$$

By matching V_o and V_i and their first-order derivatives at $r = A$, one obtains [20,37,38]:

$$V_{DH}(r) = \begin{cases} V_i(r) = -Ze^2\left(\frac{1}{r} - \frac{1}{D+A}\right), & r \leq A \\ V_o(r) = -Ze^2\left(\frac{De^{A/D}}{(D+A)}\right)\frac{e^{-r/D}}{r}, & r \geq A. \end{cases} \qquad (6)$$

In the limit when $A \to 0$, Equation (6) reduces to the screened Coulomb potential $-(Z/r)e^{-r/D}$, similar to the Yukawa potential in nuclear physics. From Equation (6), the DH model, or the DH potential V_{DH}, depends on two important parameters, as we discussed earlier in Section 1. The first one is the Debye length D, which goes to infinity when the density N equals zero for the plasma-free environment. The second one is an ad hoc parameter, i.e., the radius of the Debye sphere A, which separates the plasma-induced Debye potential V_{DH} outside the Debye sphere and the slightly modified close-in region where the atomic characteristic dominates.

The consequence of the less-attractive Debye potential V_{DH} is that all the atomic levels will experience an up-lifting in energy. Qualitatively, the change in the transition energy $\Delta\omega$ from its plasma-free transition energy ω_o for an atomic transition subject to the outside plasma depends on the decrease or increase of the difference in the relative energy shifts of the initial and final state of the transition. For the intershell transitions, although the percentage change of the orbital energy is larger for an electron with a larger principal quantum number n due to the stronger outside plasma effect, its small plasma-free orbital energy actually makes the net energy change smaller than the one for the electron with a smaller n; thus, the transition energy is redshifted. On the other hand, for the intrashell transitions, the involved electrons are from the orbitals with the same n; the change in energy is greater for the one with larger orbital angular momentum. However, additional factors such as the interplay between the electron–electron correlation and the relativistic interactions may also affect the relative energies of the initial and final states. As a result, dependent on the individual transition, the transition energy could either be blueshifted or redshifted.

Another interesting immediate consequence of the DH potential V_{DH} is that the ratio $R = \Delta\omega/\omega_0$ for the α emission line for the H-like ion depends on a single parameter, i.e., the reduced Debye length, defined as $\lambda = ZD$. Qualitatively, the energy shift of the emission line subject to the outside plasma $\Delta\omega_\alpha$ is given approximately by the difference in the energy corrections between the initial and final H-like orbitals due to the difference in the Coulomb potential and the screened Coulomb potential, i.e., $\Delta V_D = \frac{Z}{r}(1 - e^{-r/D})$. This can be estimated by the difference of the expectation values of $\Delta_{1s} = <1s \mid \Delta V_D \mid 1s>$ and $\Delta_{2p} = <2p \mid \Delta V_D \mid 2p>$, or given analytically by [21]:

$$\Delta\omega_\alpha(D) \approx \Delta_{1s}(D) - \Delta_{2p}(D) = Z^2\left[\frac{3}{4} - (1 + \frac{1}{2\lambda})^{-2} + \frac{1}{4}(1 + \frac{1}{\lambda})^{-4}\right]. \qquad (7)$$

Since the plasma-free transition energy of the Lyman-α line ω_0 is also proportional to Z^2, the ratio $R = \Delta\omega/\omega_0$ is a function of λ only, or more conveniently, R could be expressed by a simple polynomial in terms of the reduced Debye length λ for all ions with applicable nuclear charge Z. Effectively, we demonstrated a simple scaling feature for the plasma-induced transition energy shifts in terms of the reduced Debye length λ.

The theoretical results presented in this review were carried out with the multiconfiguration approaches, both non-relativistically and relativistically. The non-relativistic results were calculated with the B-spline-based configuration interaction (BSCI) approach with a complete two-electron basis corresponding to both negative and positive energy atomic orbitals. The individual one-electron atomic orbitals were generated from an effective one-electron Hamiltonian $h_o(r, D)$, i.e.,

$$h_o(r; D) = \frac{p^2}{2m} + V_{DH}(r; D), \qquad (8)$$

where p is the momentum of the electron and $V_{DH}(r; D)$ is given by Equation (6). The N-electron Hamiltonian for an atom in the plasma environment is expressed in terms of $h_o(r; D)$ as:

$$H(r_i, r_j, \cdots ; D) = \sum_{i=1,N} h_o(r_i; D) + \sum_{i>j}^{N} \frac{e^2}{r_{ij}}, \qquad (9)$$

where $r_{ij} = \mid \vec{r}_i - \vec{r}_j \mid$ represents the separation between the atomic electrons i and j. The state wave functions for the upper and lower states of the transition, ϕ^U and ϕ^L with energies ϵ^U and ϵ^L, respectively, were calculated by diagonalizing the Hamiltonian matrix with the basis set of the multiconfiguration two-electron orbitals discussed earlier and following the numerical procedure detailed elsewhere [39–42]. The energy of the emission line under the external plasma environment in terms of the Debye length D is given by the difference of the energies between the upper (U) and the lower (L) state, i.e.,

$$\omega(D) = \epsilon^U(D) - \epsilon^L(D). \qquad (10)$$

The energy of the emission line in the absence of the external plasma is given by $\omega_0 = \omega(D = \infty)$, and the transition energy shift $\Delta\omega$ is thus given by:

$$\Delta\omega(D) = \omega_0 - \omega(D). \qquad (11)$$

The details of the theoretical and calculational procedures leading to the oscillator strength were given in Section 2.4 of [39].

For the relativistic calculation of an N-electron ion with nuclear charge Z subject to the outside plasma, its N-electron Hamiltonian H_{DH} can be expressed as:

$$H_{DH} = H_{DC} + \sum_{i=1}^{N} V_d(r_i, D), \quad V_d(r_i, D) = \frac{Z}{r_i} + V_{DH}(r_i), \qquad (12)$$

where H_{DC} is the the well-known Dirac–Coulomb Hamiltonian in the absence of the plasma environment, i.e.,

$$H_{DC} = \sum_{i=1}^{N} [\, c\, \vec{\alpha} \cdot \vec{p}_i + (\beta - 1)mc^2 - \frac{Z}{r_i}\,] + \sum_{i>j}^{N} \frac{e^2}{r_{ij}}, \qquad (13)$$

where $\alpha_k = \begin{pmatrix} 0 & \sigma_k \\ \sigma_k & 0 \end{pmatrix}$ with $k = (1,2,3)$, σ_k is the Pauli 2×2 matrix, and $\beta = \begin{pmatrix} I & 0 \\ 0 & -I \end{pmatrix}$ with I the 2×2 unit matrix.

The energies E^Γ and the their corresponding state functions (ASFs) $|\Gamma PJM\rangle$ of the upper and lower states of the atomic process are derived from:

$$H_{DH}|\Gamma PJM\rangle = E^\Gamma|\Gamma PJM\rangle \qquad (14)$$

with P the parity, J the total angular momentum, M its magnetic quantum number, and Γ all other information to define the ASF uniquely. The ASFs are N-electron eigenstate wave functions, which are the linear combinations of the configuration state functions (CSFs) with the same P, J, and M, namely,

$$|\Gamma PJM\rangle = \sum_{i=1}^{n_e} C_i^\Gamma |\gamma_i PJM\rangle, \qquad (15)$$

where C_i^Γ is the expansion coefficient and γ_i represents all other information to define the CSF uniquely. The CSFs, $|\gamma_i PJM\rangle$, which form a basis set for an N-electron atomic system, are linear combinations of the Slater determinants of the atomic orbital wave functions (AOs) corresponding to the electron configurations included in the calculations. The electron correlation effects were taken into account by diagonalizing the relativistic H_{DH} with a quasi-complete basis in a revised multiconfiguration Dirac–Fock (MCDF) approach. The basis consists of the spectroscopic AOs (with $n - l - 1$ fixed nodes) and pseudo AOs (without fixed nodes). Both AOs are specified by the principal quantum number n, the orbital angular momentum l, and the total angular momentum j. The number of spectroscopic orbitals depends on the requirement of specific physical problem, i.e., the degrees of excitations of target ions, whereas the number of pseudo orbitals is determined by the desired accuracies. The atomic orbitals (AOs) were optimized using the GRASP-JT version based on the earlier GRASP2K codes [43,44]. The details of the calculational schemes have been presented elsewhere [45–47]. With the ASFs calculated with and without the outside plasma and under the dipole long-wavelength approximation, the oscillator strength of transition between atomic states can be expressed as the product of the transition energy $\omega_{\alpha\beta}$ and the square of the transition matrix element in the length gauge as:

$$g_\alpha f_{\alpha\beta} \sim \omega_{\alpha\beta} \cdot |\sum_{i,j} C_i^{\Gamma_\alpha} C_j^{\Gamma_\beta} \langle \gamma_i P_i J_i M_i | \hat{r} | \gamma_j P_j J_j M_j \rangle|^2, \qquad (16)$$

where $\langle \gamma_i P_i J_i M_i | \hat{r} | \gamma_j P_j J_j M_j \rangle$ are the dipole transition matrix elements.

For the relatively light He-like ions, the relativistic effect is small, and the results from our non-relativistic and relativistic calculations are in close agreement with each other, as shown previously [21]. As a result, the discussion in Section 3 was mostly based on our non-relativistic calculation. We should also point out that the close agreement between two very different calculation approaches (i.e., the non-relativistic BSCI approach with the spin–orbit interaction included and the relativistic GRASP2K approach with the optimized atomic orbitals in its quasi-complete basis) suggests that the electron–electron correlation was fully included in our study.

For the atomic processes involving the continuum such as the atomic photoabsorption, the detailed expressions for its cross-section σ in terms of the excitation energy and the oscillator strength (either in the length or velocity gauge) were given in detail elsewhere [39,40].

With the state wave functions generated following our discussion presented above, our calculated plasma-free photoabsorption cross-sections σ with either the length or velocity approximation for the He-like atomic systems are generally in agreement to within 1–2%. The agreement at such a level suggests again that the electron–electron correlation between atomic electrons was taken into account fully in our theoretical calculation.

3. Plasma Effects on H-like and He-like Ions

Our first attempt to apply the DH approximation to the shift of the transition energy $\Delta\omega$ subject to the outside plasma started with the Lyman-α line of the one-electron H-like ions. Our main objective was to find if the estimated $\Delta\omega$ based on the DH model would agree with the quantitatively measured redshift of 3.7 ± 0.7 eV for the H-like Al^{12+} ion from the laser-generated plasma at temperature 300 eV with its density of $(5-10) \times 10^{23}$ cm^{-3} [36]. We chose in our calculation the Debye radius as the product of the radius of the 1 s orbit a_o/Z and a size parameter η, i.e., $A = (\eta/Z)a_o$. For a H-like ion, we simply solved the one-electron Hamiltonian Equation (8) with a given D corresponding to a pair of (kT, N) and evaluated the resulting redshift. Figure 1 compares the experimental redshift to our calculated results with four different size parameters at $\eta = 0$, 1, 1.5, and 2. With the experimental density extended from $5-10 \times 10^{23}$ cm^{-3}, all our theoretically estimated redshifts with different η compared well with the observed value of 3.7 ± 0.7 eV. It also turned out that our estimated redshifts with $\eta = 0$ for the Lyman-α emission of H-like Ne^{9+} ion at 500 eV are in agreement with the theoretical result of an earlier QMIT (quantum mechanical impact theory) study by Nguyen et al. [5,6], which included the effect of total ion polarization and considered as the upper limit for the redshift. As pointed out by Nguyen et al., their limiting result was about 20% greater than the result from an earlier quantum mechanical treatment by Davis and Blaha [7] with only a partial account of the ion charge density. Interestingly, the estimate by Davis and Blaha is in agreement with our calculated redshift with $\eta = 1.0$. More discussion was presented earlier in detail in Figure 4 of [20].

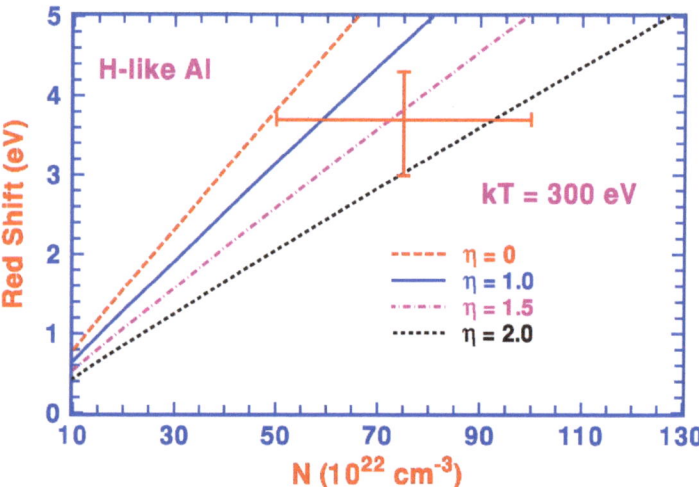

Figure 1. Comparison between the experimentally measured Lyman-α redshift of the H-like Al ion [36] with the theoretical estimations with the radius of the Debye sphere A in terms the size parameters $\eta = 0$, 1, 1.5, and 2.

Encouraged by the agreement between our theoretical results on the Lyman-α emission of the H-like ions and the experimental observation, as well as other more detailed theoretical calculations, we moved on to study the α and β emissions of the He-like ions. With the electron–electron correlations between the atomic electrons taken into account fully based on the non-relativistic and relativistic multiconfiguration atomic structure calculations outlined in Section 2, we carried out calculations for the He-like Ne, Al, and Ar ions with the radius of the Debye sphere given by $A = \eta <r>_g$, where $<r>_g = \langle 1s^2\,{}^1S|r|1s^2\,{}^1S\rangle$ is the average radius of the ground state of He-like ions and η is a size parameter. Our calculations led to (i) the plasma-free transition energies ω_0 for all three He-like Ne, Al, and Ar ions, in close agreement with the NIST values [48], and (ii) the ratio R shown in Figure 2 between the energy shift $\Delta\omega$ and the plasma-free ω_o, which follows a nearly universal curve for each size parameter η as a function of the reduced Debye length λ for all He-like ions with Z meeting the spatial and temporal criteria of the DH approximation. This is qualitatively consistent with what we already pointed out earlier in Section 2, following the quasi-hydrogenic picture based on Equation (7). For the He-like ions with relatively low Z, the relativistic interactions are small, and indeed, this was confirmed by the nearly identical values R for the He-like O^{6+} ion between the non-relativistic and relativistic results shown previously in Figure 5 of [21]. For the intermediate Z, the DH approximation did not work due to the spatial and temporal criteria, as we pointed out earlier. As Z increases further, the DH model should apply again, and the effect of the relativistic interaction was clearly shown for the heavier He-like Yb and Au ions as the values of R deviated substantially from the ones for He-like O ion, shown also in Figure 5 of [21]. Our discussion below for the plasma effects on ions with relatively low Z was mostly based on the data from our non-relativistic calculations.

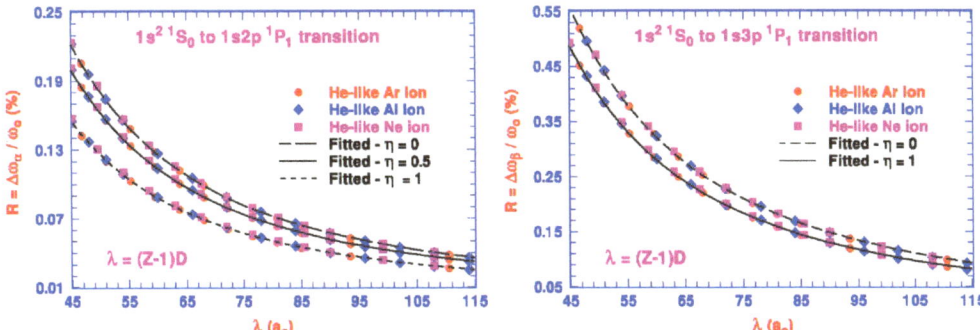

Figure 2. The universal behavior of the calculated R as functions of the reduced Debye length λ for three He-like ions and their comparison to the fit ratio R given by Equation (17) with the coefficients given in Table 1.

The nearly universal dependence of the calculated R on λ for different He-like ions shown in Figure 2 could be expressed more conveniently with a simple polynomial in terms of three numerically fit coefficients a, b, and c as:

$$R(\lambda; a, b, c) = a + b/\lambda + c/\lambda^2, \qquad (17)$$

To estimate the ratio R for other He-like ions, we first took the average of the fit coefficients a_Z, b_Z, and c_Z from the calculated R of the three ions with individual Z shown in Figure 2. Corresponding to each size parameter η, the coefficients listed in Table 1 are the average over the individually fit coefficients for the three He-like ions, i.e., $a_\eta = (a_{Ne^{8+}} + a_{Al^{11+}} + a_{Ar^{16+}})/3$ for the α and β emission lines of the He-like ions. As expected, the R values generated with the coefficients listed in Table 1 and Equation (17) agree well with the calculated R, as shown by the two plots of Figure 2.

Theoretically, the ratio R should go to zero as the energy shift approaches zero when λ or the Debye length D goes to infinity. With the two coefficients c and b several orders of magnitude larger than the small, but non-zero coefficient a, one could identify the coefficient a as the numerical uncertainty of the theoretically estimated ratio R. Based on Equation (17) and the fit coefficients listed in Table 1, it is straightforward to estimate the transition energy shifts $\Delta\omega$ following three simple steps for the α and β emission lines for any He-like ions with Z between six and eighteen. First, one starts from a specific reduced Debye length $\lambda = (Z-1)D$ with D determined from Equation (4) for a pair of plasma density N and temperature kT. Second, we proceeded to calculate the ratio R from Equation (17) for each λ with the coefficients listed in Table 1. In the third step, the estimated energy shift $\Delta\omega$ corresponding to this specific pair of N and kT is given by $\Delta\omega(N, kT) = R(\lambda)\omega_0$ with ω_0 the plasma-free transition energy. Following this simple procedure, the estimated redshifts $\Delta\omega(eV)$ for the α and β emission lines of the He-like Cl ion at 600 eV and 800 eV as functions of N in unit of 10^{23} cm^{-3} with three size parameters $\eta = 0, 0.5$, and 1 are presented in Table 2. The top plot of Figure 3 shows good agreement between our estimated redshifts listed in Table 2 for the β emission line of the He-like Cl ion and the experimental data from the recent high-resolution satellite line-free measurement [15]. Similar good agreement is also shown in the bottom plot of Figure 3 between our calculated redshifts of the α emission line of the He-like Al ion and the results of the recent picosecond time-resolved measurement [14]. It is interesting to note that the fit coefficients c listed in Table 1 are about three orders of magnitude greater than the coefficient b and seven orders greater than the coefficient a. Therefore, the estimated energy shifts should be dominated by the term c/λ^2, or proportional to the ratio N/kT. This is consistent with what is shown in Figure 3: the nearly linear dependence of the redshift $\Delta\omega$ as a function of the plasma density N at a fixed temperature kT. We should point out that this linear dependence is also consistent with the analytical expression derived by Li and Rosmej in their IS model [8,9]. On the other hand, the same reasoning based on the DH approximation that the energy shift should vary as N/kT suggests that at a constant density N, $\Delta\omega$ should vary linearly as the inverse of the temperature, i.e., $1/kT$, such as the ones shown in the bottom plot of Figure 4. This is different from the analytical expression from the IS model of Li and Rosmej [8,9], which suggests a temperature dependence of $(1/kT)^{1/2}$.

Table 1. Fit average coefficients a, b, and c in $A(e) = A \times 10^e$ for the α and β emission lines of the He-like ions with the size parameters $\eta = 0, 0.5$, and 1 (corresponding to the R in percentage).

	The α Emission Line of the He-like Ions			The β Emission Line of the He-like Ions		
η	a	b	c	a	b	c
0.0	−9.08030 (−5)	2.96178 (−1)	4.37256 (2)	−4.19761 (−4)	1.45386 (0)	1.07289 (3)
0.5	−8.89317 (−5)	2.90274 (−1)	3.92346 (2)	−4.17651 (−4)	1.44754 (0)	1.02662 (3)
1.0	−8.01969 (−5)	2.63109 (−1)	3.02430 (2)	−4.08090 (−4)	1.41677 (0)	9.26831 (2)

Table 2. The redshifts $\Delta\omega(eV)$ of the β emission line for the He-like Cl ion derived from the fit coefficients listed in Table 1 at 600 eV and 800 eV as functions of the plasma density N in (10^{23} cm^{-3}) with the size parameters $\eta = 0, 0.5$, and 1.

$N(10^{23}$ cm$^{-3})$	$kT = 600$ eV $\Delta\omega$ (eV)			$kT = 800$ eV $\Delta\omega$ (eV)		
	$\eta = 0$	$\eta = 0.5$	$\eta = 1$	$\eta = 0$	$\eta = 0.5$	$\eta = 1$
1.5	2.058	1.982	1.813	1.579	1.521	1.394
2.5	3.313	3.186	2.908	2.532	2.436	2.227
3.5	4.550	4.373	3.986	3.469	3.336	3.044
4.5	5.777	5.549	5.053	4.396	4.226	3.852
5.5	6.996	6.718	6.113	5.318	5.109	4.654
6.5	8.210	7.882	7.168	6.235	5.989	5.451
7.5	9.419	9.041	8.218	7.148	6.864	6.245

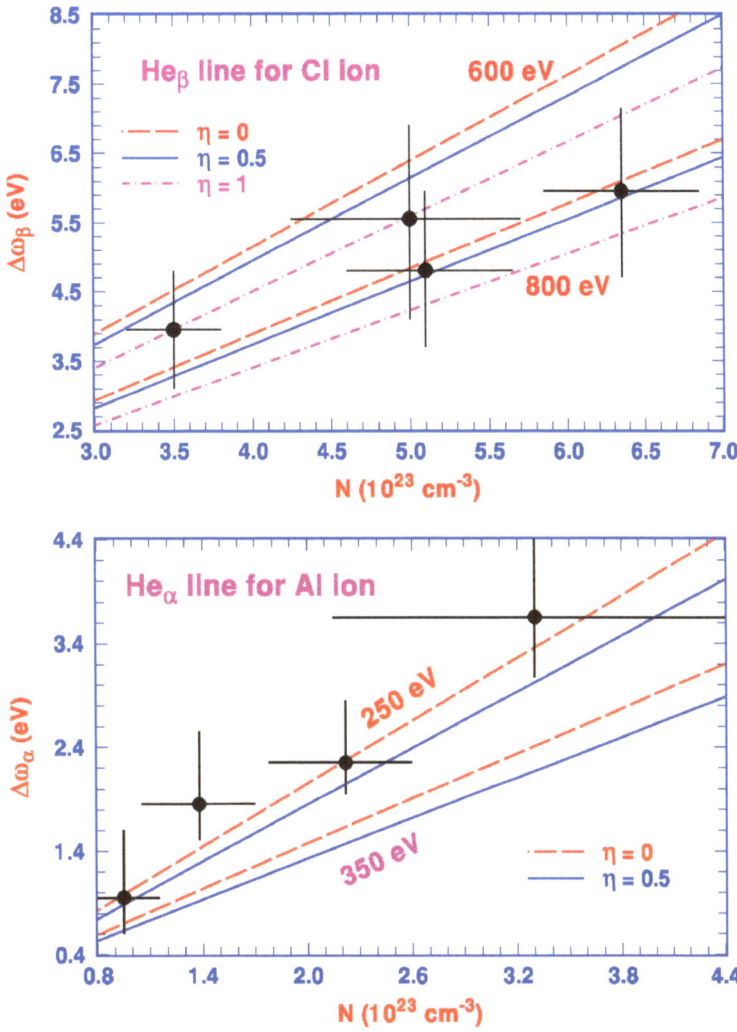

Figure 3. The comparison between the redshifts $\Delta\omega$ for the β and α emissions of the He-like ions estimated based on the DH approximation with the recent dense plasma experiments for the He-like Cl ion between 600 eV and 650 eV [15] and the Al ion between 250 eV and 375 eV [14], respectively.

As we already pointed out earlier in Section 1, both IS models were able to generate the estimated redshifts that are in agreement with only one of the recent experimental measurements [14,15], but not the other. One possibility could be due to their application of the Fermi–Dirac statistics for the outside plasma. For many other theoretical approaches, the temperature dependence would also be implicitly determined by the free electron density expressed in terms of the Fermi–Dirac distributions. In contrast, for the DH model, the electrons and ions in the outside charge-neutral plasma are treated as charged particles with no quantum mechanical interaction (such as those involving the spin of the individual particles in the solid system or inside the nucleus), and thus, the Maxwell–Boltzmann statistics is applied. This is very different from the Fermi–Dirac distribution applied in the AIS model. It is interesting to note that a statistical electron screening model was proposed very recently [13] to describe the atomic processes in warm/hot dense plasmas with a wide range of temperatures and densities. This model includes corrections for the Fermi–Dirac distribution by considering the non-equilibrium feature of the plasma electron distribution around the atomic ion caused by the three-body recombination process, which effectively broadened the phase space of the plasma electron and made the plasma electron distribution more close to the Boltzmann distribution under high temperatures. In fact, for the conditions relevant to the two recent experiments [14,15], such a sophisticated statistical model results in a plasma electron distribution almost identical to the DH model and can result in a similar conclusion as we presented earlier. Although there is no definitive quantitative measurement on the redshifts with different plasma temperatures at a fixed plasma density, the two recent experiments were carried out with a range of estimated temperatures that appeared to suggest a temperature dependence of energy shifts $\Delta\omega$ more pronounced than the fairly small variations at different temperatures obtained from the two versions of the IS model.

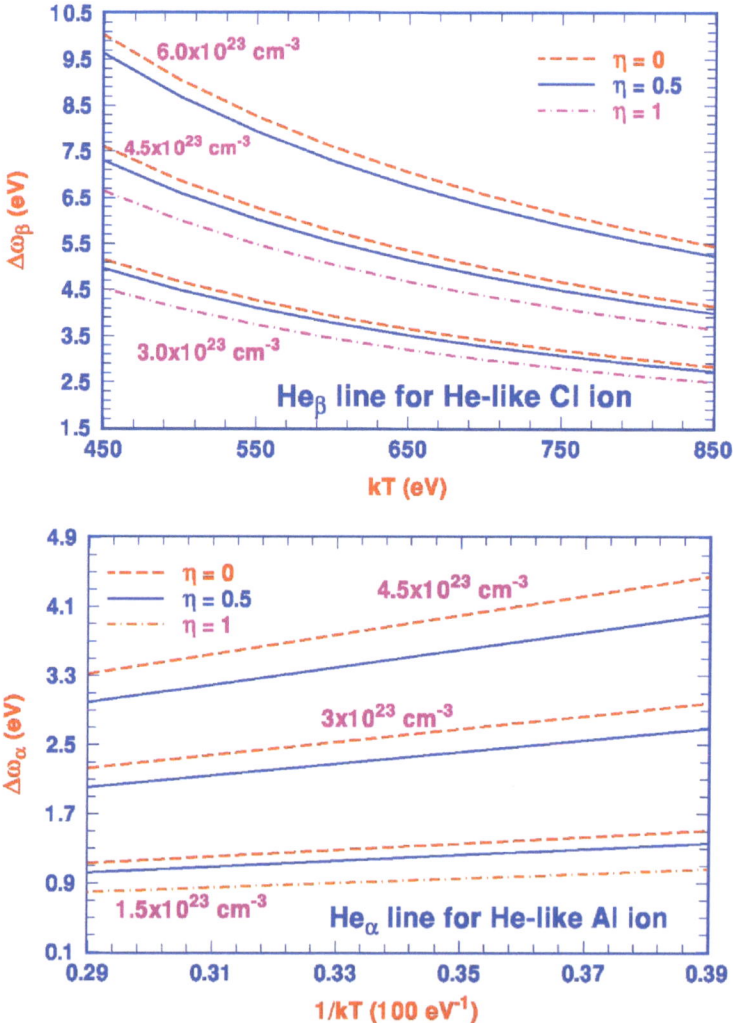

Figure 4. The theoretically estimated redshifts $\Delta\omega$ as functions of kT and $1/kT$ derived from Equation (17) at a number of plasma densities.

It is important to point out that the failure of the application of the DH model to an atomic process without taking into account appropriately the critical physical aspects of the spatial and temporal criteria should not invalidate the DH approximation as a viable phenomenological approach. For example, the DH approximation should not be applied at all to transitions involving atomic states close to the ionization threshold such as the "dip" shown near the series limit in Figure 2 of [49]. The other such example is due to the definition of the Debye length D. As we already pointed out earlier [20], some applications of the DH approximation have included an extra $(Z+1)$ factor to the plasma density in defining the Debye length. This would lead to a smaller Debye length and consequently a much stronger plasma effect. One such example is shown by the large difference between the DH calculation and the results from other calculations shown in Figure 2 of the recent work by Gu and Beiersdorfer [50] due to the fact that their Debye length was over an order of magnitude smaller than what it should be at a given temperature. Once again, we would like to emphasize that the inability of generating reliable data based on questionable applications should not be viewed as the shortcoming of the DH model.

We should also comment briefly on the size factor η. There is no good a priori theoretical prescription to determine its value other than to assume the maximum plasma effect to the atomic electron when $\eta = 0$ and the calculated $\Delta\omega$ could be identified as the upper bound for the energy shifts. Based on the spatial criterion of the DH model of keeping as much of the atomic characteristic of the transition, together with the good agreement between our estimated $\Delta\omega$ with the measured data shown in Figure 3, a reasonable compromise would be for η to be close to 0.5, but less than one. Certainly, more high-precision experiments will help refine the choice of the size parameter η.

For the effect of the outside plasma on the oscillator strength f for the α and β emissions of the H-like and He-like ions, we focus our discussion similarly to those presented above in terms of its variation as a function of the reduced Debye length λ. Specifically, we examined the ratio f_r between the change in f and its plasma-free value $f(\lambda = \infty)$, i.e.,

$$f_r(\lambda) = \frac{f(\lambda = \infty) - f(\lambda)}{f(\lambda = \infty)}. \tag{18}$$

It is interesting to note from Figure 5 that our calculated percentage variation of the ratio f_r as a function of λ for the He$_\alpha$ line exhibits a similar qualitative feature as the ratio for the energy shift R discussed earlier. In other words, this general feature could also be expressed in terms of a polynomial, i.e.,

$$f_r(\lambda; a_f, b_f, c_f) = a_f + b_f/\lambda + c_f/\lambda^2. \tag{19}$$

We also note from Figure 5 that f_r is generally a few times larger than the value of R as a function of λ, or at the same temperature kT and density N. Even with R at a fraction of 1%, the energy resolution of the current experiments such as those we referred to earlier is sufficient to measure the energy shifts with reasonable accuracy. However, it may still be difficult to experimentally measure the change in the oscillator strength even at a level of a few percent change in f_r, as shown in Figure 5.

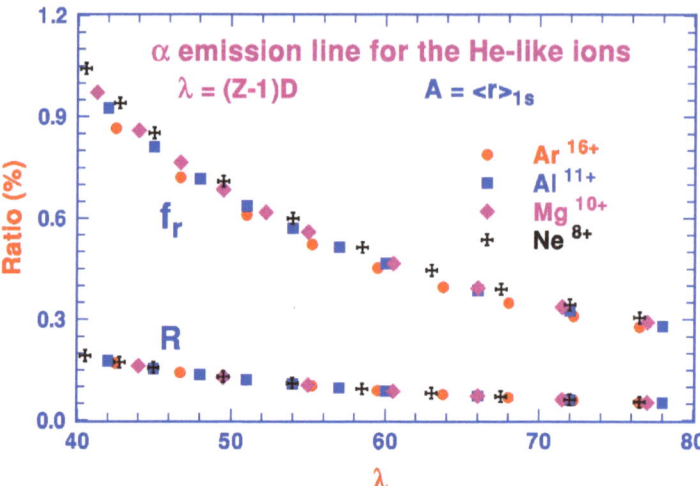

Figure 5. The percentage change of the redshifts R and the corresponding f_r of the oscillator strengths for the $1s2\,^1S \to 1s2p\,^1P$ transition of the He-like ions as a function of λ.

4. Photoabsorption

Our theoretical study on the atomic processes subject to the outside plasma actually started with the atomic photoionization from the ground states of the one- and two-electron atoms [19]. Experimentally, most of the photoionization measurements in the plasma-free environment are focused on the angular distribution of the outgoing photoelectrons, which offers more information to understand the detailed dynamics of the process either with the polarized or unpolarized incident light. In the presence of the outside plasma, the outgoing atomic electron resulting from the absorption of the incoming photon by the atom will lose its identity due to its interaction with the plasma electrons and accordingly could not be collected as the ones in the plasma-free environment. As a result, the experimental study of such a process would likely be limited to the measurement of the attenuation of the incident photon. In this section, we present our discussion of the qualitative feature of the photoabsorption process from the ground state of the one-electron atom, which meets the spatial criterion of the DH approximation.

For the hydrogen atom, the oscillator strength and its corresponding photoabsorption cross-section are proportional to the square of the dipole matrix element $<\chi_{1s}|r|\chi_{kp}>$, where χ are the solutions of the one-electron Hamiltonian h_o given by Equation (8) and $k^2 = \epsilon$ is the energy of the ionized p electron in Rydberg units with momentum k. Qualitatively, the larger the overlap between χ_{1s} and χ_{kp}, the greater the cross-section is. The radial parts of the wave function of the outgoing kp electron with slightly different energies near the ionization threshold for a plasma-free photoabsorption from the hydrogen atom are essentially the same as shown by the top plot of Figure 6 for a number of momenta k. They all reach their first local maxima and the subsequent zeros at about the same distance r and only differ from each other until they are sufficiently away from the nucleus at large r. Since the photoabsorption spectrum is dictated by the overlaps of the χ_{kp} with the 1s orbit up to a distance before the 1s orbit reaches zero (see, e.g., the bottom plot of Figure 6), only the close-in part of χ_{kp} at the relatively small r needs to be taken into account for the slowly varying dipole matrix element $<\chi_{1s}|r|\chi_{kp}>$ as the energy changes. This is consistent with the plasma-free hydrogen photoabsorption spectrum, which is known to vary smoothly near the ionization threshold [51]. We should point out that this short-range nature of the photoabsorption from the hydrogen ground state is necessary to meet the spatial criterion required in the application of the DH approximation in the presence of the outside plasma.

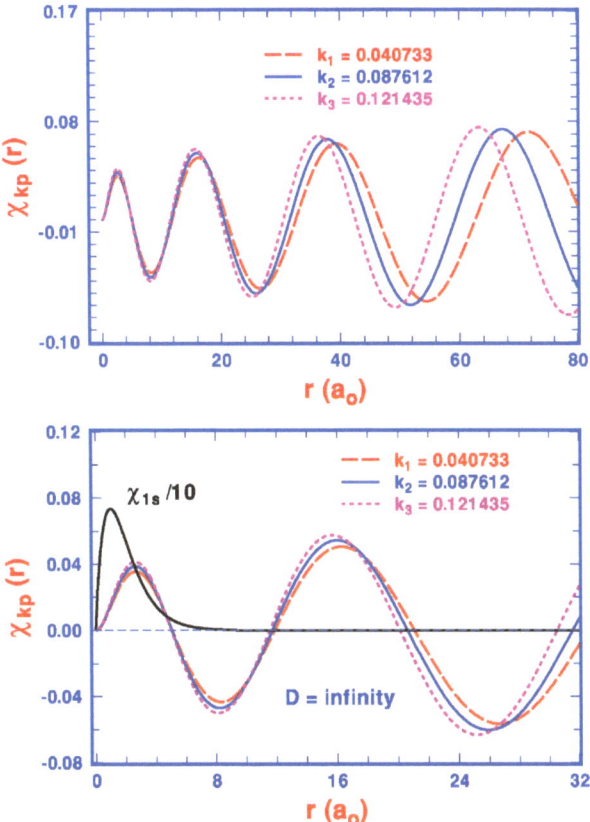

Figure 6. The radial functions χ_{kp} of the outgoing ionized p electron from the plasma-free hydrogen at a number of momenta. The bottom plot compares χ_{kp} to χ_{1s} (reduced by a factor of 10) at small r.

We limit our detailed discussion on the application of the DH approximation to the photoabsorption of the hydrogen ground state subject to the plasma environment and examine first with the Debye lengths D substantially greater than the radius of the 1s orbital to keep the plasma effect on the the orbital wavefunction χ_{1s} small. Qualitatively, one of the most outstanding features of the DH approximation is the upwards migration of the bound excited state as the Debye length D decreases. Corresponding to each bound excited state, there is a critical value of D such that the state is pushed into the continuum. In particular, we focus our discussion on the plasma-induced resonant structures in the photoabsorption spectrum from the hydrogen ground state between 989 Å and 992 Å, which are associated with the plasma-free 1s to 4p Lyman-γ line at 972.5 Å. Following the earlier plasma-free example, but with the Debye length at a critical length of 24.5 a_o given in [19] due to the outside plasma, the top plot of Figure 7 compares the radial orbital function χ_{1s} to three χ_{kp} orbitals representing the outgoing electron near the ionization threshold at three momenta $k = 0.030301, 0.034244$, and 0.037739, respectively. It is interesting to note that although the locations of their first local maxima and the subsequent zeros for the first loop from $r = 0$ for all three χ_{kp} are similar to the plasma-free ones shown in Figure 6, their magnitudes are very different as k varies. At $k = 0.030301$ and 0.037739, the magnitudes of their first loop are relatively small, indicating a minimal presence of the outgoing electron at the inner region of the atom, similar to those shown in Figure 6 for the plasma-free photoabsorption. At a slightly different energy with $k = 0.034244$, the magnitude of χ_{kp} is noticeably higher at small r, but relatively small at larger r compared to that of the two other momenta, indicating the presence of a quasi-bound radial wavefunction. As a result, its corresponding photoabsorption cross-section σ proportional to the square of the dipole matrix $|<\chi_{1s}|r|\chi_{kp}>|^2$ is substantially larger than those at the nearby k, leading to a resonant structure in the photoabsorption spectrum such as the one shown in the top plot of Figure 8 with $D = 24.5a_o$ at $kT = 600$ eV and $N = 1.973 \times 10^{22}$ cm^{-3}. Two additional narrow resonant structures due to slightly larger critical Debye lengths are also shown. It is interesting to note that the relative locations of the peak cross-sections of these three narrow resonant structures are expected under the DH approximation. This is due to the fact that the ionization threshold corresponding to a smaller Debye length is smaller than the one with a larger Debye length, and it requires less photon energy (or a longer wavelength) to "push" the $1s$ to $4p$ transition into the continuum. A similar feature due to the hydrogen $1s$ to $3p$ transition was also discussed in detail in [52]. With the electron–electron correlation between atomic electrons taken into account fully for the He atom, our study [19] also led to a similar general feature of the narrow resonance-like structure slightly above the changing ionization threshold in the photoabsorption spectrum. In spite of our theoretical understanding of the general feature of the plasma driving narrow resonances due to the migration of the excited np levels into the continuum in the photoabsorption process, unfortunately, the widths of such resonances are approximately 10,000-times smaller than the photon energy and likely not observed in the laboratory.

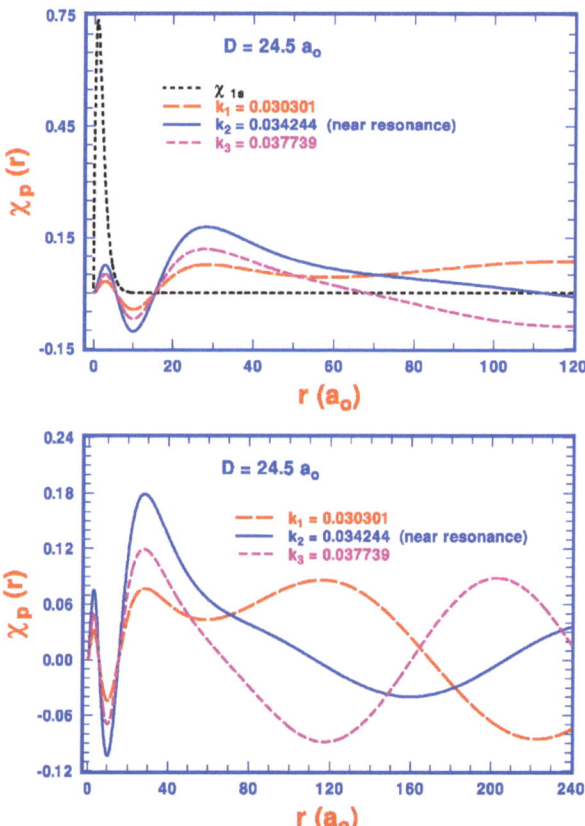

Figure 7. The top plot compares the radial functions χ_{kp} of the outgoing ionized p electron to χ_{1s} for hydrogen ground state photoabsorption subject to the outside plasma with $D = 24.5a_o$. The bottom figure shows an enlarged plot of χ_{kp} with an expanded scale up to $r = 240a_o$.

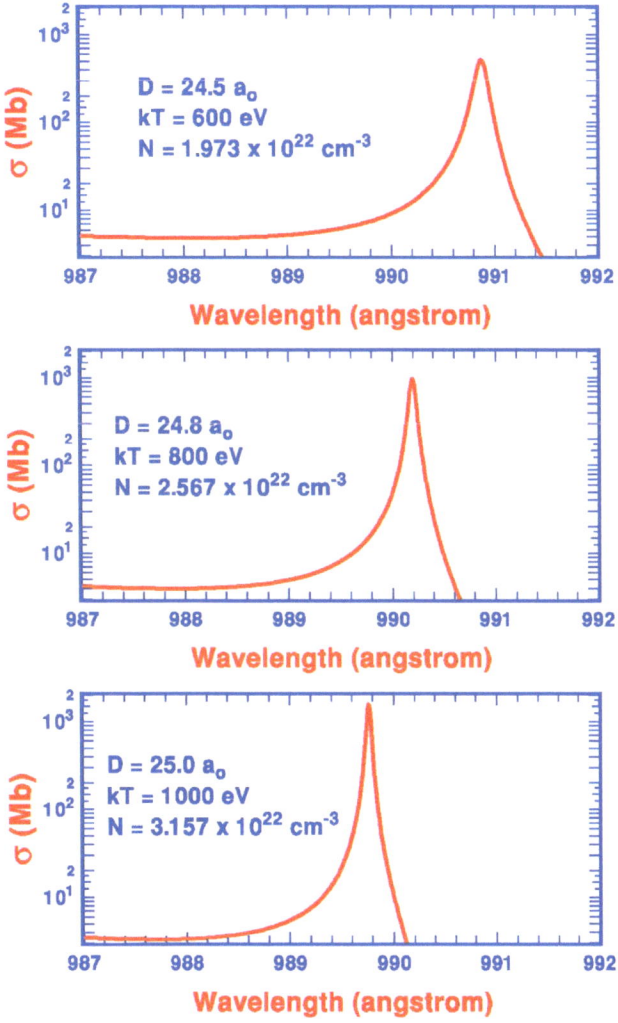

Figure 8. Hydrogen photoabsorption spectra from its ground state in terms of the cross-section σ (in units of Mb) corresponding to the $1s \to 4p$ transition as D varies around $24.5a_0$ with $A = 0$.

5. Conclusions

We presented in this paper a review of a series of our recent studies on the application of the Debye–Hückel approximation for atomic processes subject to the outside plasma environment. We focused our studies primarily on the processes that meet the all important spatial and temporal criteria for the DH model. In spite of the simplicity of the DH approximation, our theoretical results compared well with the limited data from the available experimental measurements. In addition, we identified a general scaling feature for the ratio R between the redshift $\Delta \omega$ and the plasma-free transition energy ω_0 of the α and β emission lines of the He-like ions with the nuclear charge Z between six and eighteen. More specifically, the ratio R could be expressed in terms of a simple polynomial Equation (17) of the reduced Debye length λ. Following the simple three-step procedure presented in Section 3, one could estimate the redshifts of the α or β emission lines for other He-like ions with no need for additional theoretical calculations. In fact, one such example led to good agreement between the estimated redshifts of the β emission line for the He-like Cl ion and the experimental data shown in Figure 3.

In addition to our studies on the H- and He-like ions, we were also interested in finding out if the simple scaling feature presented in Section 3 also works for ions with more electrons. Our first such study was for the 3C and 3D lines of the Ne-like ions [53]. It turned out that for the dipole-allowed 3C line, the redshifts of the transition energy and the oscillator strength follow a similar scaling feature. However, for the dipole-forbidden 3D line, due to the interplay between the relativistic spin–orbit interaction and the plasma screening effects, the simple scaling feature failed to follow [53]. Following our study for the transitions for the Ne-like ions, we extended our study to two strong dipole transitions for the C-like ions, i.e., (1) the intershell $2s^22p3d\ ^3D_1 \rightarrow 2s^22p^2\ ^3P_0$ transition and (2) the intrashell $2s2p^3\ ^3D_1 \rightarrow 2s^22p^2\ ^3P_0$ transition. For the intrashell transition, the ratio of the energy shifts to its plasma-free transition energy and the increase of the oscillator strength follows a similar general scaling property. However, due to the change of the electron correlation with respect to the relativistic spin–orbit interaction as Z varies, the decrease in the oscillator strength for the intershell transition failed to follow the scaling feature discussed earlier [54].

Finally, we should comment again on the size factor η. As we stated earlier that there is no good a priori theoretical prescription to determine its value, therefore, the most appropriate η for the theoretical calculation would depend on the good agreement between the theoretically calculated $\Delta\omega$ and the measured data such as those shown in Figure 3. The interplay between the theory and experiment could refine the value of η and offer a more reliable and accurate determination of the energy shifts of the emission lines at a given combination of plasma temperature and density. In addition, as we indicated earlier that although the temperature variation of $\Delta\omega$ at a fixed plasma density based on the limited experimental date appears to support the use of the classical Maxwell–Boltzmann statistics over the Fermi–Dirac distributions for the outside plasma, more experimental measurements are needed for such a conclusion.

Author Contributions: Conceptualization, T.-N.C., T.-K.F. and X.G.; methodology, T.-K.F., C.W. and X.G.; calculation, T.-K.F. and C.W.; writing—original draft preparation, T.-N.C.; writing—review and editing, T.-N.C., T.-K.F., C.W. and X.G.; funding acquisition, T.-K.F. and X.G. All authors have read and agreed to the published version of the manuscript.

Funding: This research was funded by the National Natural Science Foundation of China under Grant Nos. 11774023, U1930402, the National Key R&D Program of China under Grant No. 2016YFA0302104, and the Ministry of Science and Technology (MOST) in Taiwan under Grant No. MOST 109-2112-M-030-001.

Institutional Review Board Statement: Not applicable.

Informed Consent Statement: Not applicable.

Data Availability Statement: Not applicable.

Acknowledgments: T.-N.C. would also like to thank the National Center for Theoretical Science in Taiwan for its continuous hospitality. We acknowledge the computational support provided by the Beijing Computational Science Research Center.

Conflicts of Interest: The authors declare no conflict of interest.

Abbreviations

The following abbreviations are used in this manuscript:

DH   Debye–Hückel
IS   Ion sphere
AIS   Average atom ion sphere
BSCI B-spline-based configuration interaction

CSFs	Configuration state functions	
AOs	Atomic orbital wave functions	
MCDF	Multiconfiguration Dirac–Fock	
QMIT	Quantum mechanical impact theory	

References

1. Tennyson, J. *Astronomical Spectroscopy: An Introduction to the Atomic and Molecular Physics of Astronomical Spectral*; Press Imperial College: London, UK, 2005. [CrossRef]
2. Kallman, T.R.; Palmeri, P. Atomic data for x-ray astrophysics. *Rev. Mod. Phys.* **2007**, *79*, 79–133. [CrossRef]
3. Lindl, J.D.; Amendt, P.; Berger, R.L.; Glendinning, S.G.; Glenzer, S.H.; Haan, S.W.; Kauffman, R.L.; Landen, O.L.; Suter, L.J. The physics basis for ignition using indirect-drive targets on the National Ignition Facility. *Phys. Plasmas* **2004**, *11*, 339–491. [CrossRef]
4. Boozer, A.H. Physics of magnetically confined plasmas. *Rev. Mod. Phys.* **2005**, *76*, 1071–1141. [CrossRef]
5. Koenig, M.; Malnoult, P.; Nguyen, H. Atomic structure and line broadening of He-like ions in hot and dense plasmas. *Phys. Rev. A* **1988**, *38*, 2089–2098. [CrossRef] [PubMed]
6. Nguyen, H.; Koenig, M.; Benredjem, D.; Caby, M.; Coulaud, G. Atomic structure and polarization line shift in dense and hot plasmas. *Phys. Rev. A* **1986**, *33*, 1279–1290. [CrossRef] [PubMed]
7. Davis, J.; Blaha, M. Level shifts and inelastic electron scattering in dense plasmas. *J. Quant. Spectrosc. Radiat. Transf.* **1982**, *27*, 307–313. [CrossRef]
8. Li, X.; Rosmej, F.B. Quantum-number dependent energy level shifts of ions in dense plasmas: A generalized analytical approach. *Europhys. Lett.* **2012**, *99*, 33001. [CrossRef]
9. Rosmej, F.; Bennadji, K.; Lisitsa, V.S. Effect of dense plasmas on exchange-energy shifts in highly charged ions: An alternative approach for arbitrary perturbation potentials. *Phys. Rev. A* **2011**, *84*, 032512. [CrossRef]
10. Chen, Z.B. Calculation of the energies and oscillator strengths of Cl^{15+} in hot dense plasmas. *J. Quant. Spectrosc. Radiat. Transf.* **2019**, *237*, 106615. [CrossRef]
11. Glenzer, S.H.; Redmer, R. X-ray Thomson scattering in high energy density plasmas. *Rev. Mod. Phys.* **2009**, *81*, 1625–1663. [CrossRef]
12. Ichimaru, S. Strongly coupled plasmas: High-density classical plasmas and degenerate electron liquids. *Rev. Mod. Phys.* **1982**, *54*, 1017–1059. [CrossRef]
13. Zhou, F.; Qu, Y.; Gao, J.; Ma, Y.; Wu, Y.; Wang, J. Atomic-state-dependent screening model for hot and warm dense plasmas. *Commun. Phys.* **2021**, *4*, 148. [CrossRef]
14. Stillman, C.R.; Nilson, P.M.; Ivancic, S.T.; Golovkin, I.E.; Mileham, C.; Begishev, I.A.; Froula, D.H. Picosecond time-resolved measurements of dense plasma line shifts. *Phys. Rev. E* **2017**, *95*, 063204. [CrossRef]
15. Beiersdorfer, P.; Brown, G.V.; McKelvey, A.; Shepherd, R.; Hoarty, D.J.; Brown, C.R.D.; Hill, M.P.; Hobbs, L.M.R.; James, S.F.; Morton, J.; et al. High-resolution measurements of Cl^{15+} line shifts in hot, solid-density plasmas. *Phys. Rev. A* **2019**, *100*, 012511. [CrossRef]
16. Chen, Z.B.; Wang, K. Theoretical study on the line shifts of He-like Al11+ ion immersed in a dense plasma. *Radiat. Phys. Chem.* **2020**, *172*, 108816. [CrossRef]
17. Debye, P.; Hückel, E. The theory of electrolytes I. The lowering of the freezing point and related occurrences. *Phys. Z.* **1923**, *24*, 185–206.
18. Chen, F.F. *Introduction to Plasma Physics and Controlled Fusion Plasma Physics*; Springer International Publishing: Cham, Switzerland, 2016. [CrossRef]
19. Chang, T.N.; Fang, T.K. Atomic photoionization in a changing plasma environment. *Phys. Rev. A* **2013**, *88*, 023406. [CrossRef]
20. Chang, T.N.; Fang, T.K.; Gao, X. Redshift of the Lyman-α emission line of H-like ions in a plasma environment. *Phys. Rev. A* **2015**, *91*, 063422. [CrossRef]
21. Fang, T.K.; Wu, C.S.; Gao, X.; Chang, T.N. Redshift of the He α emission line of He-like ions under a plasma environment. *Phys. Rev. A* **2017**, *96*, 052502. [CrossRef]
22. Fang, T.K.; Wu, C.S.; Gao, X.; Chang, T.N. Variation of the oscillator strengths for the α emission lines of the one- and two-electrons ions in dense plasma. *Phys. Plasmas* **2018**, *25*, 102116. [CrossRef]
23. Chang, T.N.; Fang, T.K.; Wu, C.S.; Gao, X. Redshift of the isolated atomic emission line in dense plasma. *Phys. Scr.* **2021**, *96*, 124012. [CrossRef]
24. Mukherjee, P.K.; Karwowski, J.; Diercksen, G.H.F. On the influence of the Debye screening on the spectra of two-electron atoms. *Chem. Phys. Lett.* **2002**, *363*, 323–327. [CrossRef]
25. Sil, A.N.; Mukherjee, P.K. Effect of debye plasma on the doubly excited states of highly stripped ions. *Int. J. Quantum Chem.* **2005**, *102*, 1061–1068. [CrossRef]
26. Zhang, S.B.; Wang, J.G.; Janev, R.K. Electron-Hydrogen-atom elastic and inelastic scattering with screened Coulomb interaction around the n = 2 excitation threshold. *Phys. Rev. A* **2010**, *81*, 032707. [CrossRef]
27. Zhang, S.B.; Wang, J.G.; Janev, R.K. Crossover of feshbach resonances to shape-type resonances in electron-hydrogen atom excitation with a screened Coulomb interaction. *Phys. Rev. Lett.* **2010**, *104*, 023203. [CrossRef] [PubMed]

28. Qi, Y.Y.; Wu, Y.; Wang, J.G.; Qu, Y.Z. The generalized oscillator strengths of Hydrogenlike ions in Debye plasmas. *Phys. Plasmas* **2009**, *16*, 023502. [CrossRef]
29. Dai, S.T.; Solovyova, A.; Winkler, P. Calculations of properties of screened He-like systems using correlated wave functions. *Phys. Rev. E* **2001**, *64*, 016408. [CrossRef] [PubMed]
30. Lopez, X.; Sarasola, C.; Ugalde, J.M. Transition energies and emission oscillator strengths of Helium in model plasma environments. *J. Phys. Chem. A* **1997**, *101*, 1804–1807. [CrossRef]
31. Kar, S.; Ho, Y.K. Oscillator strengths and polarizabilities of the hot-dense plasma-embedded Helium atom. *J. Quant. Spectrosc. Radiat. Transf.* **2008**, *109*, 445–452. [CrossRef]
32. Kar, S.; Ho, Y.K. Photodetachment of the Hydrogen negative ion in weakly coupled plasmas. *Phys. Plasmas* **2008**, *15*, 013301. [CrossRef]
33. Shukla, P.K.; Eliasson, B. Screening and wake potentials of a test charge in quantum plasmas. *Phys. Lett. A* **2008**, *372*, 2897–2899. [CrossRef]
34. Kar, S.; Ho, Y.K. Electron affinity of the Hydrogen atom and a resonance state of the Hydrogen negative ion embedded in Debye plasmas. *New J. Phys.* **2005**, *7*, 141. [CrossRef]
35. Ghoshal, A.; Ho, Y.K. Ground states and doubly excited resonance states of H^- embedded in dense quantum plasmas. *J. Phys. B-At. Mol. Opt. Phys.* **2009**, *42*, 175006. [CrossRef]
36. Saemann, A.; Eidmann, K.; Golovkin, I.E.; Mancini, R.C.; Andersson, E.; Förster, E.; Witte, K. Isochoric heating of solid Aluminum by ultrashort laser pulses focused on a tamped target. *Phys. Rev. Lett.* **1999**, *82*, 4843–4846. [CrossRef]
37. Margenau, H.; Lewis, M. Structure of spectral lines from plasmas. *Rev. Mod. Phys.* **1959**, *31*, 569–615. [CrossRef]
38. Rouse, C.A. Finite Electronic Partition Function from Screened Coulomb Interactions. *Phys. Rev.* **1967**, *163*, 62–71. [CrossRef]
39. Chang, T.N. B-Spline Based Configuration-Interaction Approach for Photoionization of Two-electron and Divalent Atoms. In *Many-Body Theory of Atomic Structure and Photoionization*; World Scientific: Singapore, 1993; pp. 213–247. [CrossRef]
40. Chang, T.N.; Tang, X. Photoionization of two-electron atoms using a nonvariational configuration-interaction approach with discretized finite basis. *Phys. Rev. A* **1991**, *44*, 232–238. [CrossRef]
41. Chang, T.N. Application of the many-body theory of atomic transitions to the photoionization of neon and argon. *Phys. Rev. A* **1977**, *15*, 2392–2395. [CrossRef]
42. Chang, T.N.; Fano, U. Many-body theory of atomic transitions. *Phys. Rev. A* **1976**, *13*, 263–281. [CrossRef]
43. Grant, I.P. *Relativistic Quantum Theory of Atoms and Molecules*; Springer: New York, NY, USA, 2007. [CrossRef]
44. Jönsson, P.; Gaigalas, G.; Bieron, J.; Fischer, C.F.; Grant, I.P. New version: GRASP2K relativistic atomic structure package. *Comput. Phys. Commun.* **2013**, *184*, 2197–2203. [CrossRef]
45. Han, X.Y.; Gao, X.; Zeng, D.l.; Jin, R.; Yan, J.; Li, J.M. Scaling law for transition probabilities in $2p^3$ configuration from LS coupling to jj coupling. *Phys. Rev. A* **2014**, *89*, 042514. [CrossRef]
46. Gao, X.; Han, X.Y.; Zeng, D.L.; Jin, R.; Li, J.M. Broken scaling laws of the transition probabilities from jj to LS coupling transitions. *Phys. Lett. A* **2014**, *378*, 1514–1519. [CrossRef]
47. Han, X.Y.; Gao, X.; Zeng, D.L.; Yan, J.; Li, J.M. Ratio of forbidden transition rates in the ground-state configuration of O II. *Phys. Rev. A* **2012**, *85*, 062506. [CrossRef]
48. Kramida, A.; Ralchenko, Y.; Reader, J.; Team, N.A. *NIST Atomic Spectra Database (Ver. 5.8)*; National Institute of Standards and Technology: Gaithersburg, MD, USA, 2020. Available online: https://physics.nist.gov/asd (accessed on 11 October 2021).
49. Nantel, M.; Ma, G.; Gu, S.; Côté, C.Y.; Itatani, J.; Umstadter, D. Pressure ionization and line merging in strongly coupled plasmas produced by 100-fs laser pulses. *Phys. Rev. Lett.* **1998**, *80*, 4442–4445. [CrossRef]
50. Gu, M.F.; Beiersdorfer, P. Stark shift and width of x-ray lines from highly charged ions in dense plasmas. *Phys. Rev. A* **2020**, *101*, 032501. [CrossRef]
51. Marr, G.V. *Photoionization Processes in Gases*; Academic Press: New York, NY, USA, 1967.
52. Chang, T.N.; Fang, T.K.; Ho, Y.K. One- and two-photon ionization of hydrogen atom embedded in Debye plasmas. *Phys. Plasmas* **2013**, *20*, 092110. [CrossRef]
53. Wu, C.S.; Chen, S.M.; Chang, T.N.; Gao, X. Variation of the transition energies and oscillator strengths for the 3C and 3D lines of the Ne-like ions under plasma environment. *J. Phys. B-At. Mol. Opt. Phys.* **2019**, *52*, 185004. [CrossRef]
54. Wu, C.S.; Wu, Y.; Yan, J.; Chang, T.N.; Gao, X. Transition energies and oscillator strengths for the intrashell and intershell transitions of the C-like ions in a thermodynamic equilibrium plasma environment. *Phys. Rev. E* **2022**, *105*, 015206. [CrossRef]

Article

The Shapes of Stellar Spectra

Carlos Allende Prieto [1,2]

[1] Instituto de Astrofísica de Canarias, Vía Láctea S/N, E-38205 La Laguna, Tenerife, Spain; callende@iac.es; Tel.: +34-922-605200

[2] Departamento de Astrofísica, Universidad de La Laguna, E-38206 La Laguna, Tenerife, Spain

Abstract: Stellar atmospheres separate the hot and dense stellar interiors from the emptiness of space. Radiation escapes from the outermost layers of a star, carrying direct physical information. Underneath the atmosphere, the very high opacity keeps radiation thermalized and resembling a black body with the local temperature. In the atmosphere the opacity drops, and radiative energy leaks out, which is redistributed in wavelength according to the physical processes by which matter and radiation interact, in particular photoionization. In this article, I will evaluate the role of photoionization in shaping the stellar energy distribution of stars. To that end, I employ simple, state-of-the-art plane-parallel model atmospheres and a spectral synthesis code, dissecting the effects of photoionization from different chemical elements and species, for stars of different masses in the range of 0.3 to 2 $M_\odot$. I examine and interpret the changes in the observed spectral energy distributions of the stars as a function of the atmospheric parameters. The photoionization of atomic hydrogen and H^- are the most relevant contributors to the continuum opacity in the optical and near-infrared regions, while heavier elements become important in the ultraviolet region. In the spectra of the coolest stars (spectral types M and later), the continuum shape from photoionization is no longer recognizable due to the accumulation of lines, mainly from molecules. These facts have been known for a long time, but the calculations presented provide an updated quantitative evaluation and insight into the role of photoionization on the structure of stellar atmospheres.

Keywords: stars; stellar atmospheres; opacity

Citation: Allende Prieto, C. The Shapes of Stellar Spectra. *Atoms* **2023**, *11*, 61. https://doi.org/10.3390/atoms11030061

Academic Editors: Sultana N. Nahar and Guillermo Hinojosa

Received: 6 February 2023
Revised: 8 March 2023
Accepted: 13 March 2023
Published: 20 March 2023

Copyright: © 2023 by the authors. Licensee MDPI, Basel, Switzerland. This article is an open access article distributed under the terms and conditions of the Creative Commons Attribution (CC BY) license (https://creativecommons.org/licenses/by/4.0/).

1. Introduction

Starlight escapes from the atmospheres of stars, which become optically thin at different heights depending on the wavelength. The main effect causing photon absorption in stellar atmospheres is photoionization. As a result, this process reshapes the spectral energy distribution of stars, which in deeper layers must resemble black bodies, and dictates the details of how their colors change with surface temperature (or mass). In what follows, the discussion will be focused on stars with masses between 0.3 and about 2 $M_\odot$. There are of course stars with less (down to the minimum mass for hydrogen burning, about 0.07 $M_\odot$), and more mass (probably up to ~ 100 $M_\odot$), but most stars are included in this range.

Photoionization in stellar atmospheres is mainly associated with hydrogen atoms and closely related ions, in particular H^-, which is a dominant source of absorption for solar-type stars. This is not surprising since 80% of the stellar mass is usually hydrogen, and the remaining is mostly helium, with less than 2% left for the rest of the elements. Hydrogen photoionization imprints a series of discontinuities on stellar spectra, related to the various atomic levels: the Lyman ($n = 1$), Balmer ($n = 2$), Paschem ($n = 3$), Brackett ($n = 4$), Pfund ($n = 5$), etc. series, visible at wavenumbers of R_H/n^2, where R_H is the Rydberg constant, or wavelengths of 912, 3646, 8204, 14580, 22790 Å, etc. In thermodynamical equilibrium, the Saha equation shows that the ionization fraction for H is inversely proportional to the electron density, and therefore higher electron pressure, as present on a main-sequence star, burning hydrogen in its core, compared to an evolved giant star, leads to less ionization of H and more pronounced photoionization jumps.

For most stars, there is a competition between the opacity from photoionization of atomic H and H$^-$ at optical and infrared wavelenghts, and depending on this, the H series jumps become more or less prominent. In the infrared region, from $\lambda \sim 16,000$ Å onwards, free-free (inverse bremsstrahlung) typically overcomes bound-free (photoionization) for H$^-$. In the ultraviolet region, the situation becomes much more complex, since there are multiple, heavier elements that become important contributors to the opacity through ionization; usually carbon, sodium, magnesium, aluminum, silicon, and iron become very important at different wavelengths.

Reports from the 1970s to the 1990s claimed a significant mismatch between models and observations in the ultraviolet for the Sun, suggesting a missing opacity source. More recent studies (see, e.g., [1–3]) claimed to have identified the source as iron, at least in part associated with autoionizing lines missed in the earliest calculations.

Nevertheless, the far UV spectrum of solar-like stars is very hard to model, since the accumulated opacity makes photons to escape from the upper atmosphere (chromosphere, transition region, and corona), rather than the much-better-understood and easier-to-model photosphere. In those high layers, the low density, high temperatures, and the presence of magnetic fields complicate physical modeling. For moderately warmer stars, on the other hand, such as B-type stars, UV light escapes from deeper, photospheric, layers, which are simpler.

This paper, devoted to the role of photoionization in shaping stellar spectra, will first describe the data sources and codes used in our calculations. In Section 3, I dissect the various contributors to the opacity in the atmospheres of the Sun and other types of stars. Section 4 will test the models against selected high-quality spectrophotometric observations, giving us an idea of the realism of the models, as well as their limitations. Section 5 examines the sensitivity of the stellar continuum to changes in the main atmospheric parameters: surface effective temperature, surface gravity, and metallicity[1] (the fraction of heavy elements). The paper closes with a short summary and conclusions in Section 6.

2. Adopted Data

In the calculations below, we adopt classical 1D plane-parallel model atmospheres from the MARCS [4,5] and Kurucz ([3] and updates) grids—see also [6] for the most recent incarnations. The theory of stellar atmospheres is laid out in detail, for example, in the textbook [7].

The sources of atomic data adopted for the opacity and synthetic spectra computed are mostly those described in [8], with some updates. Photoionization cross-sections are from TOPBASE [9] for all the elements considered but iron, which are from [10,11]. Bound-free and free-free absorption was also considered for H$^-$, H$_2^+$, He$^-$, CH, OH, H$_2^-$, as well as collisionally induced opacity from H$_2$-H$_2$, H$_2$-He, H$_2$-H, and H-He. Rayleigh scattering on atomic hydrogen and helium, H$_2$, and the wings of Lyman alpha were also included.

Atomic line data are from the most recent files from Kurucz's website[2] updated with damping constants from [12,13]. Kurucz's website (and references therein) is also the source for the molecular line data, including H$_2$, CH, C$_2$, CN, CO, NH, OH, MgH, SiH, SiO, AlO, CaH, CaO, CrH, FeH, MgO, NaH, SiH, and VO, with the exception of Exomol data employed for TiO [14] and H$_2$O [15].

The actual calculations were performed with the latest version of the code Synspec [16,17]. The data and the code used are bundled with the Python wrapper Synple[3] [16], version 1.2.

3. Atmospheric Opacity

A star is a fairly independent entity, kept together by its own gravitational pull, and for most of its life producing its own energy through nuclear fusion in the core, where the temperature reaches tens of millions of degrees. Ionization is very high throughout the stellar interior up to the surface, where the temperature drops under $T \sim 10,000$ K, hydrogen becomes neutral, the electron density falls dramatically, and with it the material becomes transparent and radiation escapes.

The opacity at the stellar surface shapes the stellar spectrum, which is mainly due to the photoionization and inverse bremsstrahlung of the first few ionization stages of the most abundant elements, chiefly hydrogen and, in solar-type stars or cooler types, H^- and other molecules (see, e.g., [18]).

The top panel of Figure 1 illustrates the run of temperature vs. density in the atmosphere of a solar-like star. The point marked with a circle is just above the layers where the temperature of the star matches its effective temperature (T_{eff}, defined as that of a black body with the same radiative flux $F = \sigma T_{\text{eff}}^4$, where σ is the Stefan-Boltzmann constant, and from where the optical continuum is escaping. The bottom panel of the figure shows the total (blue) and photoionization (plus bremsstrahlung) (orange) opacity at the chosen point ($T = 5100$ K, $\rho = 1 \times 10^{-7}$ g cm^{-3}).

As we mentioned in the introduction, the smooth curve that dominates the continuum opacity between 4000 and 16,000 Å is due to the photoionization of H^-, while the rising continuum curve at longer wavelengths is due to H^- bremsstrahlung. In the UV region, photoionization from abundant neutral and singly ionized elements is responsible for the ragged and rapidly increasing continuum opacity. For each ion, the total opacity is the sum of the contributions from multiple levels. The height over the continuum reached by many of the lines suggests that line opacity is dominant, but most of these lines are narrow, with an FWHM of a fraction of an Ångstrom, and the shape of the stellar spectrum is, at least in the optical and near-infrared regions, dominated by photoionization.

Figure 2 dissects the continuum opacity, for the same conditions in the solar atmosphere adopted in Figure 1, into the various atomic and molecular contributors. Neutral carbon, magnesium, aluminum, silicon, and iron show up as the most important absorbers in the UV region. However, while the important role of iron at $\lambda < 3000$ Å and magnesium at $\lambda < 2500$ Å is undeniable, the dramatic Increase in opacity, not only due to photoionization but also through the accumulation of transitions at these wavelengths, shifts the layers from which UV radiation escapes higher up, and the importance of some of the ionization edges is hard to assess in practice.

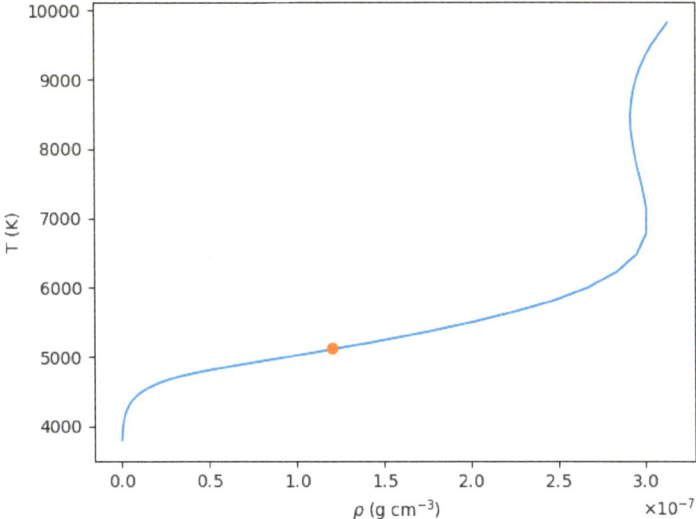

Figure 1. *Cont.*

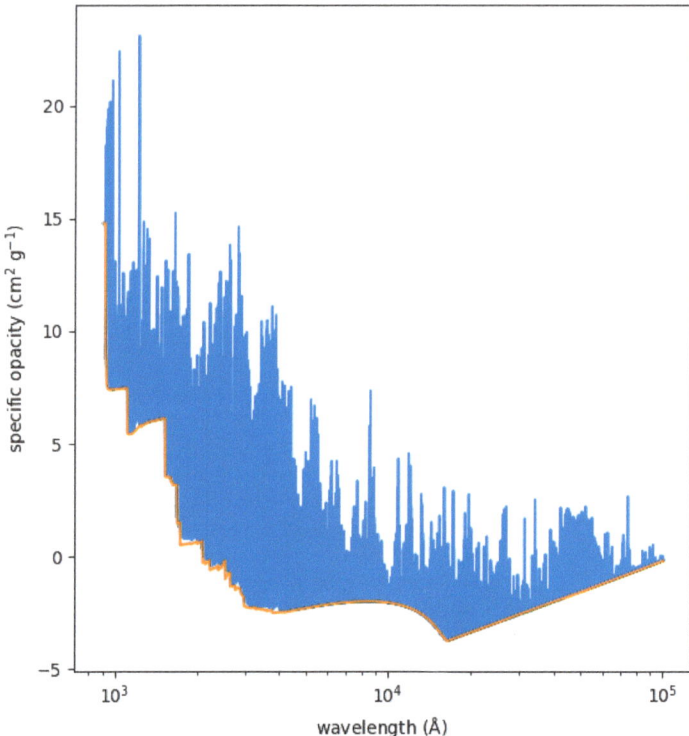

Figure 1. (**Upper panel**) Relationship between temperature and density in the atmosphere of a solar-like star ($T_{\text{eff}} = 5777$ K, $\log g = 4.437$ with g in cm s^{-2}, and the chemical abundances given in [19]) from a plane-parallel model in Local Thermodynamical Equilibrium. The indicated point is just slightly higher than optical depth unity, corresponding to $\rho \simeq 1.2 \times 10^{-7}$ g cm^{-3} and $T \simeq 5100$ K. (**Lower panel**) Continuous (photoionization plus bremsstrahlung; orange) and total opacity for the point in the T-ρ run marked in the top panel.

This picture may be incomplete, since the list of ions included in the calculations is not exhaustive, and has in practice been limited to those for which there are calculations available from the Opacity Project and the Iron Project. Furthermore, some of the opacity due to autoionization lines may be included twice as line transitions and resonances in the photoionization cross sections.

The agreement with observations (see Section 4) suggests the major contributors to opacity in stellar atmospheres have been identified, but there may be modestly or moderately important contributors missing. In a recent paper [20] it has been pointed out that the free-free opacity cross-section for negative positronium ions (the equivalent of H$^-$ for a positronium instead of a H atom) would be larger than that of H$^-$ in the infrared for the solar atmosphere. Nonetheless, the lack of knowledge of the abundance of positrons in the solar atmosphere makes it hard to assess how much relevance such a contribution would make to the total opacity.

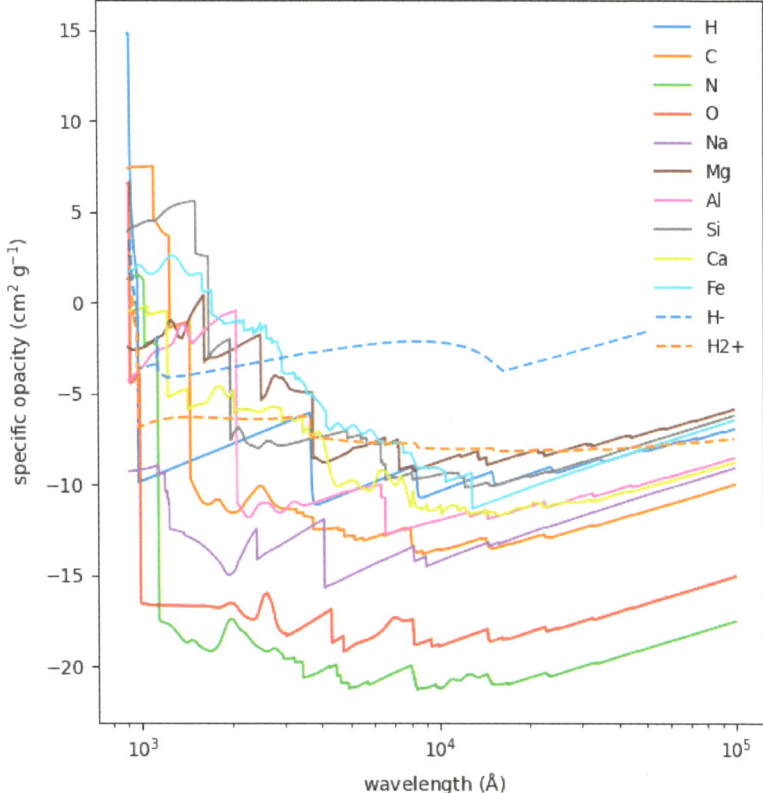

Figure 2. Bound-free Author Ok and free-free opacity associated with individual elements for $\rho \simeq 1.2 \times 10^{-7}$ g cm^{-3} and $T \simeq 5100$ K. The continuous opacity for H$^-$ and H$_2^+$ is also shown.

Figure 3 repeats the information displayed in Figure 1 for the Sun (shown here in orange), but also includes a model for an A-type star $T_{\text{eff}} = 9800$ K) and a much cooler M-type star ($T_{\text{eff}} = 3500$ K). As before, we have chosen atmospheric depths representative of the regions where the optical continuum escapes. The lower panel shows the photoionization (and bremsstrahlung) opacity for the three stars with dashed lines. For the hot (A-type, blue) and cool stars (M-type, red), the total opacity, including lines, is also shown.

The continuum of the M-type star is, like for the Sun, shaped by H$^-$ photoionization between 4000 and 16,000 Å. H$^-$ free-free opacity dominates at longer wavelengths, and the photoionization of heavier elements (and electron scattering) is most relevant in the UV. However, line absorption severely blocks much of the continuum flux. The continuum of the warmer A-type star, on the other hand, shows the characteristic shape of the hydrogenic photoionization cross-section, proportional to λ^3 [21], with discontinuities corresponding to the minimum ionization energies for each of the $n = 1, 2, 3, \ldots$, etc. levels. Note that the H$^-$ bremsstrahlung opacity shares the same slope [22].

Line absorption is progressively reduced as the surface temperature of the star increases, due to the ionization of the main line absorbers, chiefly atomic iron. It is quite obvious in Figure 3 that the lines add sharp opacity peaks on top of the continuum for the A-type star, while the lower edge of the line absorption appears detached from the continuum opacity for the cooler M-type star, indicative of a massive accumulation of lines that overlap, mainly from molecules such as CH, CN, CO, OH, and TiO.

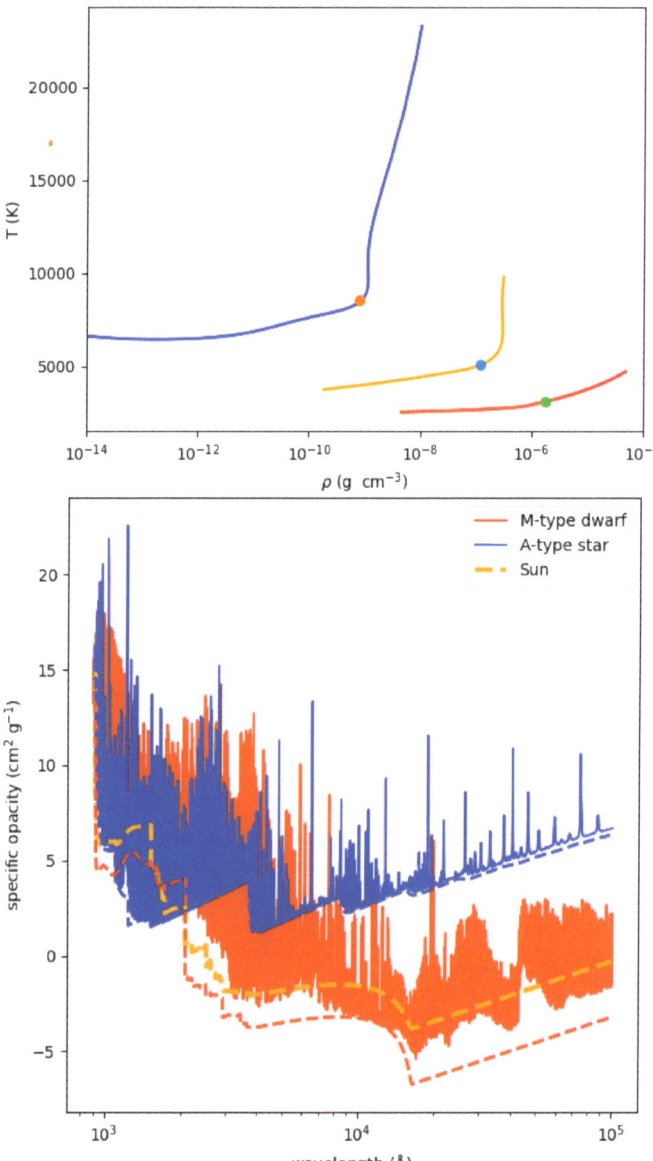

Figure 3. Similar to Figure 1 for the main sequence M-type ($T_{\text{eff}} = 3500$ K) and A-type stars ($T_{\text{eff}} = 9800$). The continuum opacity is shown with dashed lines. The total (including lines) opacity is omitted for the solar case since it is already shown in Figure 1. The representative data correspond to ($\rho \simeq 0.8 \times 10^{-9}$ g cm^{-3}, $T \simeq 8600$ K) in the case of the A-type star and ($\rho \simeq 1.8 \times 10^{-6}$ g cm^{-3}, $T \simeq 3200$ K) for the M-type star.

4. Observations

A sanity check is always necessary, for which we may use the solar spectrum. Nevertheless, it turns out that performing observations of the solar absolute flux poses notable difficulties related to its overwhelming brightness, large angular size on the sky, and the hurdles to calibrate the observations using laboratory sources or sky sources such as standard stars. This is not to say that such observations have not been made. They, in fact, keep being made with regularity, motivated by the need to understand the amount and variability of the solar radiation impacting the Earth (see, e.g., [23,24]), but to our knowledge, there is no solar spectrum of reference available with a reliable calibration over a very broad spectral range.

An alternative is offered by the various solar analogs identified over the years, largely with the motivation of performing differential determinations of chemical abundances, which benefit from the cancelation of systematic errors inherent to that type of analysis. The star 18 Sco is the nearest such star available and has been studied in detail. Broad-coverage spectroscopy from the Hubble Space Telescope is available for this target, with high-quality absolute fluxes (see, e.g., [25]).

This star is close enough and bright enough that it has been observed with an interferometer, and its angular diameter has been resolved and measured to be $\theta = 0.6759 \pm 0.0062$ milliarcseconds [26]. The Gaia parallax for this star is 70.7371 ± 0.0631 milliarcseconds, implying it is at a distance d of 14.14 ± 0.01 parsecs, and it has a radius of $R = \theta/2 \times d = 1.027 \pm 0.001 R_\odot$. The precision of the angular diameter, good to 1%, allows us to scale the model fluxes, computed at the stellar surface, and compare them to the flux measured for the star. Figure 4 illustrates this exercise, using a solar model ($T_{\text{eff}} = 5777$ K, $\log g = 4.437$, and the chemical composition from [19], [Fe/H]= 0—very close to the parameters for 18 Sco published from [26]: $T_{\text{eff}} = 5817 \pm 4$ K, $\log g = 4.448 \pm 0.012$, and [Fe/H] $= 0.052 \pm 0.005$), showing excellent agreement, except at the shortest wavelengths ($\lambda < 2300$), where multiple assumptions built in the models break down.

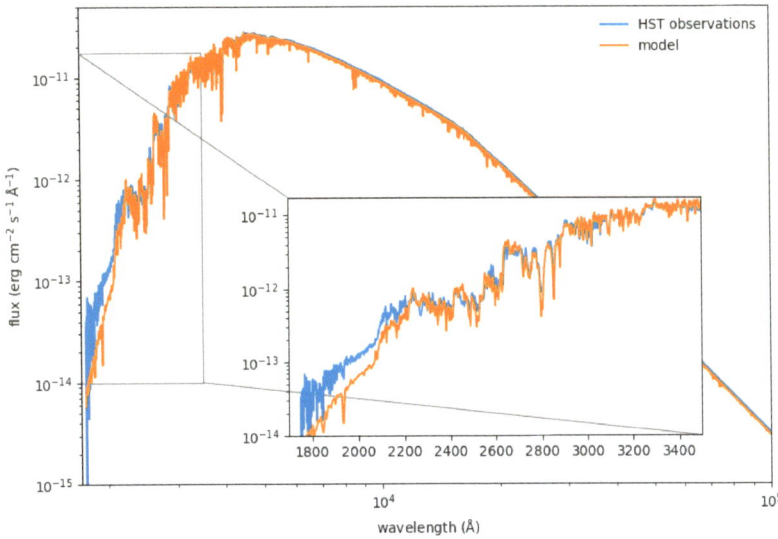

Figure 4. Observed (blue) and model (orange) spectra for the solar twin 18 Sco. The model has been scaled using the determination of the star's angular diameter from [26].

The light at wavelengths shorter than about 2000 Å escapes from very high atmospheric layers, where the classical model atmospheres we have adopted are no longer realistic due mainly to departures from Local Thermodynamical Equilibrium, the breakdown of the assumption of hydrostatic equilibrium, and the relevance of magnetic fields, which are ignored. An ad hoc chromospheric temperature increase was thought in the 1970s to solve the flux discrepancies found for the Sun at short wavelengths [27,28], but hydrostatic models cannot explain many of the observations such as the strength of CO bands [29].

A couple of features are noteworthy in Figure 4. At about 2500 Å a break is apparent, and a second one is visible near 2100 Å. These sudden reductions in flux correspond to the abrupt increases in opacity due to atomic magnesium and atomic aluminum, respectively. Other features from photoionization edges one may expect based on Figure 2 are not really visible, most likely, as pointed out above, due to the opacity enhancement shifting the region from which the continuum flux escapes to layers higher up in the atmosphere. The continuum changes in slope at about 4000 Å, longwards of which H^- photoionization dominates, and 16,000 Å, where H^- bremsstrahlung becomes the main contributor to the continuum opacity.

The brightest star in the sky, Sirius, is a good example of an A-type star, and its spectrum is included among the high-quality observations from the Hubble Space Telescope [30]. The star has a white dwarf companion that is irrelevant to our discussion. Its parallax is 379.21 ± 1.58 milliarseconds, or a distance of 2.637 pc. The angular diameter has been measured by [31] to be $\theta = 6.039 \pm 0.019$ milliarcsends, and more recently confirmed by ([32] as $\theta = 6.041 \pm 0.017$ milliarconds). Figure 5 shows the observed spectrum (blue) confronted with a model (orange) with T_{eff} = 10,000 K, $\log g$ = 4.0 and (Fe/H]) = 0 ([32] adopted for this star T_{eff} = 9845 K, $\log g$ = 4.25 and [Fe/H] = +0.5, where the two latter parameters are inherited from [33]).

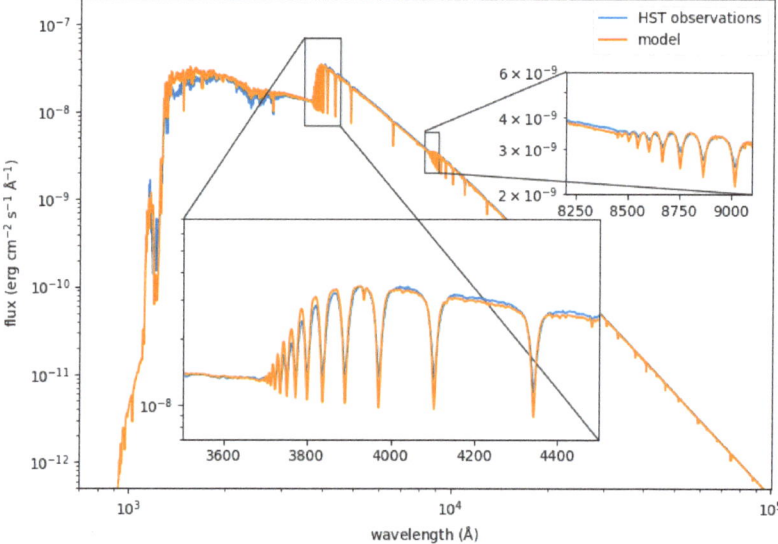

Figure 5. Observed (blue) and model (orange) spectra for the A-type Sirius A. The model has been scaled using the determination of the star angular diameter by [32].

The jumps in flux due to the photoionization of hydrogen are quite obvious at about 900 Å ($n = 1$), 3700 Å ($n = 2$), and 8200 Å ($n = 3$), becoming progressively weaker. The drop in flux at about 1300 Å is due to carbon photoionization, and there is a smaller drop at about 1500 Å caused by Si photoionization. The very strong lines at about 1200 Å are a blend of Lα (H $n = 2$ to $n = 1$ transition) with S I lines on the blue side. The cores of the H lines are deeper in the models than in the data, which could be a limitation in the models, although departures from Local Thermodynamic Equilibrium work in the opposite sense and can therefore be excluded (e.g., [34]), or an issue with scattered light in the observations.

As an example of a cool star, we have chosen Gliese 555, a well-studied red dwarf star with an effective temperature of about 3200 K. This star is one of the few M dwarfs included in the Hubble Space Telescope sample with accurate fluxes, but unfortunately is not among the short list of M dwarfs with measured angular diameters. Nonetheless, Ref. [35] have built a relation between the infrared luminosity of M dwarfs and their radii, using the stars with interferometric angular diameters, and arrived at $R = 0.310 \pm 0.013 R_\odot$ for GJ 555, which, combined with the Gaia parallax, leads to $\theta = 0.461 \pm 0.019$ milliarcseconds.

Figure 6 compares the observations with a model for $T_{\text{eff}} = 3200$ K, $\log g = 5$ and [Fe/H]= 0 ([35] give $T_{\text{eff}} = 3211$ K, $\log g = 4.89$ and [Fe/H]= +0.17), showing fair agreement. The parameters appear to be appropriate, and so is the angular diameter, but the model's imperfections are much more significant than in the cases of 18 Sco and Sirius. The complexity of the model is significantly higher due to the pervasive presence of molecules in the atmosphere of this star. As one would expect from the analysis in Section 3, and in particular from Figure 3, the shape of the spectra of this type of star is dominated by the presence of molecular bands of MgH, TiO, VO, and CaH, as well as very strong lines from low-lying levels of Na I (5900 Å and 8190 Å), K I (7680 Å), and Ca I (4227 Å).

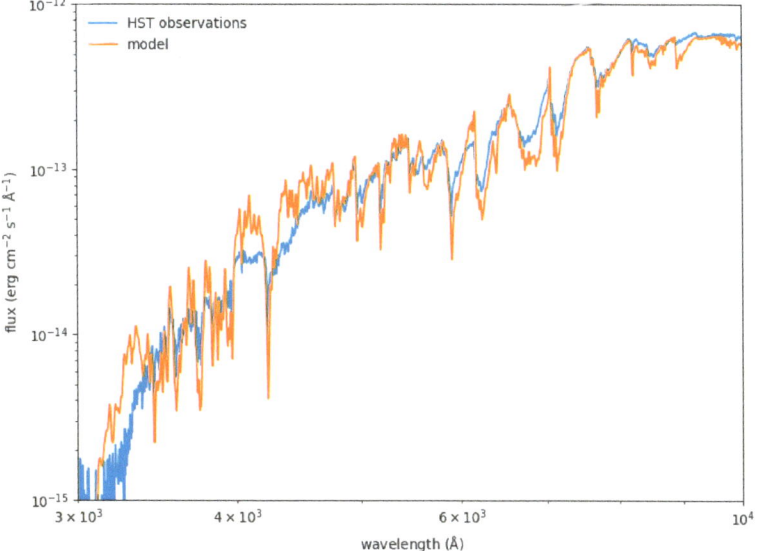

Figure 6. Observed (blue) and model (orange) spectra for the M-type dwarf star GJ 555. The model has been scaled using the estimated stellar radius from [35] and the Gaia DR3 parallax for the star.

5. Diagnostics

Most relevant for the study of stars and understanding their properties is how opacity in general and photoionization in particular vary depending upon the properties of the star.

The fundamental stellar parameters are mass and radius, plus age and chemical composition. To the zeroth-order, mass determines the fate of a star, including how long its life will be, what chemical elements it will be able to produce in its interior by nuclear fusion, and how it will die, with the most typical outcomes being a core-collapse supernova leaving behind a black hole or a neutron star or nothing for $M > 8 M_\odot$, or a white dwarf for lower-mass stars. Nevertheless, from the point of view of the spectrum of a star, the most relevant parameters are the star's surface (or *effective*) temperature, its surface gravity, and its chemical composition.

In the longest phase of the life of a star, the *main sequence*, it fuses hydrogen in its core to produce helium. In this period, the mass correlates perfectly with surface temperature: the more massive the star, the warmer the surface temperature. This is therefore the main atmospheric parameter that controls how the spectrum of the star looks, as illustrated in Figures 4–6. we now examine the impact of the other two parameters on the structure of the stellar atmosphere and the shape of the spectrum.

I have already discussed the situation for cool, M-type stars, where H^- photoionization is still the main contributor to the continuum opacity in the optical and near-infrared, but the line absorption due to molecules dominates the opacity. In what follows, we will discuss the other two warmer cases considered in our previous examples: a solar-like star and an A-type star.

In the top panel of Figure 7, we can see the run of the main thermodynamical quantities for a solar-like star with Rosseland optical depth, which is a weighted mean of the integrated opacity along the atmosphere down to a given depth, and gives a very useful reference axis when studying optical properties. There are three models shown in the figure: a reference solar-like model (in blue), another with 0.5 dex higher gravity (orange), and a third, which, in addition to the higher gravity, has a higher metal content ([Fe/H] = +0.5). The bottom panel of the figure shows the corresponding model spectra.

Our models assume that the atmosphere is in hydrostatic equilibrium

$$\frac{dP}{dm} = g, \qquad (1)$$

where P is the gas pressure (turbulent and radiation pressure are negligible for this type of atmosphere), m is the mass colum, and g is the gravitational acceleration $g = GM/R^2$. An increase in gravity compresses the atmosphere, enhancing the pressure, while keeping the fractional contribution from electrons (Pe) to it at a similar level (10^{-4}, except in the deepest layers where H begins to be ionized). The effect of this change on the continuum opacity is negligible, since the abundance of H^-, proportional to Pe, increases only mildy, while Pe/P stays nearly constant.

On the other hand, an additional increase in the abundance of the heavy elements has a profound impact on the near-UV opacity, due to the importance of iron and magnesium photoionization, and the increase in iron (and other metal) lines, and the UV flux is consequently reduced. The enhanced UV opacity cools down the outer atmospheric layers and heats the deepest ones (an effect known as *backwarming*). There is a small increase in the electron pressure, partly from the increase in the abundance of electron donors such as sodium, magnesium, calcium, and iron, which would in principle boost the optical and near-IR opacity, therefore reducing the flux at those wavelengths, but the effect observed is exactly the opposite. This is in part due to the fact that the relative enhancement of electron pressure in high atmospheric layers disappears when reaching the continuum-forming layers at the Rosseland optical depth near unity. Furthermore, an increase in continuum opacity does not necessarily imply a reduction in flux, since the model needs to self-adjust to satisfy energy conservation, which for these three models imposes that the flux integrated over all wavelengths must be the same

$$\int_0^\infty F_\lambda d\lambda = F = \sigma T_{\text{eff}}^4, \qquad (2)$$

and therefore a decrease in the UV flux has to be compensated at other wavelengths.

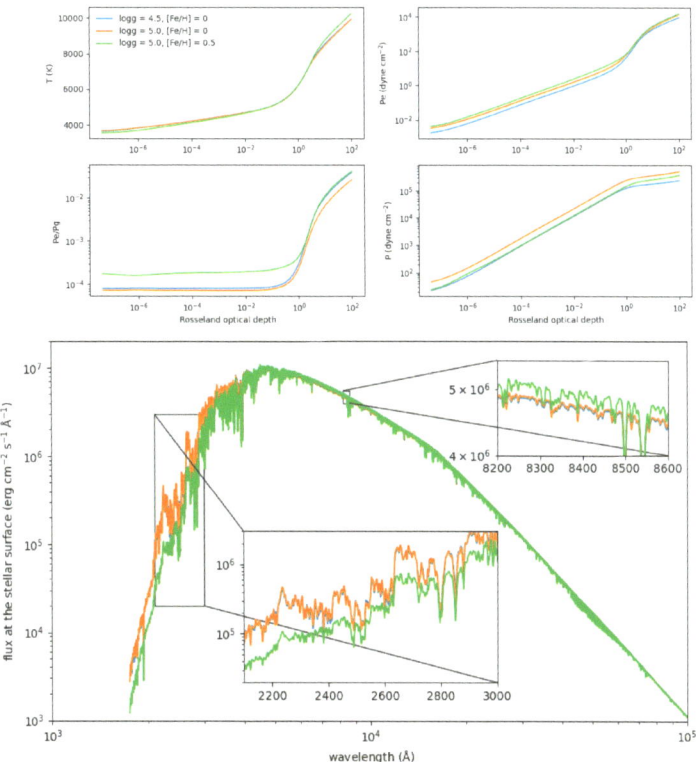

Figure 7. (**Upper panel**) Run of various thermodynamical quantities (clockwise: temperature, electron pressure (Pe), gas pressure (P) and their ratio (Pe/P)) with the Rosseland mean opacity for model atmospheres for a solar-like star ($T_{\rm eff}$ = 5777 K). (**Lower panel**) Predicted emergent flux at the atmospheric surface for the models in the upper panels (no scaling is necessary, since we are only comparing models).

The situation for an A-type star such as Sirius, illustrated in Figure 8, is different. Here, H^- plays only a minor role in the continuum opacity, and hydrogen atoms are the main contributors in the optical and near-infrared regions. An increase in surface gravity leaves the run of temperature with optical depth unchanged but compresses the atmosphere, enhancing the gas pressure and, to a lesser extent, the electron density, with an overall reduction in the electron's partial pressure. An additional increase in the abundances of the heavy elements does not change the atmospheric structure much.

In the lower panel of Figure 8, we can appreciate how the changes described affect the emergent radiative flux. The boost in pressure associated with the increase in surface gravity broadens the lines somewhat , both the hydrogen lines and other strong features. The pressure enhancement and the subsequent reduction in the electron partial pressure increases the ionization and dampens slightly the hydrogen photoionization, as clearly visible in the Balmer (3700 Å) and the Paschem (8500 Å) jumps. An enhancement in the metal abundance noticeably affects the UV absorption, in this case mainly due to (ionized) iron photoionization, reducing the UV flux, which is compensated with a slightly increase at other wavelengths to keep the integrated flux constant.

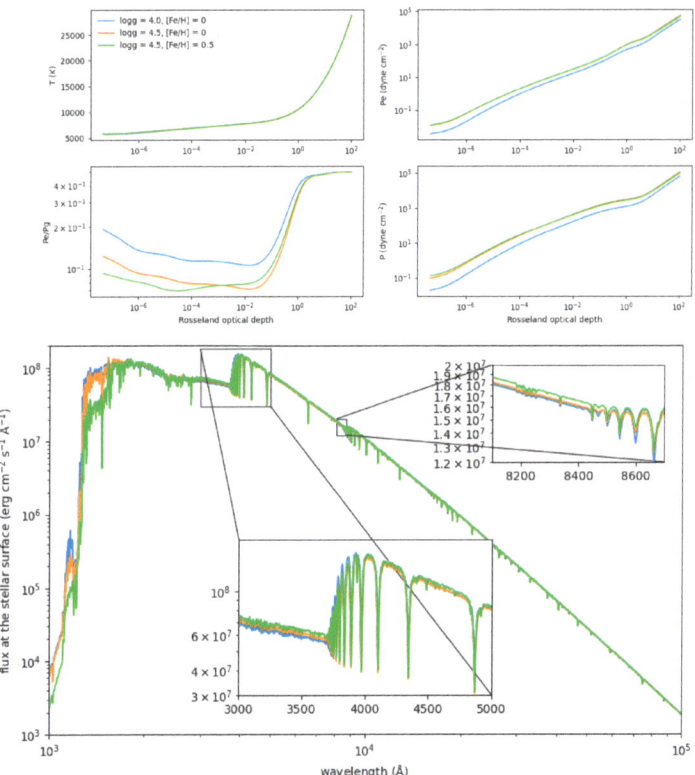

Figure 8. (**Upper panel**) Run of various thermodynamical quantities (clockwise: temperature, electron pressure (Pe), gas pressure (P), and their ratio (Pe/P)) with the Rosseland mean opacity for model atmospheres for an A-type star (T_{eff} = 9750 K). (**Lower panel**) Predicted emergent flux at the atmospheric surface for the models in the upper panels.

6. Summary and Conclusions

We have used a state-of-the-art code for computing synthetic spectra and standard plane-parallel model atmospheres to evaluate the role of photoionization in shaping the spectral energy distributions of stars.

The photoionization of atomic hydrogen or the H$^-$ ion are a dominant source of opacity in the optical and infrared regions in stars with one solar mass or more. While H$^-$ remains a dominant contributor to the optical/infrared continuum opacity, molecular line opacity becomes more important in cooler (usually less massive) stars and causes a significant redistribution of the emergent flux. This makes it harder to model stellar spectra of M-type dwarfs, which are the most common stars across the Milky Way, than warmer stars.

The ultraviolet spectra of most stars are dominated by photoionization from heavier elements (magnesium, aluminum, silicon, and iron), as well as atomic line absorption. In models of solar-type stars, the ultraviolet opacity seems to be accounted for appropriately, although a more exhaustive study is necessary to make sure that (a) no important contributors are missed (e.g., photoionization from elements with similar atomic mass or heavier than iron), and (b) no opacity sources are counted twice, (e.g., autoionization lines being included in the atomic line list and in the photoionization cross-sections).

The illustrative examples given in this paper can serve as a starting point for new, deeper, investigations, looking at the sources of opacity in various types of stars. Improving our understanding of the atmospheric opacity paves the way to refining the agreement between the observed spectral energy distributions of stars and model predictions, which are key to our ability to infer stellar properties, such as mass, radius, luminosity, chemical composition, etc., from observations.

Funding: The author acknowledges support for this research from the Spanish Ministry of Science and Innovation (MICINN) projects s AYA2017-86389-P and PID2020-117493GBI00. Funding for the DPAC has been provided by national institutions, in particular the institutions participating in the *Gaia* Multilateral Agreement.

Institutional Review Board Statement: Not applicable.

Informed Consent Statement: Not applicable.

Data Availability Statement: The opacity tables and the model spectra used in this paper have been computed with version v1.2 of Synple, publicly available from github.com/callendeprieto/synple, accessed on 13 January 2023.

Acknowledgments: I am thankful to Ivan Hubeny for his comments on an early draft of this manuscript. This work has made use of data from the European Space Agency (ESA) mission *Gaia* (https://www.cosmos.esa.int/gaia, accessed on 16 January 2023), processed by the *Gaia* Data Processing and Analysis Consortium (DPAC, https://www.cosmos.esa.int/web/gaia/dpac/consortium, accessed on 23 January 2023). This research has made use of the SIMBAD database, operated at CDS, Strasbourg, France.

Conflicts of Interest: The author declares no conflict of interest.

Notes

1. The usual convention in astronomy is used for this parameter, which prescribes that all the elements heavier than He are changed in the same ratio relative to their solar abundances, and that ratio is quantified with the parameter $[Fe/H] = \log\left(\frac{N_{Fe}}{N_H}\right) - \log\left(\frac{N_{Fe}}{N_H}\right)_\odot$, where N_X is the number density for the element X.
2. kurucz.harvard.edu accessed on 25 October 2022, file gfall08oct17.dat.
3. github.com/callendeprieto/synple accessed on 13 January 2023.

References

1. Kurucz, R.L. "Finding" the "missing" solar ultraviolet opacity. *Rev. Mex. Astron. Astrofis.* **1992**, *23*, 181–186.
2. Gustafsson, B. Opacity incompleteness and atmospheres of cool stars. *Highlights Astron.* **1995**, *10*, 579. [CrossRef]
3. Castelli, F.; Kurucz, R.L. Is missing Fe I opacity in stellar atmospheres a significant problem? *Astron. Astrophys.* **2004**, *419*, 725–733. [CrossRef]
4. Gustafsson, B.; Bell, R.A.; Eriksson, K.; Nordlund, A. A grid of model atmospheres for metal-deficient giant stars. I. *Astron. Astrophys.* **1975**, *42*, 407–432.
5. Gustafsson, B.; Edvardsson, B.; Eriksson, K.; Jørgensen, U.G.; Nordlund, Å.; Plez, B. A grid of MARCS model atmospheres for late-type stars. I. Methods and general properties. *Astron. Astrophys.* **2008**, *486*, 951–970. [CrossRef]
6. Mészáros, S.; Allende Prieto, C.; Edvardsson, B.; Castelli, F.; Pérez, A.E.G.; Gustafsson, B.; Majewski, S.R.; Plez, B.; Schiavon, R.; Shetrone, M.; et al. New ATLAS9 and MARCS Model Atmosphere Grids for the Apache Point Observatory Galactic Evolution Experiment (APOGEE). *Astron. J.* **2012**, *144*, 120. [CrossRef]
7. Hubeny, I.; Mihalas, D. *Theory of Stellar Atmospheres. An Introduction to Astrophysical Non-Equilibrium Quantitative Spectroscopic Analysis*; Princeton University Press: Princeton, NJ, USA, 2014. ISBN 9780691163284.
8. Allende Prieto, C.; Koesterke, L.; Hubeny, I.; Bautista, M.A.; Barklem, P.S.; Nahar, S.N. A collection of model stellar spectra for spectral types B to early-M. *Astron. Astrophys.* **2018**, *618*, A25. [CrossRef]
9. Seaton, M.J. *The Opacity Project*; Institute of Physics Publishing: Bristol, UK, 1995.
10. Bautista, M.A. Atomic data from the IRON Project. XX. Photoionization cross sections and oscillator strengths for Fe I. *Astron. Astrophys. Suppl. Ser.* **1997**, *122*, 167–176. [CrossRef]
11. Nahar, S.N. Atomic data from the Iron Project. VII. Radiative dipole transition probabilities for Fe II. *Astron. Astrophys.* **1995**, *293*, 967–977.
12. Barklem, P.S.; Aspelund-Johansson, J. The broadening of Fe II lines by neutral hydrogen collisions. *Astron. Astrophys.* **2005**, *435*, 373–377. [CrossRef]

13. Barklem, P.S.; Piskunov, N.; O'Mara, B.J. A list of data for the broadening of metallic lines by neutral hydrogen collisions. *Astron. Astrophys. Suppl. Ser.* **2000**, *142*, 467–473. [CrossRef]
14. McKemmish, L.K.; Masseron, T.; Hoeijmakers, H.J.; Pérez-Mesa, V.; Grimm, S.L.; Yurchenko, S.N.; Tennyson, J. ExoMol molecular line lists—XXXIII. The spectrum of Titanium Oxide. *Mon. Not. R. Astron. Soc.* **2019**, *488*, 2836–2854. [CrossRef]
15. Polyansky, O.L.; Kyuberis, A.A.; Zobov, N.F.; Tennyson, J.; Yurchenko, S.N.; Lodi, L. ExoMol molecular line lists XXX: A complete high-accuracy line list for water. *Mon. Not. R. Astron. Soc.* **2018**, *480*, 2597–2608. [CrossRef]
16. Hubeny, I.; Allende Prieto, C.; Osorio, Y.; Lanz, T. TLUSTY and SYNSPEC Users's Guide IV: Upgraded Versions 208 and 54. *arXiv e-prints* **2021**. [CrossRef]
17. Hubeny, I.; Lanz, T. A brief introductory guide to TLUSTY and SYNSPEC. *arXiv e-prints* **2017**. [CrossRef]
18. Gray, D.F. (Ed.) *The Observation and Analysis of Stellar Photospheres*, 4th ed.; Cambridge University Press: Cambridge, UK, 2022.
19. Asplund, M.; Grevesse, N.; Sauval, A.J. The Solar Chemical Composition. In *Cosmic Abundances as Records of Stellar Evolution and Nucleosynthesis in Honor of David L. Lambert: Proceedings of a Symposium Held in Austin, Texas, USA, 17–19 June 2004*; Barnes, T.G., III, Bash, F.N., Eds.; Astronomical Society of the Pacific: San Francisco, CA, USA, 2004.
20. Bhatia, A.K.; Pesnell, W.D. A Note on the Opacity of the Sun's Atmosphere. *Atoms* **2020**, *8*, 37. [CrossRef]
21. Kramers, H.A. On the theory of X-ray absorption and of the continuous X-ray spectrum. *Phil. Mag.* **1923**, *46*, 836. [CrossRef]
22. Geltman, S. Continuum States of H^- and the Free-Free Absorption Coefficient. *Astrophys. J.* **1965**, *141*, 376. [CrossRef]
23. Floyd, L.E.; Morrill, J.S.; McMullin, D.R. Solar UV Spectral Irradiance Measurements at 0.15 nm Resolution by SUSIM. In Proceedings of the AGU Fall Meeting Abstracts, San Francisco, CA, USA, 5–9 December 2011.
24. Snow, M.; McClintock, W.E.; Woods, T.N.; Elliott, J.P. SOLar-STellar Irradiance Comparison Experiment II (SOLSTICE II): End-of-Mission Validation of the SOLSTICE Technique. *Sol. Phys.* **2022**, *297*, 55. [CrossRef]
25. Bohlin, R.C. CALSPEC: HST Spectrophotometric Standards at 0.115 to 32 μm with a 1% Accuracy Goal. In Proceedings of the International Astronomical Union General Assembly, Vienna, Austria, 20–31 August 2018; pp. 449–453. [CrossRef]
26. Bazot, M.; Ireland, M.J.; Huber, D.; Bedding, T.R.; Broomhall, A.-M.; Campante, T.L.; Carfantan, H.; Chaplin, W.J.; Elsworth, Y.; et al. The radius and mass of the close solar twin 18 Scorpii derived from asteroseismology and interferometry. *Astron. Astrophys.* **2011**, *526*, L4. [CrossRef]
27. Vernazza, J.E.; Avrett, E.H.; Loeser, R. Structure of the Solar Chromosphere. Basic Computations and Summary of the Results. *Astrophys. J.* **1973**, *184*, 605–632. [CrossRef]
28. Vernazza, J.E.; Avrett, E.H.; Loeser, R. Structure of the solar chromosphere. II. The underlying photosphere and temperature-minimum region. *Astrophys. J. Suppl. Ser.* **1976**, *30*, 1–60. [CrossRef]
29. Ayres, T.R. Does the Sun Have a Full-Time COmosphere? *Astrophys. J.* **2002**, *575*, 1104–1115. [CrossRef]
30. Bohlin, R.C. Hubble Space Telescope CALSPEC Flux Standards: Sirius (and Vega). *Astron. J.* **2014**, *147*, 127. [CrossRef]
31. Kervella, P.; Thévenin, F.; Morel, P.; Bordé, P.; DiFolco, E. The interferometric diameter and internal structure of Sirius A. *Astron. Astrophys.* **2003**, *408*, 681. [CrossRef]
32. Davis, J.; Ireland, M.J.; North, J.R.; Robertson, J.G.; Tango, W.J.; Tuthill, P.G. The Angular Diameter and Fundamental Parameters of Sirius A. *Publ. Astron. Soc. Aust.* **2011**, *28*, 58. [CrossRef]
33. Cohen, M.; Walker, R.G.; Barlow, M.J.; Deacon, J.R. Spectral Irradiance Calibration in the Infrared. I. Ground-Based and IRAS Broadband Calibrations. *Astron. J.* **1992**, *104*, 1650. [CrossRef]
34. Przybilla, N.; Butler, K. Non-LTE Line Formation for Hydrogen Revisited. *Astrophys. J.* **2004**, *609*, 1181–1191. [CrossRef]
35. Mann, A.W.; Feiden, G.A.; Gaidos, E.; Boyajian, T.; von Braun, K. How to Constrain Your M Dwarf: Measuring Effective Temperature, Bolometric Luminosity, Mass, and Radius. *Astrophys. J.* **2015**, *804*, 64. [CrossRef]

Disclaimer/Publisher's Note: The statements, opinions and data contained in all publications are solely those of the individual author(s) and contributor(s) and not of MDPI and/or the editor(s). MDPI and/or the editor(s) disclaim responsibility for any injury to people or property resulting from any ideas, methods, instructions or products referred to in the content.

MDPI AG
Grosspeteranlage 5
4052 Basel
Switzerland
Tel.: +41 61 683 77 34

Atoms Editorial Office
E-mail: atoms@mdpi.com
www.mdpi.com/journal/atoms

Disclaimer/Publisher's Note: The title and front matter of this reprint are at the discretion of the Guest Editors. The publisher is not responsible for their content or any associated concerns. The statements, opinions and data contained in all individual articles are solely those of the individual Editors and contributors and not of MDPI. MDPI disclaims responsibility for any injury to people or property resulting from any ideas, methods, instructions or products referred to in the content.

www.ingramcontent.com/pod-product-compliance
Lightning Source LLC
LaVergne TN
LVHW072330090526
838202LV00019B/2386